T0296556

RADIO ASTRONOMY

INTERNATIONAL ASTRONOMICAL UNION
SYMPOSIUM No. 4

HELD AT THE JODRELL BANK EXPERIMENTAL STATION
NEAR MANCHESTER, AUGUST 1955

RADIO ASTRONOMY

EDITED BY

H. C. VAN DE HULST

Leiden Observatory

*PRINTED WITH
FINANCIAL ASSISTANCE
FROM U.N.E.S.C.O.*

CAMBRIDGE
AT THE UNIVERSITY PRESS

1957

CAMBRIDGE
UNIVERSITY PRESS

University Printing House, Cambridge CB2 8BS, United Kingdom

Cambridge University Press is part of the University of Cambridge.

It furthers the University's mission by disseminating knowledge in the pursuit of
education, learning and research at the highest international levels of excellence.

www.cambridge.org
Information on this title: www.cambridge.org/9781316612811

© Cambridge University Press 1957

First published 1957
First paperback edition 2016

A catalogue record for this publication is available from the British Library

ISBN 978-1-316-61281-1 Paperback

PREFACE

A symposium on Radio Astronomy, organized by the International Astronomical Union, was held on 25–27 August 1955 at the Jodrell Bank Experimental Station of the University of Manchester. It coincided with the tenth anniversary of this station; the sessions took place in the control building of the 250-ft. telescope which is under construction.

The symposium brought together 108 participants from: Australia, Belgium, Canada, Czechoslovakia, Finland, France, Germany, Great Britain, Italy, Japan, the Netherlands, Norway, Spain, Sweden, Switzerland, U.S.A., U.S.S.R., and Yugoslavia. The symposium committee consisted of A.C.B. Lovell, chairman; H. C. van de Hulst, secretary; J. P. Hagen, M. Laffineur, and J. L. Pawsey. About sixty mimeographed abstracts distributed beforehand or at the meeting helped to ensure an effective exchange of information.

This volume contains all but two of the papers presented. One contribution, paper 16, has been added at the editor's request. Many papers have been improved as a result of discussion at the symposium or by the inclusion of data not available in August 1955. The essential parts of the discussions have been reported.

The six parts of this volume approximately correspond to the respective morning and afternoon sessions, each starting with an introductory lecture. Purely instrumental papers or parts of papers have been omitted from this volume as they fell outside the range of topics outlined for this symposium, as also did scintillation and purely geophysical problems. It may be fairly said that this volume gives an almost complete report on all other research in radio astronomy at the time of the meeting.

The editor wishes to express his thanks to all authors for their cooperation.

H. C. van de Hulst

LEIDEN OBSERVATORY
April 1957

v

CONTENTS

POINT SOURCES: INDIVIDUAL STUDY AND
PHYSICAL THEORY

PART V

THE ACTIVE SUN

PART I

SPECTRAL LINE INVESTIGATIONS

STUDIES OF THE 21-CM. LINE AND THEIR INTERPRETATION

INTRODUCTORY LECTURE BY

H. C. VAN DE HULST

University Observatory, Leiden, Netherlands

I. HISTORY

The subject of the 21-cm. line needs no introduction to this audience. Perhaps as an introduction I may mention one moment that belongs to the pre-history of the investigations of this line. In the spring of 1944 Oort said to me: 'We should have a colloquium on the paper by Reber; would you like to study it? And, by the way, radio astronomy can really become very important if there were at least one line in the radio spectrum. Then we can use the method of differential galactic rotation as we do in optical astronomy.'

This hope has been amply fulfilled. You know the further history: 1945, first mention of the 21-cm. line in the literature on radio astronomy; 1947, laboratory measurements of its frequency; 1951, its discovery in the Galaxy; 1952, first plots of spiral structure over a good part of the galactic plane; 1953, detection of hydrogen radiation from the Magellanic clouds; 1954, discovery of high-velocity wings in the central regions; 1954, discovery of absorption effects.

2. BASIC PROBLEMS

The basic physical data seem to be complete. The statistical weights of upper and lower level are 3 and 1. The inverse transition probability is 11 million years; the rest frequency is 1420·4056 Mc./s.; there are no significant broadening effects besides Doppler effect.

The basic astronomical data are not so well known but must be found from the observational work on the 21-cm. line and from other astronomical sources. These unknowns are:

(a) The density of the atomic hydrogen gas.

3

(*b*) The state of motion of the gas (systematic and random motions).

(*c*) The temperature of the gas.

In practice it is impossible to decide how all these vary from place to place in the Galaxy. Therefore, simplifying assumptions are made, e.g. the assumption of homogeneous temperature and random motions, in order to find the density distribution (van de Hulst, Muller and Oort[1]).

3. TEMPERATURE AND OPTICAL DEPTH

The temperature occurring in the formulae is the temperature that describes the precise population ratio of the upper and lower levels. For this temperature determines how much smaller the effective (classical) absorption coefficient is than the quantum absorption coefficient. The formulae are well-known. Let the index n denote the upper level and the index m the lower level. Then the population ratio

$$N_n/N_m = 3 e^{-h\nu/kT} = 3 \left(1 - \frac{h\nu}{kT}\right). \tag{1}$$

Let B_{mn} and B_{nm} be the transition probabilities for absorption and spontaneous emission. The classical absorption coefficient is proportional to

$$N_m B_{mn} - N_n B_{nm} = N_m B_{mn} \left\{1 - 3\left(1 - \frac{h\nu}{kT}\right)\frac{1}{3}\right\} = N_m B_{mn} \frac{h\nu}{kT}. \tag{2}$$

Precisely, this quantity is $4\pi/hc$ times the classical absorption coefficient per unit length, integrated over the frequencies in the broadened line. Upon multiplication with the incident intensity, averaged over all solid angles, this gives the effective number of *absorption transitions* per unit volume per second.

As a check we may multiply it by the classical intensity of Rayleigh-Jeans at the same T. By Kirchhoff's law the product must be also the number of *emission transitions* per unit volume per second:

$$N_m B_{mn} \frac{h\nu}{kT} \cdot 2\nu^2 c^{-2} kT = 3 N_m \cdot \tfrac{1}{3} B_{mn} \frac{2h\nu^3}{c^2} = N_n A_{nm}. \tag{3}$$

It is important to note that the dependence on T cancels, as it should, for the radiation occurs by spontaneous transition.

The consequence is that *only* absorption measurements of the 21-cm. line can give information about this temperature. Self-absorption (saturation-effects) or absorption of continuous radiation will do, in principle.

4

Emission measurements on masses of gas of small optical depth, as we see in most directions, *cannot* tell us anything about the temperature.

The temperature defined above may be identified for most practical purposes with the kinetic temperature of the gas, which is defined by the Maxwell velocity distribution of the atoms. This follows from the computations of Purcell and Field[2] on the number of exchange collisions, by which radiationless transitions between the two levels occur.

So far we have made only the assumption that a local temperature exists. The determination of the temperature is simple if we make the additional assumption that the temperature is homogeneous, i.e. equal in all volume elements along the line of sight. It requires the knowledge of the saturated intensity in absolute units. The saturated intensity may be found by a rough curve-of-growth computation, if we know the intensity at two places with a very different but known ratio of the optical depths. This ratio may be computed fairly reliably from the differential galactic rotation. The absolute scale follows from a calibration of the antenna temperature, which is a separate problem. A numerical example from the Leiden data is:

highest antenna temperature in survey (at $l = 43°2$, $b = -1°0$):

$$T = 118°,$$

reduced to zero band-width and beam-width:

$$T = 124°,$$

estimated optical depth $\tau = 4\cdot2$, so gas temperature:

$$T_0 = 126°.$$

The DTM observers find a higher value, $T = 150°$. The preliminary Australian data gave even higher peaks in some directions. This may be due to calibration uncertainties or to local temperature differences.

A serious indication for the existence of local temperature differences is that an observed temperature of this order cannot easily be reconciled with the theoretical temperatures for H I regions of Spitzer and Savedoff[3] which are near $50°$. The factor is only 2 or 3 but the heat budget is drawn out of balance by several powers of ten. Kahn[4] has indicated heating by collision of clouds as the obvious extra gain; this comes only occasionally and after it the gas cools down gradually. He also has shown that the measured saturation temperature T_0 is the harmonic mean of the actual temperatures (weighted with numbers of atoms in the

5

line of sight) provided there is a rapid succession of layers with different temperatures within any small optical depth. The latter assumption is astronomically not plausible. It is more plausible that in some directions the clouds nearest the sun are thick enough to impress their temperature strongly on the saturation value. But the directions and frequencies in which we might check this are not too numerous. This is another important reason for stressing the need of accurately comparable calibrations in all regions of the sky.

A modification of the reduction is needed if there is a continuous background to the spectral line. The method to be used depends on the location and properties of the sources of the continuous radiation. Let us assume that the instrument records the intensity difference of the radiation at a frequency inside the line and outside the line.* The formulae (section 7) become simple if we make the assumption that the gas density is homogeneous across the beam. The recorded intensity difference is then unaffected by sources of continuous radiation that are nearer to us than the hydrogen gas. It is $(I_0 - I_c)(1 - e^{-\tau})$, instead of $I_0(1 - e^{-\tau})$, if the sources are beyond the gas. Here I_0 is the saturation intensity of the gas radiation, I_c the continuous intensity of the sources as measured by this antenna and τ the optical depth of the gas. Only if the sources are beyond the gas a modified reduction is needed. As the absorption measurements have already shown that the assumption of homogeneity across the beam is not fulfilled we have to use this modified reduction method with caution. Fortunately, until now the continuous spectrum necessitated only a small correction.

4. SURVEYS

Many observational results reported in this symposium have the form of surveys. They are approximations to the ideal of a scan of the intensity in three dimensions: longitude, latitude and frequency. From it we should like to obtain the hydrogen intensity distribution as a function of four dimensions: longitude, latitude, distance and velocity. The data obviously cannot be sufficient. So a simplifying assumption is made, namely that the velocity distribution is known at each point. It is assumed to be centred on the velocity corresponding to circular motion in the Galaxy and to be spread by a known distribution of random cloud velocities. This assump-

---

* This is not precisely true in Muller's receiver which operates on an A.V.C. (automatic volume control) system but, as the instrumental noise level is high, the measured noise ratio exceeds the value 1 by a small amount. This amount is recorded and is very nearly proportional to the difference that is sought.

tion makes a full reduction possible, at any rate for the small latitudes, $b < 15°$. This reduction method used in the Leiden surveys will be called the *standard method*.

5. THE STANDARD METHOD AND ITS LIMITATIONS

There are blurring effects in all three observational co-ordinates, expressed by beam-width and band-width. Additional blurring is caused by the distribution of cloud velocities in converting from frequencies to distance. In the Leiden survey the distance between half-power points in the blurring function in the Perseus arm at $l = 90°$, $r = 2·8$ kiloparsecs was $90 \times 130 \times 550$ parsecs, the first two being due to the elliptical beam, the last one mainly to the cloud velocities. Such numbers should be kept in mind in judging the extensive graphs of observational results that will be presented by Westerhout and Schmidt (papers 4 to 6). The relative success of the standard method is due to the fact that the observations give a sufficiently coarse picture of the Galaxy. In narrowing down the band-width, and even more so in narrowing down the beam-width, we may detect many more deviations from the assumptions made in the standard method than we do now. It is difficult to imagine what new reduction methods may then be needed.

In the present method the required data have been derived from the following sources.

Galactic rotation curve $\omega(R)$: inner part from observed velocities at tangent points (Kwee, Muller and Westerhout[5]); outer part and points out of plane from a model of the mass distribution (new data have been computed by Schmidt). Random cloud motions: from the observed intensities at frequencies that are forbidden on the assumption of pure circular motion and from the requirement that over-correction should be avoided. Intensity of saturated line: from highest observed intensities and estimates of optical depth at those points (absolute calibration, i.e. conversion to T_0, is not needed at this stage).

There is no guarantee that any deviations of the average local motions from circular motion will have a systematic character over the entire Galaxy. In other words, I expect that the character of such deviations will be local and erratic and that it can be studied only by long and patient observations. But I wish to mention two suggested systematic modifications. Edmondson ([6], see also paper 3) has suggested that the gas spirals out at an angle $\phi = 4°$ with circular motion. Vera Rubin[7] has made a computation based on the assumption that the gas has a motion

7

$V_s = 20$ km./sec. along a spiral arm on top of the circular motion. Such assumptions give rise to a slightly distorted form of the mass distribution obtained by the standard method. On studying the precise plots of the spiral arms as found in the Leiden surveys, however, we are so impressed by the comparative irregularity that it is difficult to assign much weight to Edmondson's point that the kink in the anticentre is drawn straight by his assumption. The centre direction would give a good check, as the circular motion gives strictly zero radial velocity, but the interpretation is confused by the large optical depth. Here observations of the deuterium line might be an important help. Also an accurate comparison of northern and southern observations of the rotational velocities would be helpful to determine the value of ϕ.

I wish to add in this connexion that systematic effects of over 1 km./sec. would also be observed if the nebular red-shift would not be due to recession but would be a cosmological distance effect that worked already inside the Galaxy.

6. HIGH-LATITUDE SURVEYS

The standard method has been successful in the sections of the Milky Way near the galactic plane, $|z| < 500$ parsecs or so, and farther away than 1 or 1·5 kiloparsecs. The study of the nearer regions requires high-latitude observations. Moreover, high latitude observations probably give information mostly on the gas in our immediate neighbourhood. Bok and his associates have concentrated on this subject. One problem they have studied in some detail is the association of neutral hydrogen gas with dark clouds. I shall not try to relate the details, which are reviewed in papers 7 to 10.

Helfer and Tatel ([8], see also paper 11) have detected at $l = 50°$ and $90°$, $b = 20°$ to $40°$, extensions to negative velocities up to 60 km./sec. The fact that similar extensions seem to be common at still higher latitudes, from $40°$ to $90°$, in the measurements made at Kootwijk (see paper 3), suggests that they are not located in the next spiral arm, but are due to a more extended medium embedding the arms.

7. ABSORPTION EFFECTS

Absorption of radiation from a continuous source by the interstellar hydrogen gas at 21 cm. gives information not otherwise available. The main asset is the higher angular resolution, which is in this case determined by the point source, if the observations are correctly interpreted.

The theory is simplest if we first make the *incorrect* assumption that the gas has a constant temperature T_h and, at a given radial velocity, a distribution in the line of sight that is constant over the solid angle Ω_b of the antenna beam. Further, let us suppose that the source is characterized by a solid angle Ω_s, an optical depth τ_s, and an average surface brightness $T_s(\mathrm{I} - e^{-\tau_s})$. Finally, let τ_1 be the optical depth of the hydrogen gas in front of the source and τ_2 the optical depth of the hydrogen behind the source, all at a certain frequency.

The antenna temperatures that might be measured by a direct radiometer are

T_{SH}: source and hydrogen,

T_S: source only, i.e. off-frequency,

T_H: hydrogen only, the so-called expected profile, which is obtained by an interpolation between off-source observations.

From the definitions we have, after a simple reduction:

$$T_H = T_h \left\{ \mathrm{I} - e^{-(\tau_1 + \tau_2)} \right\}$$

$$T_S = \frac{\Omega s}{\Omega b} T_s \left\{ \mathrm{I} - e^{-\tau_s} \right\}$$

$$T_{SH} = T_H + e^{-\tau_1} T_S - T_h \frac{\Omega_s}{\Omega_b} e^{-\tau_1} (\mathrm{I} - e^{-\tau_s}) (\mathrm{I} - e^{-\tau_2}).$$

A comparison radiometer records the difference

$$\Delta T = T_{SH} - T_S,$$

which is positive if the radiation from the solid angle $\Omega_b - \Omega_s$ is preponderant and negative if the absorption of radiation from the source predominates. The positive part may be eliminated by subtracting the 'expected profile', thus leaving

$$\Delta T - T_H = -(\mathrm{I} - e^{-\tau_1}) T_S - T_h \frac{\Omega_s}{\Omega_b} e^{-\tau_1} (\mathrm{I} - e^{-\tau_s}) (\mathrm{I} - e^{-\tau_2}).$$

This difference is always negative and gives a virtually correct idea of the absorption in front of the source, for the second term can usually be neglected, e.g. if no hydrogen is behind the source, or the source is transparent, or small.

Dropping now the assumptions made, we see at once that the last formula should be still correct if there is no homogeneity of the gas across the full beam, simply because the difference is based on the solid angle Ω_s only.

However, another theorem, that followed from the assumptions, no longer holds true, namely that $T_{SH} \geqslant T_h$ whenever $T_S \geqslant T_h$. This theorem

9

may be derived directly from the formulae above. Its physical basis may be explained as follows. Whenever a lot of hydrogen is placed in front of the source in order to repress the source radiation, the radiation by this hydrogen itself approaches the saturation value T_h in the solid angle of the source, so contributes the small amount $T_h \Omega_s / \Omega_b$ to the antenna temperature and gives insufficient compensation. However, the equivalent hydrogen that is supposed to be in the full beam of the antenna gives sufficient compensation, bringing the antenna temperature to at least T_h. It is now clear that in an inhomogeneous gas directions and frequencies may occur in which this compensation is absent, so that the source radiation is almost fully repressed without compensation by the radiation with temperature T_h from the full beam. The observed effect then is that $T_{SH} < T_h$, or $\Delta T < T_h - T_s$. The fact that such cases seem to have been observed by the N.R.L. group gives the first definite proof of the existence of a fine structure in the gas distribution, well below the antenna resolving power.

8. EXTRAGALACTIC STUDIES

The Australian radio astronomers have made a fine study of the Magellanic clouds (Kerr, Hindman, Robinson[9]) thus adding substantially to our knowledge of these companions of our Galaxy. Beyond that, no 21-cm. radiation has been detected from any object, or group of objects. The reason is the large velocity spread giving wider line profiles than in the galactic studies. In Holland, we have been un-equipped, so far, to tackle this problem seriously as the present switching interval is only 1·08 Mc./s. and the band-width is much smaller than may be profitable for such a study.

Perhaps a rough prognosis may be useful. The Andromeda nebula, M 31, assumed distance 500 kiloparsecs, is the first to be tried. The smaller angular size of M 33 is not sufficiently compensated by the smaller velocities in the line of sight in this nearly face-on nebula.

The computation of a line profile for a model nebula may be made by means of the formula

$$nl = 0 \cdot 0006 \int T^* dv$$

where v is the velocity in the line of sight in km./sec., T^* is the brightness temperature, uncorrected for self-absorption, of the nebula at this velocity at a certain point in the projected nebula, l is the length of the line of sight in kiloparsecs through the nebula at that point, and n is the average number of hydrogen atoms/cm.³ along this line. The correction for self-absorption may then be made in the usual manner. If T_h is the hydrogen or saturation temperature, then the actual brightness temperature is

$$T = T_h(1 - e^{-T^*/T_h}).$$

Finally, a weighted average of T over the full antenna beam gives the antenna temperature. A further decrease by a constant factor, due to far side lobes or antenna losses, will be neglected.

Fig. 1 shows a sample profile computed for the *Andromeda nebula* (curve *a*). An antenna beam with diameter 45′ between half-power points and 90′ outer diameter was assumed; it corresponds to the expected performance of a 25-metre telescope. It was centred on a point of the major axis 45′ = 6·6 kiloparsecs south-west from the centre. The radial velocity at this distance from the centre is ± 251 km./sec. relative to the

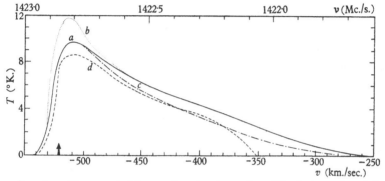

Fig. 1. Sample profile computed for an off-centre observation of the Andromeda nebula with a 25-metre telescope: (*a*) predicted profile; (*b*) without self-absorption; (*c*) without outer portions of nebula; (*d*) without outer portions of antenna beam. For detailed assumptions see text.

centre, which moves with − 270 km./sec. with respect to the sun. In order to avoid confusion with galactic hydrogen it is clearly advisable to work in the south-west portion of the nebula, which approaches us at − 521 km./sec.

The values for nl assumed are 3·2 for $R < 20′$, 1 for $20′ < R < 80′$, 0·5 for $80′ < R < 120′$ and 0·1 for $120′ < R < 160′$. As l may be about

$$4 \times 250 \text{ parsecs} = 1 \text{ kiloparsec,}$$

these values are also the approximate values of n. Circular symmetry was supposed and no reference is made to spiral arms. The circular velocities computed by Schwarzschild[10] were extrapolated to larger values of R on the basis of his model. A mesh of rectangles 5′ × 10′ in the projected nebula was made and the value of $100nl$ properly divided among the 10 km./sec. divisions of the v-scale that occur in this rectangle, thus giving the value of $0·6T^*$ in each rectangle and velocity interval. The correction for self-absorption was made assuming $T_h = 125°$. Strong saturation occurred nowhere, but mild corrections were needed in the fields close to the major axis at velocities corresponding approximately to the full rotational velocity.

It is seen in Fig. 1 that the final profile (curve a) is very asymmetric, has a peak $T_{max} = 10°$ at $v = -506$ km./sec. and a width between half-intensity points of $\Delta v = 115$ km./sec. If no saturation effects had been taken into account, curve b would have resulted with $T_{max} = 12°$. If the density in the outer parts of the nebula, beyond $R = 80' = 11·6$ kiloparsecs, is put zero, the profile changes into curve c, so the profile is not very sensitive to the density distribution in the nebula. The contribution to curve a provided by the inner circle, diameter $50'$, of the antenna pattern is given by curve d. If an antenna could be realized with nearly equal gain over this beam and no wings, curve d multiplied by $1·5$ would result, giving $T_{max} = 13°$, $\Delta v = 100$ km./sec. For a homogeneous beam of diameter $30'$ this would change to $T_{max} = 26°$, $\Delta v = 50$ km./sec., and for a homogeneous beam of diameter $10'$ to $T_{max} = 58°$, $\Delta v = 25$ km./sec.

This sample profile shows that an accuracy of $1°$ K. antenna temperature over a range of 1 Mc./s. will certainly be needed to distinguish profitably between various models of the Andromeda nebula with a 25-meter telescope. A similar accuracy in an even wider frequency range is needed for mere detection with a 7·5-metre telescope.

Another type of extra-galactic object whose 21-cm. line may be within relatively easy reach for a 25-metre telescope is formed by the nearest clusters of galaxies. A prognosis of the *Coma cluster*, situated at a distance of 25,000 kiloparsecs, has been published by Stone[11]. The main uncertainty is in the assumptions that the large mass inferred for this cluster (and other clusters) is due to neutral, inter-galactic hydrogen and that this hydrogen is atomic. Granting these assumptions and assuming a beam-width of $1°$, Stone arrives at a brightness temperature comparable with that of the Small Magellanic cloud ($24°$ K.) over a frequency range of 8 Mc./s. centred on 1389 Mc./s. The values of the frequency and frequency range correspond to the velocities $v = 6680$ km./sec. and $\Delta v = 1700$ km./sec. of the member galaxies. This velocity dispersion happens to be such that the numerical value of T equals that of nl in the formula used above.

As the Coma cluster has a diameter of $4°$ several parts of the cluster may be tested separately, for possible rotation. Similar clusters ten times farther away should still exhibit measurable temperatures, when measured with a 25-metre telescope. However, the equipment has to be specially tuned for any new cluster to be observed.

REFERENCES

[1] van de Hulst, H. C., Muller, C. A. and Oort, J. H. *B.A.N.* **12**, 117, no. 452, 1954.

[2] Purcell, E. M. and Field, G. B. *Ap. J.* **124**, 542, 1956.

[3] Spitzer, L. and Savedoff, M. P. *Ap. J.* **111**, 593, 1950.

[4] Kahn, F. D. *Gas Dynamics of Cosmic Clouds* (H. C. van de Hulst and J. M. Burgers, editors). North-Holland Publ. Co., p. 60, 1955.

[5] Kwee, K. K., Muller C. A. and Westerhout, G. *B.A.N.* **12**, 211, no. 458, 1954.

[6] Edmondson, F. K. *P.A.S.P.* **67**, 10, 1955; *A.J.* **60**, 160, 1955; also this publication papers 2 and 3.

[7] Rubin, Vera C. *A.J.* **60**, 177, 1955.

[8] Helfer, H. L. and Tatel, H. E. *Ap. J.* **121**, 585, 1955.

[9] Kerr, F. J., Hindman J. V. and Robinson, B. J. *Aust. J. Phys.* **7**, 297, 1954.

[10] Schwarzschild, M. *A.J.* **59**, 273, 1954.

[11] Stone, S. N. *P.A.S.P.* **67**, 185, 1955.

Note added in proof.

Both the radiation from the Coma cluster and from the Andromeda nebula have been observed since this paper was written:

Heeschen, D. S. *Ap. J.* **124**, 660, 1956.

van de Hulst, H. C. *Verslagen Kon. Ned. Ak. Wet. Amsterdam,* **65**, 157, 1956; *B.A.N.* (in the Press).

21-CM. OBSERVATIONS IN SYDNEY

MARTHA STAHR CARPENTER

Cornell University, Ithaca, New York, U.S.A.

The primary project in the current 21-cm. hydrogen-line work in Sydney is a study of the southern Milky Way by F. J. Kerr, J. V. Hindman, and the author. Earlier observations* made at the Radiophysics Laboratory during 1954 yielded a tentative picture of spiral structure in the southern portion of the conventional galactic plane. That picture, slightly modified by more recent results, is shown in Fig. 1.

The observational programme now in progress covers the portions of the sky which are indicated in galactic co-ordinates in Fig. 2. Observations are made simultaneously on a number of fixed-frequency channels as the aerial beam sweeps across the sky. A receiver which will have enough channels (~ 50) to delineate the whole line profile simultaneously is under development, but the present receiver has only four channels, each with a band-width of 40 kc./s. A profile is built up by repeating a given run an appropriate number of times, each time displacing the channels to a new set of frequencies. Calibration checks are made at low levels by moving the aerial to the south pole between successive galactic crossings, and at high levels by the observation of standard regions near the beginning, middle, and end of each observing period.

A 'long-run programme', in which an 8° band of galactic latitude is covered at relatively few longitudes, is alternated with a 'short-run programme' which covers a narrower band of latitude at more closely spaced longitudes. The latitude range of approximately 8° was selected in order to permit a study of features of distant spiral arms which are centred well to the south of the conventional galactic plane. Moreover, from observations of the angular widths of spiral features, important information concerning their distances may be expected.

For each track a diagram showing constant-intensity contour lines with respect to galactic latitude as the vertical co-ordinate and radial velocity as the horizontal co-ordinate is being derived. The more familiar line

---

* Reported to U.R.S.I. August 1954.

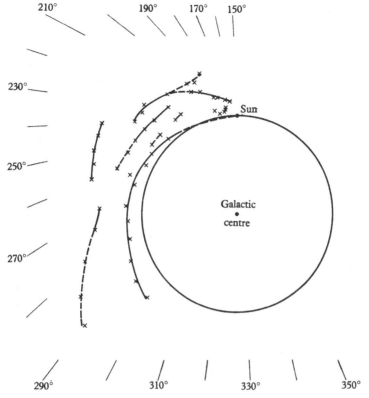

Fig. 1. Provisional diagram of galactic spiral structure based on southern observations and the rotational model of *B.A.N.* no. 452.

Fig. 2. Tracks followed in 21-cm. survey of southern Milky Way. Black: 'short-run' programme; hatched: 'long-run' programme.

profiles correspond to horizontal sections through the contours. This form of diagram is, for our method of observation, a direct presentation of observational data. Since the velocity scale is related to the distance scale, a series of such diagrams at successive longitudes leads to a three-dimensional picture of the hydrogen distribution. At present, however, distances are ambiguous inside the circle joining the sun with other points at the same distance from the galactic centre. They also depend in all regions on the galactic model chosen.

Fig. 3 gives provisional contour diagrams for the tracks which cross the galactic circle at longitudes 260°, 270°, and 275°. The corresponding scales of distance, deduced on the basis of a recent model [1] derived by Oort, are indicated. In each case at least three spiral arm peaks appear, one lying inside, the other two outside, the circle through the sun. The vertical ridge near zero velocity in the diagram for $l = 270°$ coincides with the Coalsack, but further observations at nearby longitudes are needed to confirm this identification.

Consideration of the positions of the maxima with respect to galactic longitude and distance from the galactic centre shows that each of the two outer arms tends to shift outward from the centre with increasing longitude, i.e. in this range of longitude they trail as the Galaxy rotates. All three arms indicated in Fig. 3 show in addition a strong tendency to shift southward in galactic latitude with increasing distance from the galactic centre. The outermost arm is, in fact, barely visible if we consider only the three profiles in the galactic plane. It is of interest that the Magellanic Clouds are located near these longitudes and may therefore be producing some distortion of the structure of our galaxy in this region.

Edmondson has recently suggested [2] that the directions of the mean motions in the Galaxy may be inwardly inclined by about 4° with respect to circular motions. This hypothesis can be tested by making a number of comparisons of observed radial velocities along pairs of lines of sight which lie in the galactic plane and are symmetrically placed with respect to the galactic centre. Comparisons of the *extreme* velocities in two such directions are particularly useful since these velocities are related to the actual velocities of rotation at the innermost points along the two lines of sight. However, any comparisons of regions on opposite sides of the galactic centre are complicated by inherent differences arising from the structural characteristics of the Galaxy and by artificial modifications of the line profiles due to the different aerial beam-widths with which the observations in the northern and southern hemispheres have been obtained.

A sample comparison is shown in Fig. 4, which gives part of a Sydney profile, together with two Leiden profiles for which the signs of the velocities have been reversed. The Sydney profile is for longitude 311°.2, which, on the assumption that the galactic centre is at longitude 327°.8, is symmetrical with the mean of the longitudes at which the two Leiden profiles were obtained. Edmondson's theory predicts that, for $\phi = 4°$, the numerical

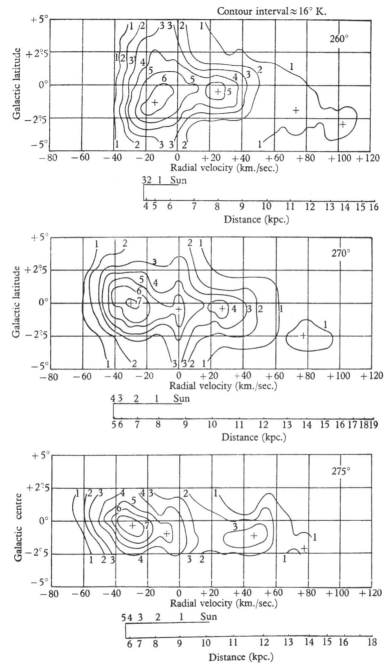

Fig. 3. Sample contours obtained at longitudes 260°, 270° and 275°. The distance scales are based on the model of *B.A.N.* no. 452.

value of the southern 'cut-off' velocity should in this case be 29 km./sec. greater than that of the northern, but no such difference appears. Other comparisons have also failed to show the predicted differences. Conclusive evidence for or against the hypothesis awaits, however, a systematic comparison over a wide range of longitudes. This would amount to a rotational determination of the apparent direction of the galactic centre.

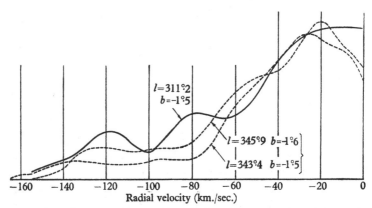

Fig. 4. Comparison of portions of line profiles referring to inner part of the Galaxy. The Sydney profile at $l=311°2$ is compared with the Leiden profiles at $l=345°9$ and $343°4$ plotted on a reverse velocity scale.

The author is indebted to the Commonwealth Scientific and Industrial Research Organization for a research grant in support of her participation in this programme over the past year.

REFERENCES

[1] van de Hulst, H. C., Muller C. A. and Oort, J. H. *B.A.N.* **12**, 117–49, no. 452, 1954.
[2] Edmondson, F. K. *P.A.S.P.* **67**, 10–11, 1955.

DEVIATIONS FROM CIRCULAR MOTION AND THE IMPORTANCE OF SOUTHERN HEMISPHERE 21-CM. OBSERVATIONS

F. K. EDMONDSON

Goethe Link Observatory, Indiana University, Bloomington, Indiana, U.S.A.

(1) The suggestion of systematic deviations from circular motion by an angle ϕ, constant throughout the Galaxy, was first presented (Edmondson, 1955) [1] as an *ad hoc* way to recompute distances so as to make the long circular arm discovered by the Leiden observers (van de Hulst, Muller and Oort, 1954) [2] spiral in. The logarithmic spiral motions postulated in this model are not necessarily along the spiral arms, in contrast to the assumption made by Mrs Rubin (1955) [3].

(2) The value $\phi = 4°$ was based on the discrepancy between the longitude of the centre derived from differential motions and from direct determinations (Oort, 1952) [4], but this is highly uncertain. This value of ϕ implies motion along the arms if the Perseus arm and the Sagittarius arm are portions of the same arm.

(3) It was shown later (Edmondson, 1955) [5] that the apparent kink in the anti-centre direction is straightened out when distances are computed with $\phi = 4°$ in place of the assumption of circular motions. Diagrams illustrating this have been published elsewhere (Edmondson, 1955) [6], and will not be reproduced here.

(4) There is very little difference between the predicted radial velocities toward the galactic centre from the circular orbit and spiral orbit models for distances less than that to the centre. This is illustrated in Figs. 1 *a* and 1 *b* which have been computed using the rotational velocities from *B.A.N.* no. 458. At and beyond the centre the spiral model predicts negative velocities, in contrast to the zero velocities expected if the orbits are circular. McClain's observations (1957) [7] of the galactic centre source (I.A.U. 17S2A) show a small, but well defined, secondary maximum at -40 km./sec. A similar maximum is less clearly shown on the Leiden profile (Kwee, Muller, and Westerhout, 1954) [8] at $l = 328°3$, $b = -1°4$. McClain (1955) [9] has pointed out: (*a*) that this corresponds to the

maxima at -80, -70, and -60 km./sec. shown on the Leiden diagrams for longitudes $321°1$, $323°6$ and $325°8$; and (b) that at $l = 330°8$ it merges with the more intense low velocity part of the profile but still shows up as a slight asymmetry. These figures are consistent with $\phi = 5°$ if we are observing a spiral arm about 3 kiloparsecs beyond the centre.

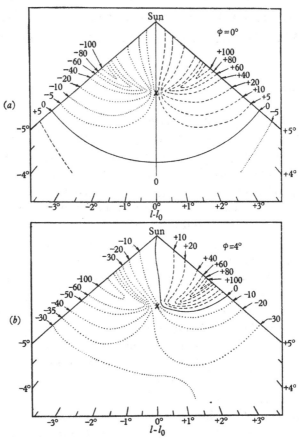

Fig. 1. Contours of equal radial velocity in the galactic plane computed from the rotational velocities in *B.A.N.* no. 458. × marks the position of the galactic centre; (a) circular orbits, $\phi = 0°$; (b) spiral orbits, $\phi = 4°$.

These four points are suggestive, but are not conclusive evidence that the proposed deviation from circular motion exists. Such evidence could be provided by a detailed comparison of values of V_{max}, as defined by the Leiden observers, in the northern and southern hemispheres. If $\phi = 4°$ the southern hemisphere V_{max} should be systematically larger by

$$30 \cos (l - l_0) \text{ km./sec.,}$$

20

disregarding signs. The difference will be reduced at those longitudes where the southern hemisphere V_{max} are observed too small because there is no hydrogen in the line of sight at the distance corresponding to V_{max}. It will be increased where the northern hemisphere velocities are observed too small. Hence, comparison of profiles at a single longitude is not a sufficient basis for any firm conclusions regarding the presence or absence of this effect.

The Australian profile for $l = 311°2$ (Martha S. Carpenter, 1957) [10], while in general appearance similar to the two Leiden profiles ($l = 343°4$ and $345°9$) with signs reversed, has a cut-off velocity slightly larger than the Leiden cut-off velocity for the same value of $| l - l_0 |$, and the difference corresponds to $\phi \sim 1°$. Unfortunately, the longitude chosen for this comparison is too close to the galactic centre to give a reliable determination, even if the previously mentioned effects were absent. It is quite important that many more such comparisons should be made over a wide range of longitude, and V_{max} should be computed from the southern profiles following the method of *B.A.N.* no. 458. If there is a systematic difference, it will give an accurate determination of the amount of deviation from circular motion.

It should also be possible, in principle, to use radial velocities of distant supergiants to check whether the relationship between radial velocity and distance from the sun follows the circular orbit model or the spiral orbit model. Unfortunately, the accuracy with which distance determinations can be made at the present time is not quite sufficient for this problem. The southern hemisphere 21-cm. observations can give a quicker and more certain answer.

REFERENCES

[1] Edmondson, F. K. *P.A.S.P.* **67**, 10–11, 1955.
[2] van de Hulst, H. C., Muller, C. A. and Oort, J. H. *B.A.N.* **12**, 136, no. 452, 1954.
[3] Rubin, V. C. *A.J.* **60**, 177, 1955.
[4] Oort, J. H. *Ap. J.* **116**, 237, Table 1, 1952.
[5] Edmondson, F. K. *A.J.* **60**, 160, 1955.
[6] Edmondson, F. K. *Sky and Telescope*, **14**, 321, 1955.
[7] McClain, E. F. This symposium, p. 80, 1957.
[8] Kwee, K. K., Muller, C. A. and Westerhout, G. *B.A.N.* **12**, 214, no. 458, 1954.
[9] McClain, E. F. Private communication, 1955.
[10] Carpenter, M. S. This symposium, p. 18, 1957.

PROGRESS REPORT ON 21-CM. RESEARCH BY THE NETHERLANDS FOUNDATION FOR RADIO ASTRONOMY AND THE LEIDEN OBSERVATORY

G. WESTERHOUT

University Observatory, Leiden, Netherlands

From November 1953 to August 1955, about 2500 line profiles have been measured with the Kootwijk receiver, under the supervision of Ir C. A. Muller. His receiver is of the well-known frequency-switching type. Two pairs of channels in the second i.f. amplifier, about 5 Mc./s., have bandwidths of 36 kc./s. The components of one pair are 1080 kc./s. apart; the second pair is shifted 500 kc./s. with respect to the first. Each pair of channels, combined with the continuously variable second local oscillator and with the fixed, crystal-controlled pair of first local oscillators, which are also 1080 kc./s. apart, gives a switching system. The power at a certain frequency is compared with that at a 1080 kc./s. higher frequency during half the switching period and with that at a 1080 kc./s. lower frequency during the other half by switching between the two first local oscillators. The signals are fed through an integrating network with a time constant of 54 sec. At the outputs of the two pairs are two separate recorders, each giving one-half (or more) of a line profile (see Fig. 1). A continuous frequency calibration of the second local oscillator provides 10 kc./s. markers on the records, accurate to 1 in 10^6. The limiting sensitivity is $0.7°$ K., with an overall noise figure of 6·3 (8·0 db.). The 7·5 m. Würzburg aerial has a beam-width between half-power points of $1°9 \times 2°7$.

The programme, finished during a nearly continuous one-and-a-half years operation, was mainly concentrated on a study of the large-scale structure of the galactic system. It consisted of the investigation of a strip around the galactic equator, where line profiles were obtained, and special studies based on these profiles were made, as summarized in points (1) to (6) below. Some additional investigations are mentioned in points (7) to (10).

(1) Line profiles from $l = 320°$ to $l = 42°5$, spaced $2°5$ in longitude, at latitude $-1°5$ and corresponding profiles at latitudes of approximately

$-5°5$, $-3°5$, $+0°5$, $+2°5$ and $+4°5$ along the same declination circles. A study of these profiles, combined with the measures mentioned in point (4), is reported in the contribution by M. Schmidt to this Symposium (paper 6).

(2) Line profiles from $l = 45°$ to $l = 110°$, $b = +10°$ to $-10°$, spaced $2°5$ in longitude and $2°5$ in latitude.

(3) Line profiles from $l = 115°$ to $l = 220°$, $b = +10°$ to $-10°$, spaced $5°$ in longitude and $2°5$ in latitude. At selected points the latitude interval was somewhat extended. The results of the study of the profiles in points

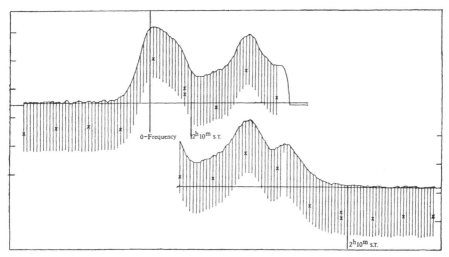

Fig. 1. A double line profile at $l = 67°5$, $b = +2°5$, as recorded simultaneously on two recorders. The vertical lines are 10 kc./s. frequency markers. The temperature of the main top is $64°$K.

(2) and (3) are to be found in the next paper. Fig. 2 shows the line profiles at or near the galactic equator that are contained in programmes 1-3.

(4) About 200 drift-curves were obtained by M. Schmidt, at fixed frequency and position, in the longitude range $l = 340°$ to $35°$. These served to study the latitude distribution in the central regions.

(5) The positive velocity ends of the line profiles in the region $l = 320°$ to $l = 40°$ were studied by K. K. Kwee, C. A. Muller and G. Westerhout[1]. The circular velocity as a function of the distance from the galactic centre was determined with a hitherto unattained accuracy. Long wings of very low intensity, extending to radial velocities of 200 km./sec., were discovered close to the centre and are discussed in the same paper. They justify the hypothesis that the hydrogen in the region $R < 2.5$ kiloparsecs is in a highly turbulent state.

(6) M. Schmidt made a study of the regions where the optical thickness of the neutral hydrogen may be assumed to be very high: (a) the centre, where radial velocity due to galactic rotation is zero and the density is high; and (b) galactic longitudes $l = 40°–45°$, where the line of sight is almost tangent to a circle through the sun, around the centre, and passes through the Orion spiral arm over a considerable distance. A new value of the antenna temperature at infinite optical thickness, T_0, was obtained by means of the formula $T = T_0(1 - e^{-\tau})$. After correction for aerial pattern, a minimum value $T_0 = 125°$K. was found for both regions. Although it is realized that local variations in the temperature are quite likely, a constant $T_0 = 125°$K. was used throughout all the reductions.

It should be noted that the absolute value of this temperature is based on the noise figure quoted. This noise figure in turn was determined by assuming that the noise increase when the aerial is pointed into a nearby pine-wood corresponds to $280°$K. It is felt that this value is probably correct to within 10 %. The intensity scale was checked daily by measuring the main top of the profile at $l = 50°$, $b = 0°$, the height of which corresponds to $100°$K.

(7) A study of the Taurus dark regions was made by Dr F. D. Kahn, to obtain information about the relation between dust and interstellar gas. The preliminary results based on four selected points and published in *B.A.N.* no. 452 [2] indicate that the regions of very strong obscuration

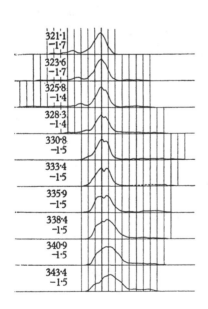
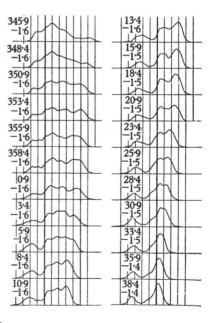

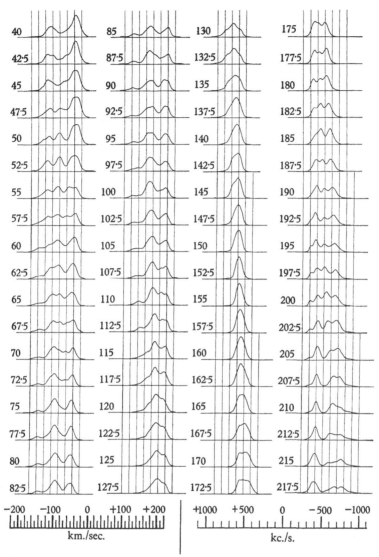

Fig. 2. Line profiles, in the galactic plane. From $l = 321°1$ to $38°4$, the latitudes are given in the figure, from $l = 40°$ to $l = 217°5$, the latitude is $0°$. The vertical lines are at 20 km./sec. intervals.

do not have a striking excess of 21-cm. emission. Most of the gas in the very dense clouds may be molecular.

(8) E. Raimond studied two regions of $10° \times 10°$ each, centred around the ζ Persei association ($l = 129°$, $b = -15°$) and the Lacerta aggregate ($l = 67°$, $b = -14°$), which have been investigated optically by Blaauw[3] and Blaauw and Morgan [4].

25

No conspicuous features were found in the ζ Persei region, although there is a slight indication that the line of maximum intensity bulges out towards the association in a velocity-longitude plot.

In the Lacerta region an outstanding maximum shows up between $l = 62°5$ and $70°$, $b = -12°5$ and $-20°$. It appears both in the line profiles and in a plot of optical depth, corrected for cloud velocities, against velocity and latitude. The relative velocity of the centre of this mass ($l = 67°$, $b = -16°$) with respect to the aggregate is -17 km./sec. At some

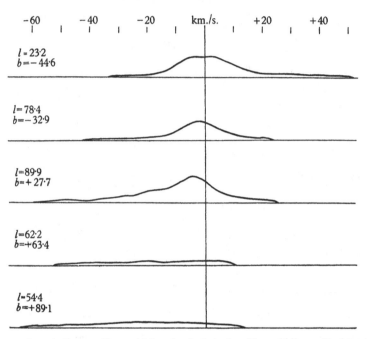

Fig. 3. Sample line profiles at high galactic latitudes. The middle profile is at the celestial North Pole, the bottom profile at the galactic North Pole.

points, there are clearly visible bulges of the lines of equal optical depth, away from the aggregate. If the dimensions of this hydrogen region are taken as $65 \times 65 \times 65$ parsecs, at a distance of 500 parsecs, we find a mean density $\bar{n}_H = 1$ atom/cm.³ and a total mass of the order of 5000 solar masses. Photographs show hardly any indications of dark clouds in this region.

(9) A first general impression of the hydrogen distribution at high latitudes was obtained from thirty line profiles made at latitudes $| b | > 35°$, evenly distributed over the visible part of the sky.

Most of these profiles show long wings of very low intensity (Fig. 3). Twelve have a clearly outspoken top at about the zero frequency, with maximum temperatures between 10°K. and 26°K. The remainder, although usually reaching maximum intensity at the zero frequency, has the appearance of a very broad low-intensity spectrum ($T_{max} \leqslant 9°$K.).

There appears to be a tendency for the long wings to extend farther to the negative velocity side. The mean maximum velocities are -42 km./sec. and $+24$ km./sec. There is no definite correlation with position.

Integration of the eighteen profiles with $T_{max} < 9°$K. shows that the mean total number of hydrogen atoms in a column of 1 cm.^{2} cross-section is $\bar{N}_H = 3 \times 10^{20}$ cm.$^{-2}$. Assuming a length of 1000 parsecs for the line of sight we find an average density $\bar{n}_H = 0.1$ cm.$^{-3}$. Separating the positive and negative velocities, we find that 70 % is moving towards us.

From these first, very preliminary results, it may be concluded that a continuous stream of atomic hydrogen is moving towards the galactic plane. To attain the observed velocities, it must come from considerable heights above the plane. Some of the hydrogen may have reached these heights in high-velocity clouds, which have dispersed on their way up.

(10) From measurements of the maximum intensity of line profiles close to the horizon, L. Woltjer and G. Westerhout made an estimate of the extinction at 21 cm. The transmission coefficient at the zenith was found to be $p = 0.9915 \pm .0008$, corresponding to an extinction of

$$0.037 \text{ db.} \pm 0.003.$$

Expressed in magnitudes this gives an extinction coefficient

$$k = 0.^m 009 \pm .001.$$

This figure is in good agreement with data given by Van Vleck [5].

All profiles reduced since the publication of *B.A.N.* no. 452 (1954) are corrected for the effect of extinction, and for the effect of the continuous radiation from the air and the ground, which influences the automatic volume control system of the receiver.

The work described above has been published in *B.A.N.* no. 475[6].

The observations were made possible through the support of the Netherlands Organization for Pure Research (Z.W.O.)

The 21-cm. equipment of the Netherlands Foundation for Radio Astronomy has recently been moved from the Kootwijk transmitting station to the Radio Astronomical Observatory, Dwingeloo. We are greatly indebted to the Netherlands Post and Telegraph service, who rendered us hospitality for so many years.

The new, azimuthally mounted 25-metre paraboloid reflector at the Dwingeloo Observatory is approaching completion; two additional, equatorially mounted Würzburgs are also being installed.

It is a pleasure for me to express the thanks of all of us in the 21-cm. research group to Ir C. A. Muller, whose fine engineering work—almost wholly a one-man job—has made it possible to obtain the exciting results reported here.

REFERENCES

[1] Kwee, K. K., Muller, C. A. and Westerhout, G. *B.A.N.* **12**, 211, no. 458, 1954.
[2] van de Hulst, H. C., Muller, C. A. and Oort, J. H. *B.A.N.* **12**, 117, no. 452, 1954.
[3] Blaauw, A. *B.A.N.* **11**, 405, no. 433, 1952.
[4] Blaauw, A. and Morgan, W. W. *Ap. J.* **117**, 256, 1953.
[5] van Vleck, J. H. *M.I.T. Radiation Lab. Series*, no. 13 (S. Silver, ed.), p. 641, 1951.
[6] Muller, C. A., Westerhout, G., Schmidt, M., Raimond, E., van de Hulst, H. C. and Ollongren, A. *B.A.N.* **13**, 151, no. 475, 1957.

Discussion

Greenstein: Optical observations by G. Münch, in unpublished investigations, show that most of the discrete clouds in front of stars 1000 parsecs from the galactic plane give systematically negative velocities from interstellar K. This is an unexpected observation, because it had been thought that clouds at great heights above the galactic plane should be in the process of expulsion from the plane by high-temperature expansion processes. The importance of study of the high-velocity and high-velocity-dispersion 21-cm. gas at considerable heights above the galactic plane cannot be overstressed.

A 21-CM. LINE SURVEY OF THE OUTER PARTS OF THE GALAXY

G. WESTERHOUT

University Observatory, Leiden, Netherlands

In a strip 20° wide around the galactic equator, extending from $l = 340°$ to $l = 220°$, line profiles were measured at 620 points. The interval in latitude was $2°.5$, the longitude interval was $2°.5$ from $l = 340°$ to $110°$, and $5°$ from $l = 115°$ to $215°$. At some longitudes the latitude range was extended to $\pm 15°$.

A one-dimensional correction was made for the finite width of the antenna pattern. As the intensity changes slowly along the galactic equator, relative to the half-width of the antenna pattern, only a correction perpendicular to the equator was applied.

The results appeared to be influenced by the fairly large band-width used. Therefore a correction for the blurring effect of the band-width was also made.

The profiles were reduced to optical depth at zero random cloud velocity, following the procedure described in *B.A.N.* no. 452. In the correction for random cloud velocities a dispersion of 6 km./sec. was used throughout, although it was realized that this dispersion may change from point to point. An investigation by Mr Pottasch showed that the dispersion might range from $5\cdot5$–9 km./sec., whilst the velocity distribution function, especially in the regions near the sun, resembles a Gauss curve instead of the exponential function found by Blaauw.

Many of the unpermitted wings show extensions of very low intensity, reaching as far as 25–40 km./sec. from the zero frequency. These may be due to systematic velocities, but may also be part of the velocity distribution function. A form which might better represent the observed distribution is for example $ae^{-v^2/2\sigma^2} + be^{-v/\eta}$ where $b \ll a$ and $\sigma \ll \eta$. It was not found necessary in this stage to investigate this possibility much closer.

An investigation by A. Ollongren and H. C. van de Hulst showed that for a programme of this size the Eddington approximation for removing the blurring effect of the cloud velocities, although not applicable at first sight, is, when slightly modified, a satisfactory and not too laborious reduction method. A Fourier analysis of this correction procedure shows that a velocity dispersion of 6 km./sec. fits the observations best. If the profiles are Gaussian in shape, negative intensities arise inevitably with

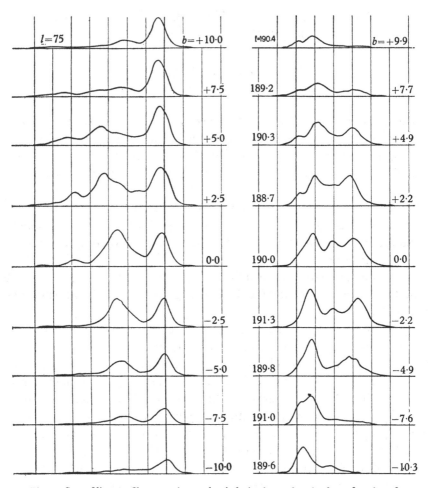

Fig. 1. Sets of line profiles at various galactic latitudes, at longitudes 75° and 190°. The vertical lines have intervals of 20 km./sec.

the present correction method. However, they occur mainly at the forbidden side of the profiles, and have been corrected for by keeping the sum of the ordinates constant. It was found from the reduction of artificial profiles that this should be done by distributing the sum of the negative ordinates evenly over the neighbouring steep slope.

30

On the basis of a new model of the galactic system, recently prepared by M. Schmidt and based on the latest dynamical data available, the frequency scale was transformed into a distance scale and the hydrogen densities were determined as a function of distance from the sun. The velocity–distance relation at latitudes $\neq 0°$ starts to deviate from the one at $b = 0°$ towards higher latitudes. For the region between $l = 135°$ and $160°$, a velocity-distance relation cannot be established.

The hydrogen densities were plotted in planes perpendicular to the galactic plane, at all longitudes for which this was possible. Some examples of the final isodensity contours are given in Figs. 1 and 2.

The instrumental constants may be expressed in terms of distance. In the Perseus arm, for example at $l = 75°$, $r = 3500$ parsecs, the receiver band-width represents a distance of 570 parsecs, while the resolving power of the aerial corresponds to distances of 115 and 165 parsecs, perpendicular to each other and to the line of sight. Details much smaller than 500 parsecs in the line of sight and 130 parsecs in a direction perpendicular to it cannot be detected. This is clearly illustrated by Figs. 1 and 2.

A smaller band-width would show up smaller details in the direction of the line of sight, but the position of the hydrogen clouds causing these details would be indeterminate for three reasons. (*a*) The random velocity would give it an uncertainty of at least 450 parsecs (corresponding to 6 km./sec.). (*b*) Systematic velocity deviations give an uncertainty which may be even greater. (*c*) The aerial resolving power would still set the limits mentioned above.

For a large-scale survey of regions far from the sun, the band-width of 7·5 km./sec. is therefore quite suitable, with the aerial used.

In the present survey, systematic velocity deviations from a purely circular velocity may well distort the shape of the equidensity lines. Nothing definite can be said about this problem so far. It may be that, for example, at $l = 75°$, $r = 3·2$ kiloparsecs, $z = +0·5$ kiloparsec, the trunk sticking out of the main body of the Perseus arm is due to large scale motion of the part of this arm at $r = 3·6$ kiloparsecs, $z = 0·4$ kiloparsec. The same applies to an extension at $l = 80°$, $r = 1·4$ kiloparsecs, $z = 0·2$ kiloparsec.

The hydrogen densities were projected on the galactic plane to give a picture consistent with M. Schmidt's representation. These projected densities are given in Fig. 4.

At several places, it is clear that a systematic distortion of the density contours in the line of sight influences the picture. This is particularly noticeable in the region close to the anti-centre. In this region the velocity–distance relation is very steep, and small deviations from circular motion

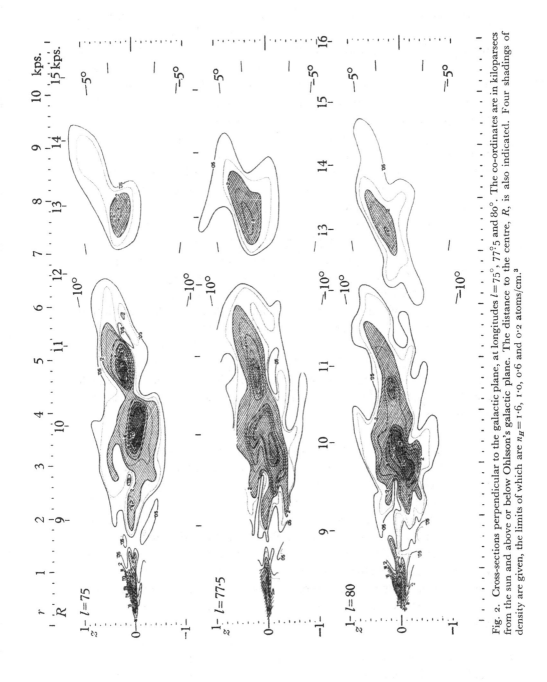

Fig. 2. Cross-sections perpendicular to the galactic plane, at longitudes $l = 75°$, $77°.5$ and $80°$. The co-ordinates are in kiloparsecs from the sun and above or below Ohlsson's galactic plane. The distance to the centre, R, is also indicated. Four shadings of density are given, the limits of which are $n_H = 1·6$, $1·0$, $0·6$ and $0·2$ atoms/cm.^{3}

correspond to large distance variations. Also a slightly larger random velocity than assumed tends to widen the spiral arms considerably. This explanation may also be correct for the directions around $l = 82°5$.

In the future it may be possible to make an estimate of the variation in random cloud velocities, and to find some indication of systematic motions,

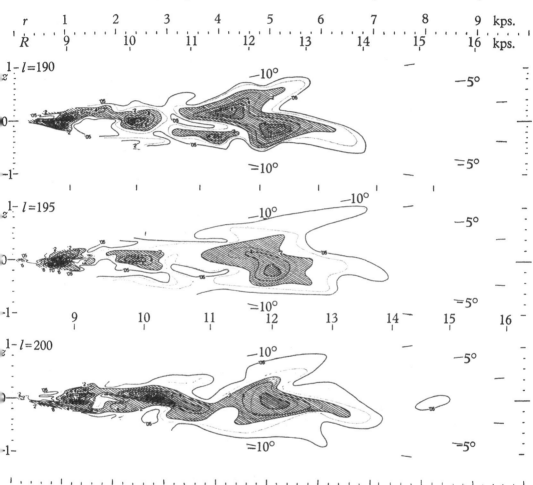

Fig. 3. Cross-sections perpendicular to the galactic plane, at longitudes $l = 190°$, $195°$ and $200°$. For explanation see Fig. 2.

from detailed comparisons of different regions in the galactic system. It should be realized, however, that these deviations usually show up only in investigations like the present one, where large regions may be inter-compared.

A striking feature of most cross-sections is the tilt of the arms. This tilt averages approximately $+1°5$ and is opposite to the tilt found in the central parts. It is due to the wrong position of Ohlsson's plane. From the present measurements, supplemented with measures of the southern hemisphere, a very accurate determination of a new plane may be made. Real deviations from the mean plane may be found at several places, for example around $l=80°$, $r=3$ kiloparsecs, and in the arms at $l=75°$, $r=3·5$ kiloparsecs and at $l=190°$, $r=4$ kiloparsecs.

The representation of the results in the form of a projection on the plane $b=0°$ is not quite adequate, as some arms that partly overlap in the r-coordinate, but are distinctly apart in latitude, tend to blend in this picture.

This contribution may therefore be concluded by a brief description of the features of the outer parts of the galactic system, as seen from a combination of all available observations in a three-dimensional model. All distances are given with respect to the galactic centre, the sun being situated at $R=8·2$ kiloparsecs; the longitudes are counted in the normal manner from the sun, the anti-centre being at $l=147°5$.

(1) *The Orion arm.* At $l=45°$ part of the Orion arm extends inwards to distances $<8·2$ kiloparsecs. A branch of it gradually moves outward, lying at a mean distance of $8·5$ kiloparsecs between $l=52°5$ and $l=65°$. Between $l=65°$ and $l=72°5$ this branch turns sharply outward to $R=8·9$ kiloparsecs and dies out there. The position of the main body of the arm cannot be determined, as its velocity varies around the zero velocity. A $l=80°$ it splits up again, and a very dense part of it runs about 100 parsecs above the plane at $R=8·6$ kiloparsecs. The rest of the arm joins this branch at $l=105°$. From the profiles in the anti-centre region it is clear that the sun is situated at the inner edge of the arm. From $l=170°$ onwards, the mean distance of the arm is between $8·7$ and 9 kiloparsecs.

(2) *The Perseus arm* may be followed from $l=340°$ to $l=220°$. Between $l=340°$ and $l=15°$ it is split up in two parts. The innermost runs at $R=9$ kiloparsecs, 400 parsecs below the plane, the outer at approximately 200 parsecs below the plane to $l=0°$, and from there on moving to about 200 parsecs above the plane at $R=11·5$ kiloparsecs. At $l=15°$ both parts come together at $R=10·5$ kiloparsecs, in the plane. From there onward it lies on the average in the galactic plane. The region between $l=15°$ and $l=50°$ is very complicated. At $l=45°$ the distance is 11 kiloparsecs, to decrease continuously to 10 kiloparsecs at $l=80°$. Between $l=75°$ and $l=82°·5$ the arm splits up and a fairly dense part goes out to $10·8$ kiloparsecs again. From $l=85°$ onwards, the mean distance of the arm is $10·5$ kiloparsecs. The mean height in the entire region is between

100 and 250 parsecs above the plane. At $l = 170°$ the arm may be traced again and is seen to move outward from 11 to 12·8 kiloparsecs at $l = 215°$. It lies in the plane here and tends to a negative height from $l = 210°$ onwards.

(3) *Between the Perseus arm and the Orion arm,* at two places strong intermediate arms spring up. One is at $R = 9·5$ kiloparsecs, running from $l = 45°$ to $l = 70°$. It is a very dense arm in places and does not have any connexion with the Perseus or Orion arms in the region investigated. The connexion at $l = 65°$ on the projection map is spurious. Contrary to the Perseus arm, this arm lies approximately in the plane.

At $l = 175°$ an arm originates at $R = 9·7$ kiloparsecs, of which it is not clear whether it branches off from the Orion arm or from the Perseus arm. It is an arm of the same size as the Orion and Perseus arms and has still to be named. It runs gradually outward to $R = 10·5$ kiloparsecs at $l = 220°$, and lies between 0 and 150 parsecs above the plane.

(4) The arm previously called *outer arm* begins to show at $l = 55°$, $R = 12$ kiloparsecs. It has a considerably smaller density than the other main arms, but is definitely an arm by itself. Partly due to the tilt of Ohlsson's plane, it runs at a mean height of 400 parsecs up to $l = 65°$, the height then gradually decreasing to about 200 parsecs. It can be traced as a separate arm to $l = 107°5$, where the distance is 14 kiloparsecs. A branch splits off at $l = 85°$ and moves outward to more than 15 kiloparsecs at $l = 105°$. Beyond this longitude the densities become too small to give reliable equidensity curves, but hydrogen can be traced farther than 20 kiloparsecs from the centre at $l = 130°$. It is very striking that at the other side of the anti-centre, up to $l = 180°$, no hydrogen is detected beyond $R = 14$ kiloparsecs. A second outer arm originates from the Perseus arm at $l = 12°5$, $R = 12·5$ kiloparsecs, 600 parsecs above the plane, and runs outward to about $R = 14$ kiloparsecs at $l = 35°$. It bends sharply inward again to rejoin the Perseus arm at $l = 45°$. Its mean height, between 500 and 1000 parsecs above the plane, is very peculiar.

A detailed account of this investigation has been published in *B.A.N.* no. 475[1].

REFERENCES

[1] Westerhout, G. *B.A.N.* **13**, 201, no. 475, 1957.

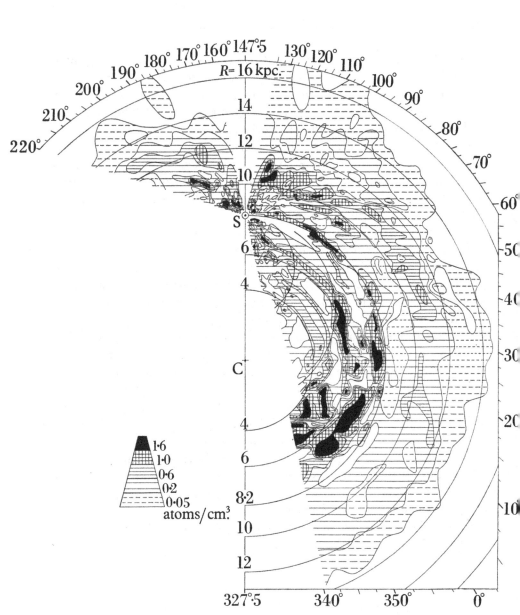

Fig. 4. Maximum hydrogen densities projected on the galactic plane.
This Figure represents the combined results described in papers 5 and 6.

THE DISTRIBUTION OF ATOMIC HYDROGEN IN THE INNER PARTS OF THE GALAXY

M. SCHMIDT

University Observatory, Leiden, Netherlands

The determination of the distribution of hydrogen from 21-cm. observations in parts of the Galaxy, which are nearer to the centre than the sun, is seriously handicapped by the fact that the observed radial velocity of the hydrogen clouds determines only the distance to the galactic centre. So two possible values of the distance to the sun correspond to one value of the frequency. We have used as a criterion to separate the contributions from the two regions the latitude distribution of the radiation.

About 200 drift curves, with paraboloid and frequency fixed, were obtained with the Kootwijk receiver at frequency intervals of about 40 kc./s. at longitudes $l = 340°$, $345°$, ..., $35°$. Further, line profiles are available in this region at $2\frac{1}{2}°$ longitude intervals from a survey supervised by C. A. Muller and G. Westerhout. The profiles at latitudes $+\frac{1}{2}°$, $-1\frac{1}{2}°$, $-3\frac{1}{2}°$ were used in the present investigation. Finally, some preliminary measurements of the continuous radiation at about 21-cm. wave-length were used. Ohlsson's pole at $12^h\ 40^m$, $+28°$ was used in all reductions.

All the drift curves and line profiles were corrected for the effects of antenna pattern and band-width of the receiver.

To explain the remaining part of the reductions we shall first assume that the effects of the continuous radiation and the dispersion of cloud velocities may be neglected. We may then convert all the intensities into optical depths. For the separation of the contributions from every two corresponding points on the line of sight we need the linear distribution of hydrogen perpendicular to the galactic plane. This may be obtained from frequencies for which the two contributing points are close together or coincident, i.e. from frequencies near the maximum frequency. The distance for these frequencies is known and the latitude distribution may be converted to a linear distribution vertical to the galactic plane. This may be done for all longitudes so that the vertical distribution can be studied over a range of distances from the galactic centre. Now the separations may be carried out. For each longitude and frequency the optical depth is read at four or more different latitudes. The longitude and frequency determines, with the known curve of rotational velocities, the distances of the contributing points.

Therefore the vertical linear distribution, determined before, may be converted to a latitude distribution at each of the two contributing points. The latitudes of the centres of the distributions as well as the corresponding maximum optical depths are unknown. Each of the optical depths read at four different latitudes may now be expressed as a sum of two contributions, each of which involves the known latitude distribution, an unknown mean latitude and an unknown maximum optical depth. The four unknowns may be solved from these equations. The separation has then been carried out.

However, as mentioned before, the effects of the continuous radiation and the peculiar cloud velocities has been neglected here. The effect of the continuous radiation is difficult to deal with, since it depends on the yet unknown hydrogen densities. The difficulty was solved by a preliminary separation, carried out in intensities instead of in optical depths. The four equations then include the intensity of the continuous radiation originating from behind the hydrogen region considered. These intensities were obtained from a model of the space distribution of continuous galactic radiation. The solution of the four equations mentioned before, then yielded a preliminary model of the distribution of hydrogen. This model was used only to give the conversion of measured intensities into optical depths.

A further difficulty is formed by the peculiar cloud velocities. The study of the vertical distribution indicates that the thickness of the gas layer between the points of half the maximum density is about 220 parsecs throughout the inner parts of the Galaxy. Now the general mass density and, therefore, the vertical force will increase towards the centre. This implies an increase in peculiar velocities which undoubtedly will not be restricted to the vertical components of the peculiar velocities. From an unpublished model of the distribution of mass in the Galaxy the rate of increase of the dispersion of cloud velocities was estimated. The values of the average cloud velocities used are given in Table 1. All measurements were then corrected for the effect of peculiar cloud velocities.

The final reduction proceeds as in the simplified case described above. The resulting relative distribution of hydrogen vertical to the galactic plane is given in Fig. 1. It does not seem to depend on the distance from the galactic centre. The distance between half-density points is 220 parsecs.

The mean latitudes resulting from the separations may be converted into vertical distances or heights z. These values were smoothed in rings of constant distance to the galactic centre. The smoothed values $\bar{z}$ are shown in Fig. 2. The separations were now carried out again with the smoothed z values, and the optical depths re-determined. These were reduced to hydrogen densities, which are plotted in Fig. 4 of paper 5. It is seen at

Table 1. *The assumed run of the average peculiar cloud velocity η with distance R from the galactic centre*

R	2	3	4	5	6	7	8·2	kiloparsecs
η	14·8	11·6	9·5	8·1	6·9	6·0	5·0	km./sec.

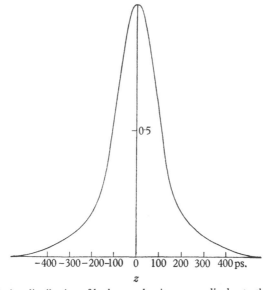

Fig. 1. The relative distribution of hydrogen density perpendicular to the galactic plane.

once that the hydrogen is concentrated in arms, as it is in the outer parts of the Galaxy. The Orion arm appears at $l=40°$ at 4·5 kiloparsecs distance and may be followed to the other side of the Galaxy, except for a break near $l=30°$. The next arm is the Sagittarius arm. It runs near the sun about 1·5 kiloparsecs inside the sun's circle and there contains the O-associations of Morgan, Code and Whitford (1953) [1]. It may also be followed to the other side of the Galaxy. There the structures of both this arm and the Orion arm seem to become very complicated, however. Further arms or parts thereof may be noted; they are tangential to the line of sight at $l=6°$, $l=359°$ and perhaps at $l=346°$.

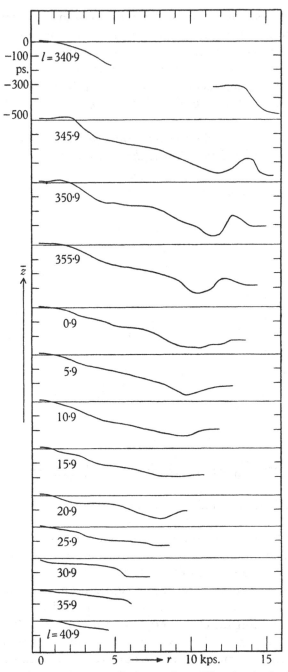

Fig. 2. The mean heights $\bar{z}$ of hydrogen above Ohlsson's galactic plane as function of distance r for different longitudes.

The Sagittarius arm, the arm at $l = 6°$ and the possible arm at $l = 346°$ fit well with the secondary maxima in the rotational velocities found by Kwee, Muller and Westerhout (1954) [2].

Both the Orion arm and the Sagittarius arm are seen to be inclined in the sense that they are trailing.

A detailed account of this investigation has been published in *B.A.N.* no. 475 [3].

REFERENCES

[1] Morgan, W. W., Code, A. D. and Whitford, A. E. *Ap. J.* **118**, 318, 1953.
[2] Kwee, K. K., Muller, C. A. and Westerhout, G. *B.A.N.* **12**, 211, no. 458, 1954.
[3] Schmidt, M. *B.A.N.* **13**, 247, no. 475, 1957.

Discussion

De Vaucouleurs: The *z*-component of the velocity dispersion among H II-regions in the Large Magellanic Cloud appears to be of the order of 5 km./sec. (Cf. *Publ. Ast. Soc. Pacific*, **67**, 397, 1955.)

PROGRESS REPORT ON THE PROJECT IN RADIO ASTRONOMY AT THE G. R. AGASSIZ STATION OF HARVARD OBSERVATORY

B. J. BOK

Harvard College Observatory, Cambridge, Mass., U.S.A.

Two years ago I reported at the Jodrell Bank symposium on the initiation of the Agassiz Station project in Radio Astronomy. At that time the 24-ft. antenna was under construction and the electronic equipment for observation of the 21-cm. line of neutral hydrogen was being built by Harold I. Ewen, who is the co-director of the project. The first successful observations were obtained in the autumn of 1953 and reported early in 1954 (see Bok[1]). Since then two papers have been published giving the results of the first studies by Heeschen[2] and Lilley[3]; a report on the equipment and the basic programme was given by Bok and Ewen[4]. Earlier this year, at the Princeton Meeting of the American Astronomical Society, Bok[5] drew heavily upon the Agassiz Station results in a paper entitled 'Gas and Dust in Interstellar Clouds', in which an attempt was made to blend the results of radio and optical research. We note also at this point a joint paper by Heeschen and Lilley[6] in which attention was drawn to the important role of Gould's Belt in the distribution of neutral hydrogen in the vicinity of the sun.

I. INSTRUMENTS*

For the present our activity is limited entirely to research on the 21-cm. line of neutral hydrogen. The receiver built by Ewen has a band-width of 15 kc./s., which at 1420 Mc./s. is equivalent to a resolution in radial velocity of the order of 3 km./sec. The equipment that is presently under construction will have a band-width of 5 kc./s., with an equivalent resolution of 1 km./sec. in radial velocity. It will have twenty channels, the output of which is recorded by an electrically operated typewriter. It will be

---

* As no space could be allowed within this symposium publication for description of telescopes and electronic equipment, this section has been greatly condensed.

used mostly with the 60-ft. reflector now being built for us by D. S. Kennedy and Company of Cohasset, Massachusetts. This reflector is constructed of welded aluminium tubing covered with extended aluminium mesh measuring 0·95 cm. on a side. The focal ratio is 0·35 and the equatorial mounting will give full hemispherical sky coverage. The expected accuracy of positioning and following is 1'.

2. OUTLINE OF THE AGASSIZ STATION PROGRAMMES

From the start, it has been our aim to take the best advantage of the high resolution in frequency of our equipment as is demonstrated by the published papers of Heeschen [2] and Lilley [3] and by the three Harvard papers to follow. The emphasis in our work to date has been primarily on regional surveys as detailed as can be undertaken with our equipment, and on parallel radio and optical research. It is becoming more and more apparent that radio and optical astronomy are not separate or loosely related fields, but rather that the techniques of radio and optical research neatly supplement each other in the study of the physical universe, and notably so for the study of the distribution of hydrogen, neutral and ionized, in our Milky Way system. Heeschen has followed this approach in his study of the 21-cm. profiles for fields in the section of Sagittarius and Ophiuchus, and Lilley similarly in his study of the anti-centre section, which established clearly that neutral hydrogen is present in increased amounts in the large complexes of cosmic dust. A subsequent paper by Bok, Lawrence and Menon [7] presented further evidence on the relation between neutral hydrogen and cosmic dust. The 21-cm. observations do not indicate a further increase in the amount of neutral atomic hydrogen in the densest dark spots in Taurus and Ophiuchus when the relevant signal strengths are compared with the strengths observed for positions that are less markedly affected by cosmic dust, though still inside the dust complex.

The three studies on which we report in the following papers (8 to 10) are concerned with various aspects of research in the fine structure of the distribution of neutral hydrogen in the Milky Way system. In Matthews' paper, which is a continuation of his paper presented at the Berkeley Meeting of the American Association for the Advancement of Science [8], an attempt is made to unravel some of the details of spiral structure for the section between galactic longitudes $l = 60°$ and $130°$. In his research special emphasis is placed on those directions for which the relevant optical data are established with highest precision. In Menon's research,

43

the primary aim is to study the distribution of neutral hydrogen for the Orion nebula and its surroundings, in the hope of establishing a model hydrogen nebula that will explain the appearance of the 21-cm. profiles over this area of the sky. In Lawrence's paper, as in fact in Matthews' and Menon's papers, a preliminary investigation is made of the relation between the clouds that produce the optical interstellar absorption lines (clouds that may have very small angular dimensions) and the neutral hydrogen clouds that are observed by 21-cm. technique (clouds which must have considerable angular dimensions to become detectable with present-day equipment).

Much of the current observational programme at the Agassiz Station deals with the section of the Milky Way between $l = 330°$ and $60°$.

David S. Heeschen and Campbell M. Wade are studying the 21-cm. profiles for intermediate latitudes $+10°$, $+15°$ and $+20°$ for positions $5°$ apart in galactic longitude between $l = 335°$ and $50°$. The primary aim of this research, which is well on the way toward completion, is to study the variation with galactic longitude of the widths, mean radial velocities and total intensities of the profiles, and also to investigate the presence of extended wings for some of the profiles. The variations in the 21-cm. parameters are being correlated with the known optical features of the section, particularly with variations in the interstellar absorption.

A related programme calls for profiles at low galactic latitudes between $l = 335°$ and $l = 35°$; this programme is by David S. Heeschen and William E. Howard III. The primary purpose of this programme is to provide data for a preliminary, rather exploratory, study of the structure and motions of the inner regions of the Galaxy. In the first phase of this study, Heeschen and Howard are concentrating on the section between $l = 335°$ and $l = 350°$. This region contains a number of known H II regions and many stars with observed interstellar absorption lines. A preliminary analysis of our profiles corroborates the existence of a spiral arm interior to the Sagittarius arm, as found by Kwee, Muller and Westerhout [9].

To round out the research for this section of the Milky Way, May K. Kassim is embarking on a programme for the section between $l = 35°$ and $60°$, $b = -10°$ and $+10°$, at positions spaced $2°.5$ apart. This research has the dual purpose of investigating the fine structure of the spiral features (curvature of arm, distribution of H I across the arm, expansion effects produced by OB aggregates) and of studying effects associated with the obscuration in the Great Rift.

The initial results obtained by Lawrence (paper 10) have encouraged Heeschen to initiate, jointly with Frank D. Drake, a programme for the study of possible correlation between 21-cm. profiles and the interstellar lines observed by Adams [10]. Twenty-three centres, all but one of them outside the zone $b = -10°$ to $+10°$, have been selected in which there are two or more stars on Adams' list within the area covered by the antenna beam of our 24-ft. reflector. The average radial velocities and total intensities of the 21-cm. profiles will be compared with the corresponding quantities for the interstellar absorption lines. The observations to date have indicated several regions of special interest. One of these is the region of the Lacerta aggregate, which has been mentioned by Westerhout (see paper 4). A second region, that of the Pleiades cluster, is now being investigated by Heeschen and Drake. They find that the 21-cm. profile here is double-peaked; one peak has the radial velocity of the Pleiades cluster, while the radial velocity of the second peak corresponds to the mean radial velocity of the interstellar lines observed in this region.

Papers 8 (Matthews) and 9 (Menon) describe at length the programme for the section $l = 60°$ to $130°$ and the Orion section. To supplement Matthews' programme, Robert J. Davis is observing positions separated by 2° in galactic latitude for $b = -10°$ to $+10°$ at $l = 120°$ and also a strip at $b = +3°$ with positions between $l = 110°$ and $l = 130°$ separated by 2° in galactic latitude. At the request of George H. Herbig of Lick Observatory, Menon is studying the profiles for NGC 2244, NGC 2264 and surroundings.

Two features of the Agassiz Station programmes remain to be stressed. The first is that, with the completion of the above programmes, we shall have available high-resolution 21-cm. profiles for most of this galactic belt between $l = 335°$ and $l = 180°$. It is our intention to publish in the not-too-distant future the derived mean profiles for all positions and thus make our material widely available for further analysis. The second feature of our present programme is that we consider it exploratory to the extent that we intend to observe all regions of special interest again and intensively with the 60-ft. reflector and new electronic equipment now under construction.

Our present programme is limited to some extent by the nature of our electronic equipment. Our strength seems to lie in the very high frequency resolution provided by Ewen's equipment, which is in essence a very fine comparison radiometer. At the moment we are not equipped for studies of the continuous background, or for relatively broad-band research on galaxies or inter-galactic matter. We hope to enter research in these and related fields in the near future.

45

3. ACKNOWLEDGMENTS

A project in radio astronomy, and notably in the 21-cm. field, is expensive. The Agassiz Station project could not have been initiated without substantial support from a friend of the G. R. Agassiz Station and from the National Science Foundation. Continuing support from these same two sources, as well as through funds toward basic maintenance and operation provided by the Harvard Observatory Council, is proving to be essential for the operation of the programmes outlined above. The National Science Foundation has provided most of the funds for the construction of the 60-ft. reflector and a special gift from a friend accounts for the remainder. Two special gifts and a grant from the Research Corporation have given us the necessary funds for the construction of the new electronic equipment.

A successful project in 21-cm. research requires more than funds alone. The enthusiastic encouragement which we have received from the organizations and individuals that have given us financial support and from the Observatory's Visiting Committee has been a constant source of inspiration. We have received much assistance already from some of the large electronic industries and without prospects of continued help of this nature we could hardly hope to remain successful. Our present group of reports provides in itself adequate proof that our relations with D. S. Kennedy and Company have been far more intimate than the average one between purchaser and manufacturer.

In concluding this introductory statement, it does not seem out of order for me to express my appreciation to my immediate associates in the Agassiz Station project. To Harold I. Ewen, who as co-director of the project remains a valued friend and advisor, goes the credit for the excellence of our equipment. To John A. Campbell, the project-engineer, my associates and students join me in expressing our heart-felt thanks for his continued and unceasing effort to keep the equipment in tip-top running condition. I cannot mention here all my students, but it would be unfair to close without a special word of thanks to A. Edward Lilley, now at the Naval Research Laboratory, and to David S. Heeschen, now back with us as the resident Agassiz Station Radio Astronomer; Heeschen and Lilley saw the project through its initial difficult period and we all realize our indebtedness to them for their efforts on its behalf.

REFERENCES

[1] Bok, B. J. *J. Geophys. Res.* **49**, 192, 1954.
[2] Heeschen, D. S. *Ap. J.* **121**, 569, 1955.
[3] Lilley, A. E. *Ap. J.* **121**, 559, 1955.
[4] Bok, B. J. and Ewen, H. I. *A.J.* **59**, 318, 1954.
[5] Bok, B. J. *A.J.* **60**, 146, 1955.
[6] Heeschen, D. S. and Lilley, A. E. *Proc. Nat. Acad. Sci.* **40**, 1095, 1954.
[7] Bok, B. J., Lawrence, R. S. and Menon, T. K. *P.A.S.P.* **67**, 108, 1955.
[8] Matthews, T. A. *P.A.S.P.* **67**, 112, 1955.
[9] Kwee, K. K., Muller, C. A. and Westerhout, G. *B.A.N.* **12**, 211, no. 458, 1954.
[10] Adams, W. S. *Ap. J.* **109**, 354, 1949.

47

REPORT ON 21-CM. OBSERVATIONS
BETWEEN $l=60°$ AND $l=135°$

T. A. MATTHEWS

Harvard College Observatory, Cambridge, Mass., U.S.A.

Observations of the 21-cm. hydrogen profiles have been taken between galactic longitudes $l=60°$ and $l=135°$, with the 24-ft. radio telescope at the George R. Agassiz Station. The beam-width of the antenna is approximately $1°7$ between the half-power points. The electronic equipment used was a D.C. comparison radiometer, which has a signal channel with a frequency band-width of 15 kc./s. between the half-power points. At galactic latitude $b=0°$ the observations are generally spaced $2°5$ apart, although a few gaps of $5°$ exist. At $b=+15°$ and $b=-15°$ the centres are $5°$ apart. A strip in latitude at $l=100°$, and other centres at various latitudes (mainly at $l=75°$ and $l=87°5$) have also been taken, but little reference will be made to this material in the present paper.

I. SPIRAL STRUCTURE IN THE GALACTIC PLANE

One of the most striking features on all the scans is the abundance of small details. Each large maximum (a spiral arm) has superimposed on it several of these. See, for example, the reduced profiles for the regions shown in Figs. 2, 3 and 5. Unfortunately, many of these details are near the limit of detectability of the equipment so that little can be said about them. Their appearance, however, agrees with observations of the optical interstellar lines, which are multiple for stars in both the Orion and Perseus arms.

A preliminary report giving the general features of spiral structure in the galactic plane has been published (Matthews [1]). Fig. 1 has been taken from that article. The velocities of the larger details in the galactic plane have been plotted against galactic longitude, and lines have been drawn to indicate how the details might run. These lines suggest that the details are continuous from one region to the next. This is probably true for the Orion arm, but for the Perseus and Distant arms the lines indicate, rather, the general trends of the various large details. There is a satisfying agreement between this and a similar plot made by Westerhout [2]. The few differences will probably be resolved when the reductions now in progress have been completed. The plot in Fig. 1 is still preliminary, and a final result will have to await complete reduction of all the observations, in a manner which will be suggested below. The analysis now in progress gives evidence that the final plot will be more detailed, and that many of the features do, indeed, occur over large intervals of longitude, even for the more distant arms.

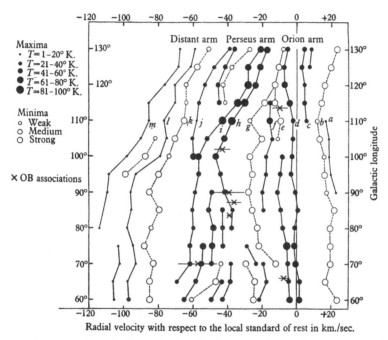

Fig. 1. 21-cm. observations of hydrogen in the galactic plane. The velocities of the larger details in the galactic plane, both maxima and minima, are plotted as a function of longitude. Lines have been drawn to suggest how the features change with longitude. The velocities of seven OB associations, deduced from the observed radial velocities, are also plotted together with their probable errors.

Little can be said now from these observations about the tilt of the spiral arms, or about their structure away from the plane, although both subjects are of great interest, as has been shown by Helfer and Tatel [3] and by Westerhout [4]. The observations for low galactic latitudes at $l = 100°$ are being reduced now and will be ready for discussion shortly. Further observations have been taken this summer at the Agassiz Station by Mr R. J. Davis to investigate these problems between $l = 110°$ and $l = 130°$.

Two features of Fig. 1 are of particular interest. The first is a branch of the Orion arm, which has definitely separated from the main arm to more negative velocities at $l = 100°$, and can be traced to smaller longitudes. This branching has already been noted by Oort, van de Hulst and Muller[5] and corresponds closely to the distribution of OB aggregates as given by Morgan, Whitford and Code [6]. The branch does not stop at $l = 115°$ as the diagram suggests. |Observations above the plane show that it continues there to at least $l = 130°$. The line of maxima which runs from $l = 60°$ to $l = 100°$ at velocities between -4 km./sec. and -12 km./sec. seems to be due to the pronounced maxima at these velocities, which occurs at latitudes near $b = +5°$ in this longitude interval according to Helfer and Tatel[3].

The second interesting feature is the double line of maxima that either branches off from the negative velocity side of the Perseus arm as suggested by Fig. 1, or arises near the arm as suggested by Westerhout[2]. It first becomes distinct near $l = 110°$. If the observed velocities of this arm, or branch of an arm, are interpreted as being due to distance, then we find that it is inclined at an angle of about 25° to the radius vector from the galactic centre. This assumes that the hydrogen has no systematic velocity, and explains the observed velocities through galactic rotation. Its slope will be greatly modified, however, if the hydrogen possesses any peculiar radial motion with respect to the local standard of rest, since the effects of galactic rotation are rapidly decreasing to zero for longitudes greater than 110°. It is interesting to note that the velocities of this line of maxima are, for successive longitudes, closely parallel to those of the Perseus arm, but shifted by about -20 km./sec. If we may assume that the hydrogen giving rise to this feature has the large peculiar velocity of -20 km./sec., then it is at the same distance as the Perseus arm over the interval from $l = 115°$ to $l = 135°$, and possibly to even greater longitudes according to Westerhout's plot (Westerhout[2]). If the latter assumption is correct, we have the case of a large mass of hydrogen moving with a considerable peculiar velocity, whereas the only other direct evidence for large peculiar motions pertains to small masses of hydrogen having positive velocities. Since velocities if due to galactic rotation are negative in this longitude interval, all points in Fig. 1 with positive velocities must be due to hydrogen having a systematic velocity of recession. The large positive velocities shown in the diagram refer to very small amounts of hydrogen, while the small positive velocities are associated with larger masses of hydrogen. This is in general agreement with the results for optical interstellar lines (Whipple[7]; Blaauw [8]).

2. FINE STRUCTURE OF THE 21-CM. PROFILES

All the analysis of these observations that has been carried out shows that the profile of a spiral arm is usually made up of more than one significant part. Many of the scans show several very obvious maxima; for example, the profile for the Orion arm at $l = 114°$, $b = +4°$ (Fig. 3) shows two well-separated, approximately equal, maxima. The same profile also shows several definite, but less pronounced, details: one between the Perseus and Orion arms, and several more at higher negative velocities. These details

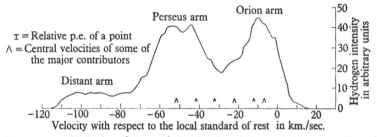

Fig. 2. The mean profile for $l = 97°$, $b = -1°$ (LF 5). The relative probable error indicated is the probable error of one point relative to another point 3 km./sec. away. The inverted V's indicate the central velocities of the Gaussian curves used to analyze the observed profile.

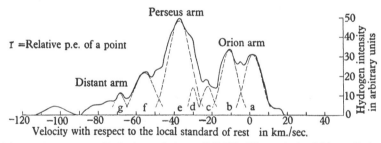

Fig. 3. The mean profile for $l = 114°$, $b = +4°$ (LF 6). The analysis of this profile has been made according to the method described in Section 2.

repeat on profiles taken several months apart, and similar details are found on scans for centres five or more degrees away. The velocities of the details may differ by two or three km./sec. in two regions that are close together, and the intensities may be somewhat different in the two regions: but the similarity in appearance of the details, together with the small change in velocity, often make their common origin probable. Details such as these suggest that the frequency band-width used (15 kc./s. between the half-power points, which is equivalent to 3·2 km./sec.) is small enough to show real details of structure in the hydrogen. The fine details are by no means fully revealed on profiles taken with a 15 kc./s. band-width; this is clearly demonstrated by Fig. 4.

4-2

The 5 kc./s. band-width profile is taken from the only pair of scans obtained with the present Harvard equipment using the narrow band-width. Additional observations should be taken to check the detailed features of the profile. This will be possible with the new equipment under construction.

To assess the reality of the individual features in the 5 kc./s. band-width profile of Fig. 4, the average probable error of the difference in intensity between points a band-width apart was computed from the differences between the two scans that form the mean profile. The average probable error for a point computed in this manner is $\pm 0\cdot53$ units for the 5 kc./s.

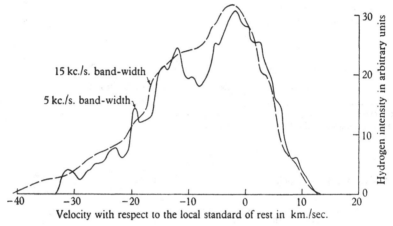

Fig. 4. Mean profiles for $l = 100°$, $b = +15°$. The dashed-line profile is the mean of two pairs of scans taken with a 15 kc./s. receiver band-width at the half-power points. The solid-line profile is the mean of one pair of scans taken with a 5 kc./s. band-width. There is an uncertainty of about 5 % in the factors used to reduce the pairs of scans to a common intensity scale.

band-width profile, and $\pm 0\cdot52$ units for the 15 kc./s. band-width profile. The low value for the 5 kc./s. band-width profile was secured mainly because its scans were taken with an integration time four times that used for the 15 kc./s. band-width scans.

The reality of each of the individual features on the 5 kc./s. band-width profile was evaluated from a consideration of the mean profile and the probable error of a point, and also from the agreement between the original scans in each feature. The evaluation shows: that the features at $V_{LSR} = +2\cdot5$, $-2\cdot0$, $-12\cdot0$, $-14\cdot5$, $-19\cdot5$, $-23\cdot0$ are definitely real; that those at $V_{LSR} = +8\cdot5$, $+6\cdot0$, $+0\cdot5$, $-9\cdot5$, $-31\cdot0$ are probably real; and that those at $V_{LSR} = +4\cdot5$, $-5\cdot5$, $-27\cdot0$ are of doubtful reality.

Yet another piece of corroborative evidence for the fine structure is given by the observations of optical interstellar lines. Dr G. Münch has very kindly given the author the results of his observations on interstellar sodium and calcium lines for fourteen stars between $l = 90°$ and $l = 106°$. Adams[9] has observed one star in this region. All these observations are from stellar spectra having dispersions greater than 5 Å/mm. Sixteen of the lines which are present in the spectra of these stars are due to clouds in the Orion arm. Their velocities with respect to the local standard of rest are mostly between -2.5 km./sec. and -6.5 km./sec., with a few velocities outside this range. The computed standard deviation for all sixteen lines is 3.6 km./sec.; the value of σ is strongly influenced by one discordant velocity.

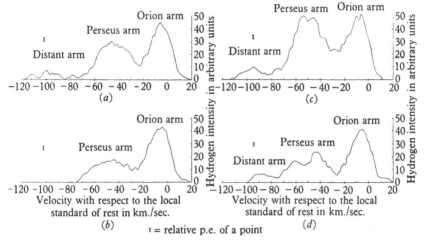

Fig. 5. (a) The mean profile for $l = 102°6$, $b = -2°7$. (b) The mean profile for $l = 102°6$, $b = -3°8$. These two profiles include the region of the h and χ Persei clusters. (c) The mean profile for $l = 100°0$, $b = -2°1$. (d) The mean profile for $l = 99°9$, $b = -4°2$.

If the broadest reasonable gaussian curve is used to represent the Orion arm both at $l = 97°0$, $b = -1°3$ (LF 5, Fig. 2) and at $l = 102°6$, $b = -2°7$ (Fig. 5a), which are representative curves for the longitude interval between 95° and 106°, we find that the values of σ are 7·1 km./sec. and 7·0 km./sec. respectively. One might expect that the values of σ from the optical and radio data would be approximately equal. If we assume that at least one of the main contributors to the 21-cm. profile has a dispersion of about 3 km./sec., then the analysis of the Orion arm profiles for these two regions is greatly altered. The small, but not insignificant, contributors which were needed when a $\sigma = 7$ km./sec. was used become much more prominent. The Orion arm is now composed of three roughly equal components.

The velocities of the optical interstellar lines in the spectra of stars included in the antenna beam pattern generally agree quite well with the velocities of the maximum of the Orion arm profile, or with some of the more prominent fine details. This is particularly true for the clouds in the Orion arm since both types of observations refer to the integrated effect of the whole arm.

Dr Münch[10] has pointed out that the optical interstellar lines for the Perseus arm, which are well separated from the components belonging to the Orion arm, are very complex at longitudes $85°$ and $102°.5$. The observed velocities occur over the whole velocity interval covered by the 21-cm. profile of the arm (see also van de Hulst, Muller and Oort[11]). The stars observed by Münch around $l = 102°.5$ are all members of the h and χ Persei clusters. Although the exact location of the double cluster in the Perseus arm is still unknown, it is probable that the observed spread of velocities in the optical interstellar lines can be attributed to random motions of the clouds of the order of 10 km./sec. If so, these random velocities would explain the featureless character of the 21-cm. profiles for the Perseus arm at the position of the h and χ Persei clusters. The profiles are shown in Figs. $5a$ and $5b$, and can be compared with the profiles at $l = 100°$ given in Figs. $5c$ and $5d$.

A comparison between all the available profiles at $l = 100°$ and those at $l = 102°.5$ (many more than are shown in Fig. 5) shows that the variation with latitude is quite similar at the two longitudes for (a) the maximum intensity of the Perseus arm, (b) the intensity at $V_{LSR} = -39$ km./sec., which is the velocity of the h and χ Persei clusters, and (c) the total area under the profile of the Perseus arm. These facts indicate that the h and χ Persei clusters have at present only a very slight influence on the gross characteristics of the neutral hydrogen in the Perseus arm. The absence of any H II region near the double cluster (Shajn and Hase[12]) further strengthens this belief.

An examination of the Orion arm profiles in Fig. 5 shows how the fine details persist, and also change, over a limited region. The persistence of these details, and the manner in which they change, over an area of the sky, coupled with the foregoing discussion, suggest that it is not out of order to consider the observed 21-cm. profiles as the sum of a limited number of simple gaussian curves. Each gaussian curve can be considered as being produced by a hydrogen 'cloud' or grouping of 'clouds'. The maximum intensity and dispersion of each gaussian curve is determined by fitting their sum to the observed profile.

It is important to remember that the true curves for an individual hydrogen 'cloud' may not be gaussian, and they may well be asymmetrical. For a first attempt, however, we may make the foregoing assumptions and see if the results are reasonable. The profile for LF 6 has been analyzed in this manner and the results for the larger details are indicated in Fig. 3. The Orion arm can be almost completely represented by two such curves, as indeed the original profile itself suggests; for curve (a), $\sigma = 4 \cdot 6$ km./sec., for curve (b), $\sigma = 4 \cdot 3$ km./sec. The small detail between the two maxima in the Orion arm seems to be almost negligible. Several small details on or near the profile of the Perseus arm appear, however, to have more importance. The profile for the Perseus arm is more difficult to analyze. It can be represented rather well by one major contributor having a $\sigma = 5 \cdot 9$ km./sec., with two or three much smaller components.

The quality of performance of the equipment necessary to obtain the observations discussed in this paper was due mainly to the efforts of Mr J. C. Campbell.

REFERENCES

[1] Matthews, T. A. *P.A.S.P.* **67**, 112, 1955.
[2] Westerhout, G. Unpublished communication, 1955.
[3] Helfer, H. L. and Tatel, H. E. *Ap. J.* **121**, 585, 1955.
   Helfer, H. L. and Tatel, H. E. Paper read at the Princeton Meeting of the A.A.S., 1955.
[4] Westerhout, G. This symposium publication, paper 5.
[5] Oort, J. H., van de Hulst, H. C. and Muller, C. A. *Kon. Nederl. Akad. Wetenschappen, Amsterdam*, **61**, no. 8, 1952.
[6] Morgan, W. W., Whitford, A. E. and Code, A. D. *Ap. J.* **118**, 318, 1953.
[7] Whipple, F. L. Centennial Symposia (*Harvard Obs. Mono.* no. 7), p. 109, 1948.
[8] Blaauw, A. *B.A.N.* **11**, 459, no. 436, 1952.
[9] Adams, W. S. *Ap. J.* **109**, 354, 1949.
[10] Münch, G. *P.A.S.P.* **65**, 179, 1952.
[11] van de Hulst, H. C., Muller, C. A. and Oort, J. H. *B.A.N.* **12**, 117, 1954.
[12] Shajn, G. A. and Hase, V. F. *Pub. Crimean Astrophysical Observatory*, **10**, 152, 1953.

Discussion

Westerhout: The situation in Matthews' Fig. 1 near $l = 80°$, where he finds four different details in the Perseus arm running closely parallel, may be explained from the Leiden observations. There it is shown that at $l = 75°$, $b = +2°$ a branch with higher negative velocity branches off from the Perseus arm and runs parallel to it until it joins up again at $l = 85$, $b = +2°5$. Parts of this positive-latitude arm stick through the plane $b = 0$ at several places and give rise to the great number of little maxima found by Matthews.

From the Leiden observations it is also clear, that a series of maxima that occurs at the negative-velocity side of the Perseus arm near $l = 110°$, if ascribed to galactic rotation, is due to a small extra arm which runs at a latitude of approximately $+2°$.

55

A 21-CM. STUDY OF THE ORION REGION

T. K. MENON

Harvard College Observatory, Cambridge, Mass., U.S.A.

An analysis of the profiles of the 21-cm. radiation from neutral hydrogen promises to be of great importance for a study of the internal motions of specific regions of the Galaxy. The two factors which influence the shape of the profiles are the velocity distribution and the density distribution of the neutral hydrogen atoms in the line of sight. The velocity distribution is essentially determined by three factors (1) galactic rotation, (2) the random motions of the gases, and (3) local peculiar motions as, for example, expansion. In the plane of the Galaxy in any specific direction the isolation of a region of particular interest is made difficult because of the superposition of the radiation along the entire line of sight. Hence regions at intermediate galactic latitudes are more suitable for study of internal motions than regions on or near the galactic equator. Also, for the study of peculiar motions, regions with small galactic rotation terms have the distinct advantage that any prevailing preferential motion will be clearly indicated by the profiles. The Orion region satisfies most of the above requirements. The galactic latitude of the section under consideration falls between $-10°$ and $-25°$, and at the mean galactic longitude of $170°$ the galactic rotation term in radial velocity amounts to 7 km./sec. at a distance of 500 parsecs. Moreover the Orion region contains many features of considerable interest like the Orion Nebula, the Orion Association, the great arc of ionized hydrogen and many smaller H II regions. The great arc of Barnard (1895) [1] forms part of an almost elliptical ring of emission nebulosity with dimensions $14 \times 12°$. At the distance of 500 parsecs for the Orion Association these dimensions are of the order of 120×105 parsecs. It is of interest to note that the major axis of this ellipse is parallel to the galactic equator. This ellipticity could presumably be caused by galactic rotation, by a galactic magnetic field with lines of force along the spiral arms, or by the rotation of the whole mass itself. Further investigation is necessary to decide which of the above effects is most important.

56

I. OPTICAL PROPERTIES OF THE ORION REGION

Observations of interstellar absorption lines for about forty stars of the Orion region are included in Adams' list (1949) [2]. This list gives radial velocities and the intensities of the components of the H and K lines of calcium. Most of the O and B stars in Adams' list for the Orion region are relatively nearby. Upon close examination of the velocities of the interstellar components, one is struck by the considerable number of components with excessive negative velocities. This fact was recognized early by Sanford and Adams. Later Whipple (1948) [3], Blaauw (1952) [4], and Schlüter, Schmidt and Stumpff (1953) [5] have also commented on this anomaly. Whipple (1948) in his analysis of Adams' data (in an attempt to find evidence for the cloud structure of the interstellar medium) grouped components with about the same velocities in stars close together in the sky and assumed that those components are produced by the same cloud. This is an extremely simplified model. For example, of the two stars θ^1 Orionis and Bond 619 which are less than 3' apart, θ^1 Orionis has four interstellar components with velocities $-15\cdot5$, $-5\cdot8$, $+0\cdot5$, $+11\cdot9$ km./sec. and intensities 2, (?), 3, 4 respectively, whereas Bond 619 has only one interstellar component with a velocity of $+2$ km./sec. and intensity 4b. In the majority of stars the strongest interstellar component has a positive radial velocity while the faintest one has generally the largest negative velocity. Also, Adams found that almost invariably the strongest component of a complex line is the one which corresponds to pure galactic rotation. This fact, combined with the anomaly noted above, seems to suggest that the components with excessive negative velocities may be produced close to the individual stars and hence have a distinctly different origin from the strong positive components. This would be the case if the negative velocity components were formed near the boundary regions of the Strömgren spheres of the individual stars—a conclusion which is strongly supported by the work of Schlüter, Schmidt and Stumpff (1953) [5], who have plotted the distribution of early type stars and the distribution of large negative velocity components. Recent theoretical studies by Oort (1954) [6] and by Spitzer and Oort (1955) [7] furnish a mechanism for acceleration of interstellar clouds in such boundary regions.

The anomalous position of η Orionis has been pointed out by Whipple (1948) [3] and Adams (1949) [2]. The spectrum of the star shows double interstellar H and K lines of unequal intensity. The stronger component, which has a negative velocity of $-10\cdot2$ km./sec., gives a large residual when corrected for galactic rotation very different from that for other stars of the same region; but the fainter component, which has a positive velocity of +6·8 km./sec., gives an accordant value. This peculiarity of the region around η Orionis is reflected in the 21-cm. profiles as we shall show below.

Using Sharpless' (1954) [8] tabulation of early-type stars belonging to the Orion Association, a plot has been made of the distribution of stars earlier than B4 (Fig. 1). This diagram shows that practically all stars

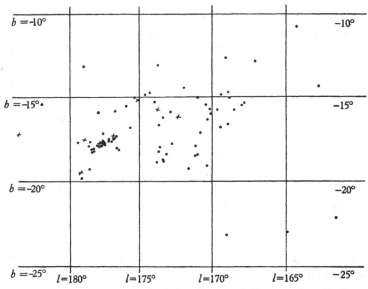

Fig. 1. Distribution of early-type stars in Orion. Crosses: O–Bo stars; dots: B1–B3 stars.

belonging to this group are distributed in a strip between $l = 166°$ to $l = 178°$ and $b = -19°$ to $b = -14°$. There are a dozen O–Bo stars, 16 B1 stars and 57 B2–B3 stars within this strip. We note that the Orion Association extends twice as far in galactic longitude as in latitude and hence it possesses a decidedly flattened shape (Pannekoek, 1929) [9].

2. THE 21-CM. OBSERVATIONS FOR THE ORION REGION

For a detailed study of the distribution of neutral hydrogen in the Orion region, profiles of the 21-cm. line were obtained at 3° intervals in galactic latitude and longitude for the region $l = 160°$ to $182°$ and $b = -25°$ to $-10°$. Fig. 2 shows the reduced profiles for most of the region. The velocities are with respect to the local standard of rest. The observations were made with the equipment of the G. R. Agassiz Station as described by Lilley (1955) [10], using a beam-width of 1°7 and band-width of 15 kc./s. The profiles have not been corrected for band-width or random velocity. The band-width correction was found to be so small as to be negligible. The random velocity corrections were generally not made for the profiles since it was not certain what value of random velocity would be appropriate for any particular region under consideration. However, some of the profiles of particular interest were corrected for an assumed average random motion to study the effect of such a correction on the profiles.

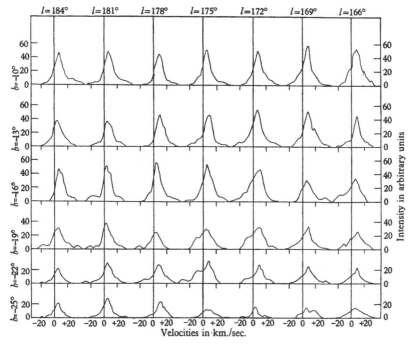

Fig. 2. Observed 21-cm. line profiles in the Orion region.

The most striking characteristic of the 21-cm. profiles of the Orion region is the rapid variation in the shape of the profiles from position to position. The profiles for positions north of $b = -16°$ are, in general, sharp with small irregularities in the wings.

The profiles for a region bounded by $l = 172°$ to $178°$; $b = -19°$ to $-22°$ show a marked asymmetry on the negative velocity side. If these profiles are corrected for random motions, assuming any reasonable value of the random velocity, the correction is such that the resultant profiles have two peaks, one on the negative velocity side and the other on the positive velocity side. The principal peaks of all the profiles show positive velocity. We note here that the galactic rotation effect for the Orion section is positive. Hence any subsidiary peak on the negative velocity side definitely indicates a real motion towards us and cannot be an effect of distribution of atoms along the line of sight. As pointed out earlier, the interstellar absorption lines for η Orionis ($l = 172°$, $b = -19°$) also indicate a strong negative component. This negative component is distinctly different from the excessive number of negative velocity components found in the region, since the latter are all weak in intensity. It is surprising to note that the region, where we have the maximum amount of hydrogen moving towards us, is comparatively free of emission nebulosities or early-type stars.

Maximum asymmetry is shown by the profiles for the section $l = 172°$ to $175°$; $b = -19°$ to $-22°$. Confining our attention to this small section, we can compute from the area of the profiles the total number of neutral hydrogen atoms along the line of sight in a cylindrical column 1 cm.^{2} in area. This value is of the order of 1.7×10^{21} atoms. Now, on the assumption that the whole contribution in the line of sight is from a column 100 parsecs in length we can derive the space density of neutral hydrogen atoms. This density proves to be of the order of 5 atoms/cm.^{3}. Before we can derive the total density, we must have an idea of the ionization equilibrium in the region. This is difficult because there are no reliable estimates of the emission measures for the region available. However, for the restricted region under consideration, H_α photographs show only extremely feeble emission. An order of magnitude check on the derived densities is obtained as follows: The star σ Orionis ($l = 174.5°$, $b = -16°$) has one of the best defined associated Strömgren spheres. The diameter of the sphere is approximately 17 parsecs. Taking the value of $s_0 N_H^{\frac{2}{3}}$ for an O9.5 star from Strömgren's tables (1948) [11] we find that for the observed dimensions the value of N_H is of the order of 10 atoms/cm.^{3}. Considering the uncertainties in the data, the agreement is good. There are obviously local variations in density over the region, but 21-cm. observations combined with the above optical estimate indicates a well-established lower limit to the average density of about 5 atoms/cm.^{3} and a maximum average density of some three times this value.

For the restricted region considered above the total mass represented by the approaching portion of the profiles is found to be about 3200 solar masses for a region 26×26 parsecs and the comparable mass with positive velocity is about 6000 solar masses. These values represent only rough lower limits. The velocities of the positive and negative components are $+12$ km./sec. and -5 km./sec. with respect to the local standard of rest.

The galactic rotation effect is about 7 km./sec. for this region and, correcting the zero velocity line for this velocity, we find that the two peaks have velocities $+5$ km./sec. and -12 km./sec. respectively.

Before we consider the interpretation of these observations, there are a few further points of interest to be noted in the profiles. In photographs of this region (*Ross Atlas*, no. 34) it is seen that the absorption increases with decreasing galactic latitude. Along with this increase in absorption, the half-widths of the 21-cm. profiles decrease quite clearly. This is not surprising since the presence of any large amounts of dust within the gas may presumably inhibit to a certain extent any large-scale internal motions of the gas. In other parts of the Galaxy we have noted that the sharpest 21-cm. profiles are often associated with regions of high local obscuration (Lilley, 1955) [10].

3. INTERPRETATION OF OBSERVATIONS

When the radial velocities of the peaks of the profiles are plotted against longitude it is found that for the latitude strip at $b = -19°$ the variation is almost sinusoidal with a maximum at $l = 167°5$ of $+9·2$ km./sec. and a minimum at $l = 179°5$ of $+2·8$ km./sec. There is also an indication of such a behaviour at $b = -16°$. The average radial velocity of all the profiles for the Orion region is $+6·6$ km./sec. Hence the semi-amplitude of the radial velocity variation is about 3 km./sec. It is also interesting to find that the points $l = 179°5$, $b = -19°$ and $l = 167°5$, $b = -19°$ coincide with the edges of the H II ring. More observations are under way to determine the plane of maximum variation of radial velocity. The present observations can be successfully interpreted if we assume that the whole gaseous mass contained inside the H II ring is in rotation about an axis perpendicular to the plane of the Galaxy, passing through the point $l = 173°5$, $b = -19°$ and with a linear velocity of rotation of about 3 km./sec. Preliminary calculations indicate that the observed ellipticity of the H II ring is consistent with the hypothesis of rotation of the whole gaseous mass. Detailed calculations are in progress to check on the consistency of having a gaseous cloud of the estimated mass, size and shape rotate at the above rate and the study is being extended to include possible magnetic effects from the galactic field.

On the basis of Strömgren's theory we can estimate the dimensions of the H II regions associated with the various groups of stars at the densities estimated above. Let us first take the giant H II ring. The obvious symmetry of the ring precludes the possibility of its excitation being due to any arbitrary distribution of exciting stars. The only small homogeneous group of stars is the Trapezium cluster and considering the nine stars which are close together all of spectral type O9·5 or B0, the observed diameter of 120 parsecs for the ring requires only a density of about 1·6 atom./cm.³ This is certainly too low a density to be reconciled with the 21-cm. observations. For a density of 5 atom./cm.³ the diameter of the H II region comes out to be about 56 parsecs. Moreover, if the Trapezium Cluster were the source of excitation, then practically all the hydrogen inside such a region should be ionized. This certainly is not the case since we observe no marked decrease in H I intensity at the centre of the region. The only valid conclusion appears to be that the density near the OB stars is so high that practically no radiation escapes outside these dense regions, to be available for ionization. Greenstein's (1946) [12] investigation of the Orion nebula, the appearance of the Horse Head nebula, as well as the estimate made earlier for the region near σ Orionis, tend to support the conclusion of very high density in these few regions (Seaton, 1953) [13]. If this is true, then we have to look for a different mechanism for the excitation of the giant H II ring.

Returning to the 21-cm. observations, let us consider a spherical, expanding region. The line profile for the centre of such a region should contain two peaks separated by the maximum positive and negative velocities of expansion. Since the density distribution is not necessarily uniform in all directions, the velocity of expansion will also not be the same in all directions. From aerodynamic considerations (see Oort (1946) [14]; Burgers (1946) [15]), it follows that the velocity will be greatest in regions of low density and smallest in regions of high density. As we move away in the sky from the centre of the region, the separation of the peaks should become smaller and the two peaks should coalesce when the separation is smaller than the half-width due to random motions alone. Let us assume a random velocity of 6 km./sec. We predict then for an expanding region with a diameter of 90 parsecs and at a distance of 500 parsecs that we may expect single peaks farther than 6° from the centre of the expanding region. The distance could be considerably less than 6° if the velocity were not uniform in all directions. Present observations indicate that the centre of this expansion is approximately at $l = 173°5$, $b = -18°5$. Further observations are being undertaken to locate the centre more accurately. The fact that we do not observe double-peaked profiles at distances more than about 3° from this centre indicates that either the diameter of the expanding region is smaller or the velocity is not uniform in all directions. We shall take as an approximation a diameter of 85 parsecs. We can estimate the maximum distance of the centre of expansion from the observation of the interstellar components in the spectrum of η Orionis. This star is one of the nearest in the association and is estimated to be at a distance of 360 parsecs. Hence it seems likely that the centre of expansion is on the inner edge of the spiral arm, in which case the higher density in the spiral arm might slow down the velocity of recession. If we take the maximum observed velocity as 12 km./sec. and a radius of 42·5 parsecs we get a value of 3·3 million years for the interval since the expansion began.

The following table gives approximate positions of some centres of interest:

Object	l	b
Orion nebula	$176°5$	$-18°$
Horse Head nebula	$174°5$	$-15°$
Centre of H II ring	$173°5 \pm 2°$	$-18° \pm 2°$
Centre of OB association	$172° \pm 2°$	$-21° \pm 1°$
Centre of 21-cm. expansion	$173°5 \pm 1°$	$-18°5 \pm 1°$

We note from the above tabulation that the three latter centres are probably the same, whereas the Orion nebula is not directly connected with the centre of expansion. In this connexion it is interesting to recall the discovery by Blaauw and Morgan (1954) [16] that the motion of AE Aurigae and μ Columbae indicates that the two stars may have had a common origin in Orion. They found that these two stars move with the same speeds—127 km./sec.—in opposite directions away from the Orion region. They give the point of origin of the two stars as $l = 174°8$, $b = -17°7$, but the uncertainties in the observational data are such that the point of origin might well be moved down to $l = 173°5$, $b = -18°5$. They estimate that the stars must have left their point of origin about $2·6 \times 10^6$ years ago. This estimate is in good agreement with that made above from the radio observations. Though Blaauw and Morgan were inclined to believe that the point of origin might be close to the Orion nebula, it is more logical to attribute the motion of the two stars and the expansion of the gases to one and the same cause and hence expect the two centres to coincide.

We seem to have the following picture: there exists a region of expansion of neutral hydrogen associated with a grouping of early-type stars, all of these features being bounded by a giant H II ring. The observed intensity of 21 cm. radiation is incompatible with the idea that the whole region is one of high ionization. But there are many small regions of high ionization around the early-type stars. However, the giant H II ring cannot be the edge of a Strömgren sphere since in such a case the interior of the sphere also must be completely ionized. Hence the giant H II ring may be part of a thin shell, which is excited by an unknown mechanism, but with direct stellar radiation playing only a subsidiary role.

Two different mechanisms have been put forward to explain the origin of stellar associations and their associated expansions. The first mechanism was proposed by Öpik (1953) [17]. According to Öpik's theory the giant H II ring is the remnant of the expanding shell of a supernova. The initial large velocities of a supernova explosion are resisted by the inertia of the surrounding interstellar gas and the expansion is decelerated in proportion to the mass of interstellar gas displaced from the volume occupied by the shell. His estimates indicate that the observed radius should correspond to a present velocity of expansion of 10 km./sec. and an interval of about $1 \cdot 5 \times 10^6$ years since the supernova explosion. These estimates are very approximate. During the process of expansion the compressed shell is supposed to have condensed into stars. The stars by this mechanism would have the velocity of expansion of the shell at the time of their formation. The different velocities of expansion of the stars in our case could presumably be due to their birth at different epochs of the expanding shell. One consequence of this mechanism is that the ages of the stars and the interval since the supernova outburst need not be precisely the same.

In this connexion it is interesting to recall a recent suggestion by Ambartsumian (1955) [18] that the origin of stellar associations is probably a single body—a proto-star. This proto-star is supposed to divide and form a trapezium system which in turn gives rise to an association.

In the second mechanism, proposed by Spitzer and Oort (1954, 1955) [6,7], the initial expansion is set off by the formation of a hot star in the middle of dense clouds of dust and gas. The resultant differences in pressure in the H II and H I regions give rise to various aerodynamical phenomena, for example the production of shock waves at the boundary of the H I and H II regions, and provide energy for the acceleration of H I clouds. It is supposed that the resultant conditions are sufficient for the formation of expanding groups of stars. However, Spitzer and Oort (1955) [7] point out that this mechanism cannot be responsible for the acceleration of the two stars AE Aurigae and μ Columbae already discussed.

Again, according to this theory, 21-cm. profiles should show for the Orion region maximum hydrogen with negative radial velocity in the direction of the early-type stars. But, as pointed out earlier, observations indicate that this maximum occurs in a region devoid of early-type stars. Also the presence of the early-type stars within giant symmetrical H II ring is not to be expected from the Spitzer–Oort theory, since in their theory the stars are supposed to be formed at the boundary of H II regions. However, the observations of interstellar absorption lines indicate that the Spitzer–Oort mechanism may provide the acceleration for the small clouds responsible for their appearance. But it seems unlikely that a similar mechanism is responsible for the whole range of expansion phenomena of neutral hydrogen observed at 21 cm.

The author wishes to express his sincere gratitude to Dr B. J. Bok for his constant encouragement during the course of the above investigation. This study was made possible by a grant from the National Science Foundation.

REFERENCES

[1] Barnard, E. E. *Pop. Astr.* **2**, 151, 1895.
[2] Adams, W. S. *Ap. J.* **109**, 354, 1949.
[3] Whipple, F. L. *Centennial Symposia* (Camb. H.C.O.), Sec. I, 8, 1948.
[4] Blaauw, A. *B.A.N.* **11**, 405, no. 433, 1952.
[5] Schlüter, A., Schmidt, H. and Stumpff, D. *Z. Ap.* **33**, 194, 1953.
[6] Oort, J. H. *B.A.N.* **12**, 177, no. 455, 1954.
[7] Spitzer, L. and Oort, J. H. *Ap. J.* **121**, 6, 1955.
[8] Sharpless, S. *Ap. J.* **119**, 200, 1954.
[9] Pannekoek, A. *Publ. Astr. Inst. Amsterdam*, no. 2, 1929.
[10] Lilley, A. E. *Ap. J.* **121**, 559, 1955.
[11] Strömgren, B. *Ap. J.* **108**, 242, 1948.
[12] Greenstein, J. L. *Ap. J.* **104**, 414, 1946.
[13] Seaton, M. J. *Liège Symposium*, p. 452, 1953.
[14] Oort, J. H. *M.N.R.A.S.* **106**, 159, 1946.
[15] Burgers, J. M. *Kon. Ned. Akad. Wet.* **49**, 589, 1946.
[16] Blaauw, A. and Morgan, W. W. *Ap. J.* **119**, 625, 1954.
[17] Öpik, E. *Irish Astr. J.* **2**, 219, 1953.
[18] Ambartsumian, V. A. *Observatory*, **75**, 72, 1955.

RADIO OBSERVATIONS OF INTERSTELLAR NEUTRAL HYDROGEN CLOUDS

R. S. LAWRENCE*

National Bureau of Standards, Boulder, Colorado, U.S.A.

The detailed relationship between optical interstellar absorption lines and 21-cm. observations is investigated in this paper.

Dr Guido Münch, of the Mount Wilson and Palomar Observatories, provided the list of six intermediate-latitude stars shown in Table 1. The spectra of these stars all show complex absorption lines due to interstellar Ca II. The 21-cm. line is measurable in four of the six regions, although the peak intensity is low in each case. It is noteworthy that for the first two stars on the list the radio velocity agrees closely with the velocity of an intense optical component. In view of the great difference in angular resolution, the failure to find correspondence in every case is not surprising.

Table 1. *Comparison of 21-cm. and optical interstellar lines*

Star	Galactic co-ordinates l	b	Type	Mag.	Radial velocity (km./sec. relative to the sun) 21 cm.	Optical components of Ca II (in order of decreasing intensity)			
HD 219188	$52°$	$-51°$	B2n	6·9	-9	$-7,$	$+18,$	$(-29?)$	
HD 215733	$54°$	$-37°$	B2	7·7	-10	$-26,$	$-11,$	$-44,$	-57
HD 203664	$30°$	$-28°$	B3n	8·3	$+5$	$-8,$	$+66,$	$(-28?)$	
HD 119608	$289°$	$+42°$	cB	7·3	-10	$-3,$	$+18,$	$(-11?$	$+33?)$
HD 91316	$204°$	$+54°$	cB0	3·8	—	$-13,$	$-4,$	$+16$	
HD 93521	$150°$	$+64°$	B3n	6·9	—	$-11,$	$-53,$	$+5,$	-34

The full text of this paper has been published in *Astrophysical Journal*, **123**, 30, 1956.

Discussion

Hagen: We have observed a local cloud at high latitude ($l = 328°8$, $b = 53°7$) with velocity of recession of 100 kc./s. and brightness temperature 20°, covering an area with diameter 3°.

* This work was done at Harvard College Observatory, Cambridge, Mass.

21-CM. MERIDIAN PLANE SURVEYS

H. E. TATEL

Department of Terrestrial Magnetism, Carnegie Institution of Washington,
Washington, D.C., U.S.A.

The meridianal surveys[1] at 21-cm. of interstellar hydrogen gas have been continued. Surveys at galactic longitudes of 50° [1], 60° [2], 80° [2], 90° [1], 110° and 200° have been completed while partial surveys have been made at longitudes of 20°, 205°, and 210°.

These surveys of apparent antenna temperature as a function of frequency are carried out at one longitude for a series of latitudes. The

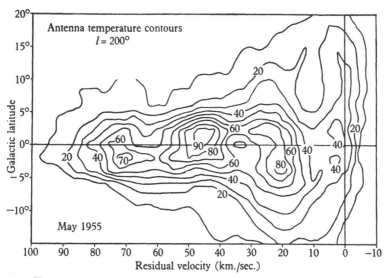

Fig. 1. Isotherms at wave-length of 21 cm. for galactic longitude of 200°.
Temperature scale is approximate.

latitude interval is from two to five degrees depending upon the structural detail. The 'slit' width used is 12 kc./s. It is instructive to plot the data on a cartesian system in which the abscissa is the Doppler shift and the ordinate is the galactic latitude. Contours of constant apparent antenna temperature are drawn. These plots show distinct maxima associated with the

67

spiral arm structure as first pointed out by the Australian [3] and Dutch [4] workers. These particular plots are useful in that the original data are entered with no adjustments. On the other hand, they give distorted pictures of the spatial configuration of the gas. In order to improve the presentation certain assumptions have to be made such as random cloud velocities and the variation of galactic rotation with distance. These will not be considered here.

By making plots of this nature certain simple properties of the galactic gas distribution appear. One interesting aspect is that these plots give an excellent presentation of the solid angle of a given gas concentration which is related, to a first approximation, to any particular contour grouping.

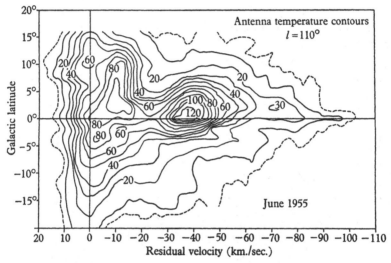

Fig. 2. Isotherms at wave-length of 21 cm. for galactic longitude of 110°.
Temperature scale is approximate.

A more sophisticated analysis in which self-absorption is taken into account will shift the true gas centres with respect to the apparent temperature system here under consideration. Near the sun these shifts should not vary enough with direction to affect the gross conclusions, as the apparent temperature gradients do not vary rapidly with longitude. The gas concentrations near the sun have a small Doppler shift and subtend large angles compared to the more distant concentrations. As these near gas structures are examined in different longitudes, from 50° to 210° (Figs. 1, 2), it is seen that the structure changes.

First there is the gas which at longitude 60° extends to latitudes of ± 20° and greater, with appreciable concentrations. At $l = 200°$ (Fig. 1) the peak

68

temperature of the distribution has dropped to almost half and at 210°, which is difficult to observe because it is so close to our horizon, there are only the most vestigial traces of this concentration, particularly at negative latitudes. A preliminary survey at 20° longitude indicates that this local gas has a high concentration in this direction. Thus, it appears that the sun is in a gas concentration which is denser in the region from $l = 50°$ to $110°$ and far less dense in the direction $l = 210°$. This local gas concentration is distinct and apart from what we have been calling the local arm.

Inspecting once again the gas concentration of the local arm, there is a suggestion that the main body of this gas is inclined out of the galactic

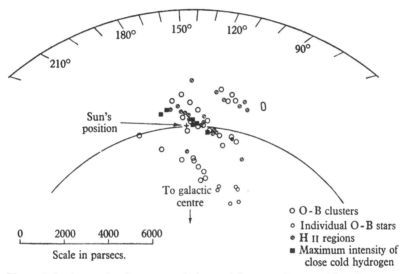

Fig. 3. Galactic map showing star associations and the approximate position of apparent antenna temperature maxima of the cold hydrogen in the near arm.

plane. At $l = 110°$ (Fig. 2) it is 4° out of the plane and fairly well separated from the local cloud. At 200° (Fig. 1) there is no doubt, for there are four major concentrations instead of the three found between $l = 60°$ to $110°$. While we have not traced these four concentrations in intermediate longitudes, for example between $l = 170°$ to $190°$, there is good evidence for their existence in the equatorial traces made at these longitudes for this region. The traces show four well-defined peaks instead of the three as found at $l = 60°$ to $110°$. As this near arm gas concentration becomes more separated from the local cloud, its Doppler shift is also increasing. If the position of greatest temperature is mapped (Fig. 3) assuming that its distance from the sun is a function of the galactic rotation constants and the sine of twice the galactic longitude, then the distance from the sun of this arm is seen to increase as l both increases and decreases from $l = 70°$. We have one point at $l = 50°$ and the data for $l = 20°$ are at least consistent with this hypothesis. Thus, the near arm, which apparently coincides in position with the star associations of Morgan, Whitford, and Code, has an inclination with respect to the circle through the sun about the galactic centre.

The values of the Doppler Shift presented by Helfer and Tatel in *Ap. J.* **121**, 585, 1955, should be increased by a non-linear factor equal to 7% at 100 km./sec. I am indebted to Messrs Raimond and Westerhout of Leiden, who called this to my attention.

REFERENCES

[1] Helfer, H. L. and Tatel, H. E. *Ap. J.* **121**, 585, 1955.
[2] *Carnegie Inst. of Wash. Year Book*, no. 55, 80, 1955–1956.
[3] Christiansen, W. N. and Hindman, J. V. *Aust. J. Sci. Res.* A, **5**, 437, 1952.
[4] Hulst, H. C. van de, Muller, C. A. and Oort, J. H. *B.A.N.* **12**, 117, no. 452, 1954.

THE MEASUREMENT OF THE DISTANCE
OF THE RADIO SOURCES

R. D. DAVIES AND D. R. W. WILLIAMS

Jodrell Bank Experimental Station, University of Manchester, England

I. INTRODUCTION

In radio astronomy it is becoming increasingly important to know the distance of the radio sources. An identification with astronomical objects observed optically is then more readily obtained and this in turn may allow further investigation of the mechanism of radio emission. A measurement of the distance of sources will also resolve the problem of their distribution in space, showing which are galactic and which are extra-galactic. Furthermore the surface area and absolute luminosity can be estimated from a knowledge of the distance and angular size of a source.

Early attempts to measure distances by the method of parallax by Ryle [1] and Mills [2] served to place the major sources outside the solar system at a distance greater than $\frac{1}{2}$ parsec. This paper describes a completely new method, which involves a measurement of the absorption spectrum of the source produced by interstellar neutral hydrogen at a wave-length of 21 cm. Three different methods have been used to obtain a distance from these data. First, the very existence of absorption at the frequency of a spiral arm in the direction of a source places it within or beyond that arm. Secondly, the absorption line width of a source enables its distance to be determined in a way analogous to that used in optical astronomy. Thirdly, if the distribution of hydrogen along the line of sight in the direction of the source is known, the distance can be determined from the known amount of hydrogen between the sun and the source. The last two methods are useful in the galactic centre and anti-centre regions where the Doppler displacement relative to the sun is small.

Absorption measurements have been made on the four intense sources in Cassiopeia, Cygnus, Taurus and Sagittarius. Cygnus A lies beyond the limits of the Galaxy while the other sources are within it. Cassiopeia A appears to have a distance of at least 2·5 kiloparsecs which is five times greater than the optical determination. The Sagittarius source which some observers have identified with the galactic nucleus is at a distance of 3 kiloparsecs. An identification is suggested.

2. THE OBSERVATIONS

Preliminary results [3] indicated that the equipment had to be made more sensitive to give higher accuracy on the absorption in the Cassiopeia and Cygnus sources and to permit observations on weaker sources. The improved equipment consists of a radio spectrometer used in conjunction with a focal plane paraboloid 30 ft. in diameter illuminated by a dipole feed and reflector. The beam-width to half-power points is $1°45$ in the H-plane and $1°65$ in the E-plane. The aerial mounting is alt-azimuth and in order to be able to follow a source to within a quarter of a beam-width an autofollow mechanism is employed in the altitude and azimuth driving circuits. It is thus possible to obtain a spectrum. The accurate aerial position calibration for all parts of the sky was made through observation of the sources on total power.

The receiver employs the comparison principle in order to obtain the required stability. It is switched at 30 c./s. between the line frequency and a comparison band 1 Mc./s. on one side of the line. A new principle, whereby the line is observed continuously, involves having a second pair of receiving bands, one on the line frequency and a comparison band on the other side of the line. This 'double comparison technique' gives an improvement in sensitivity of $1·5$ db. In order to observe the line profile a variable frequency is added to the frequency-controlled first local oscillator. The noise factor of the receiver was measured to be $8·5$ db. using an argon noise source. This was checked by comparing the receiver noise temperature with the observed sun temperature. The receiver band-width was 40 kc./s. and the observing time constant was 30 sec. which gave peak-to-peak fluctuations of $7°$ K.

A total power receiver is required to measure the intensity of the radio sources in the continuum. A rotating capacity switch which has a switching ratio of 20 db. gives the desired stability. The measurements were made using the broad-band ($1·1$ Mc./s.) section of the receiver. The temperature of the sources was obtained by using narrow-band (40 kc./s.) and comparing Cassiopeia with the hydrogen-line intensity at $l = 50°$ which was taken as $100°$ K.

In order to determine the absorption of the continuous emission of a source, it is necessary to observe the spectrum not only in the direction of the source but also in adjacent directions. The changing spectrum of the

72

hydrogen near the source can then be taken into account in order to obtain the real emission spectrum in the precise position of the source. Accordingly, spectra were taken on a grid of eight points displaced 3° from the source either side in right ascension, declination, and at the four diagonal points. Spectra were obtained in sets of three across the source and were repeated three times. Two spectra are then obtained for each source, one of which is the average of all the points surrounding the source and

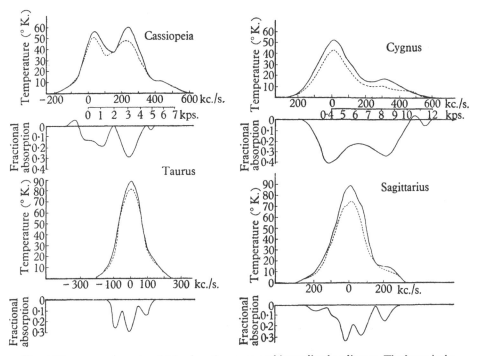

Fig. 1. The averaged spectra obtained on the source and immediately adjacent. The latter is the uppermost curve in each case. The fractional absorption spectra are plotted in the lower set.

represents the emission spectrum of the hydrogen if the source were not present, and the other is the average of the spectra taken on the source. The absorption spectrum for a source is obtained by subtracting these two spectra. Further, the fractional absorption can be derived from a knowledge of the temperature of the source obtained using the receiver on total power. A 10 % correction to the observed source temperature is required to allow for reception at the image frequency. The averaged spectra and the fractional absorption spectra for each of the four sources studied (Cassiopeia A, Cygnus A, Taurus A and Sagittarius A) are given in Fig. 1.

73

3. THE POSITION OF THE CYGNUS AND CASSIOPEIA SOURCES RELATIVE TO THE SPIRAL ARMS OF THE GALAXY

The concentration of neutral hydrogen into spiral arms within the Galaxy is now well established [4]. The exact position of these spiral arms depends upon an accurate knowledge of the rotational velocity in all regions of the Galaxy. In the present study the circularly symmetrical Oort model is used in conjunction with the hydrogen-line profiles to establish the distance of the arms in the direction of the various sources. In Cygnus the first peak comes from hydrogen near the sun and at 4 kiloparsecs while the second peak is from hydrogen at 8 kiloparsecs. In Cassiopeia the arms are at 0·5 and 3·0 kiloparsecs with a third faint arm at 5·5 kiloparsecs, while in Taurus and Sagittarius there is insufficient Doppler displacement to assign a distance to the hydrogen.

The following simple picture is used to describe the observed absorption results. It is assumed that the source itself blocks out only a negligible amount of neutral hydrogen from behind it. Let there be a source lying between two spiral arms. Fig. 2 a represents the spectrum observed very close to the source; in fact, it is the spectrum in the direction of the source if it were not there. Fig. 2 b is the absorption spectrum of the source due to its continuous emission traversing the nearer spiral arm. Then the observed spectrum in the direction of the source will be that in Fig. 2 c. The absorption spectrum of the source, which contains all the required information, can be obtained by subtracting spectrum Fig. 2 c from Fig. 2 a. It can be seen that if a source lies beyond or within a spiral arm it will exhibit absorption at the frequency characteristic of that arm. Thus the Cygnus source which shows absorption within both arms must be extra-galactic while the Cassiopeia source lies between the outer two arms at 3·0 and 5·5 kiloparsecs. This distance for Cygnus is consistent with the optical identification of the source with colliding galaxies at 34 megaparsecs. The Cassiopeia result is at variance with the distance of 500 parsecs derived by Baade and Minkowski [5] for a supernova remnant in this direction. On the above interpretation of the radio results the nearest that this source could be placed is at 2·5 kiloparsecs, on the inner side of the second arm.

Hagen, Lilley and McClain [6] have studied the Cassiopeia source with a band-width of 8 kc./s. and found fine structure within the two absorption peaks, one line within the first peak and two within the second. It is not certain whether these narrow absorption lines are due to individual gas clouds. They may rather be due to groups of clouds because the density of clouds in the line of sight is probably about 10 per kiloparsec within an arm [7]. Whichever interpretation is given, these results do not affect the above conclusions, for these individual clouds or groups of clouds can only belong to the spiral arm within whose velocity limits they lie. The hydrogen line emission profiles may be considered as the probability distribution of cloud velocities within the arms and as the profiles fall almost to zero between the arms it is very unlikely that a cloud with a velocity characteristic of one arm should belong to the other.

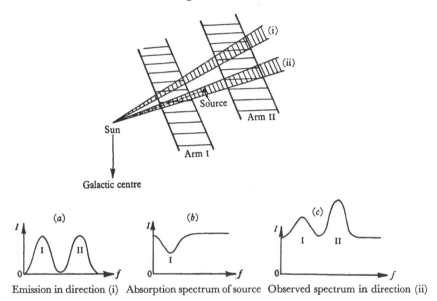

Emission in direction (i)   Absorption spectrum of source   Observed spectrum in direction (ii)

Fig. 2. A schematic diagram of the production of absorption of a source by neutral hydrogen.

4. THE TAURUS SOURCE

The Taurus A measurements enable the total number of hydrogen atoms/ cm.^{2} in the line of sight between the sun and the source to be determined by the method described in the above section. The kinetic temperature of the gas was assumed to be 125° K. Since the distance of this source, which is identified with the Crab nebula, is known to be 1·1 kiloparsecs [8] the proportion of hydrogen in front of and behind this distance may be estimated: 30 % of the hydrogen is between the sun and the source and 70 % is beyond it. This information is of some value because in this part of the Galaxy ($l = 153°$, $b = -1·5°$) there is too little Doppler diplacement to find the distribution of hydrogen by the conventional method. The published survey of the Galaxy [4] gives the hydrogen distribution only up to $l = 135°$ and beyond $l = 160°$. The published density contours were extrapolated across this region and the percentage of hydrogen up to and beyond a distance of 1·1 kiloparsecs was found to be precisely that obtained in the present work. This suggests that there may be some justification in extrapolating the distribution in (4) across the region $l = 135°$ to 160°.

5. THE SAGITTARIUS SOURCE

This source lies very close to the galactic centre, where its distance cannot be measured by the method used for the Cassiopeia and Cygnus sources. Two new methods have been devised to deal with sources in the centre and anti-centre regions in particular, although in principle they may be applied to any source.

(a) *The equivalent width of the source absorption line.* In optical astronomy the intensities of interstellar absorption lines are known to give a statistical measure of the distance of stars. In a recent study Beals and Oke [9] compared the measured distance of a large number of stars up to 2·5 kiloparsecs with the equivalent width of their absorption lines and found a linear relationship. The distance of individual stars could be measured to an accuracy of at least 25 % using the derived linear relation. This error was largely due to the non-uniform distribution of absorbing gas in the direction of the stars.

This method can be used to measure the distance of radio stars by calculating the equivalent widths of their 21-cm. absorption lines. The equivalent width, H, is defined by

$$H = \int_{-\infty}^{+\infty} \frac{I_0 - I_v}{I_0}\, dv \text{ km./sec.,}$$

where I_0 is the intensity of the unabsorbed continuum from the source and I_v is the intensity of absorbed radiation at a velocity v. It follows from this definition that the equivalent width is independent of the receiver bandwidth. Table 1 gives the distances and derived equivalent widths of the absorption lines for the four sources.

The distance assigned to the Cassiopeia source is the shortest possible consistent with the above interpretation, which brings it closest to the optical determination. The distance given in the case of Cygnus A is the

Table 1

Source	Distance (parsecs)	Equivalent widths (km./sec.)
Cassiopeia	2500	13 ± 2
Cygnus	8000	36 ± 3
Taurus	1100	8 ± 3
Sagittarius	?	14 ± 3

distance to the outer arm of the Galaxy in that direction, which is the path-length to the source lying in neutral hydrogen. The data for the three sources in Cassiopeia, Cygnus and Taurus along with the origin, enable a linear line-width against distance calibration curve to be drawn as in the optical case. The linear relation may be written

$$r = 220\,H \text{ parsecs.}$$

The optical result for the interstellar calcium K line was

$$r = 34 \cdot 8\,K \text{ parsecs}$$

which holds at least to a distance of 2·5 kiloparsecs. The hydrogen absorbing line is about six times weaker than the calcium line and the linear relationship for hydrogen may be expected to hold to 15 kiloparsecs before saturation effects come into play. This calibration curve becomes increasingly accurate for more distant sources but is liable to some error for the nearer sources because of the known non-uniform distribution of neutral hydrogen. This method is applicable to all sources but is particularly useful in cases where galactic rotation is small.

At first sight it is unexpected that the equivalent width of a line should increase linearly with distance since Oort [10] has shown theoretically that clouds are likely to have optical depths of the order of unity, and therefore where there are a number of clouds in the line of sight saturation effects are likely to come into play. The fallacy in this argument is in expecting the absorption spectrum of each cloud to be the same, whereas in fact each cloud has, in addition, its own peculiar line-of-sight motion, thus spreading the absorption spectrum and making the total equivalent width nearly equal to the sum of the individual equivalent widths. The above explanation was derived from Wilson and Merrill [11] who clarified a similar anomaly in the optical case.

This method is now applied to find the distance of the source in Sagittarius (17S2A). From the calibration curve its equivalent width of 65 kc./s. corresponds to a distance of 2·9 kiloparsecs.

(b) *The distribution of hydrogen in the direction of the source.* The observed absorption spectrum of a source allows an estimate to be made of the number of hydrogen atoms per unit cross-section in the line of sight to the source. Then, if the distribution of the hydrogen in this direction can be determined by some means, the distance of the source is known.

For the Sagittarius source a useful estimate of the distribution can be made by examining the emission spectra of neutral hydrogen from the inner parts of the Galaxy. A rigorous solution is not attempted here but a workable result comes from an analysis of the observed intensity of the low-frequency cut-off end of the spectra taken between galactic longitudes 350° and 35°. This leads to an estimate of the number of hydrogen atoms at points where the line of sight makes a tangent to a circle about the galactic centre. The locus of such points is a circle centred midway between the sun and the galactic centre. The number of hydrogen atoms at a given radius on that locus is then taken to be the number at the same radius on the line of sight to the centre of the Galaxy.

In a preliminary survey, a number of spectra were taken with an 18 kc./s. band-width at galactic latitude $-1°5$. The narrow band-width was necessary to delineate more clearly the arm structure shown in the Dutch profiles [12]. The profiles show a sharp rise to a high temperature when the line of sight is tangent to a spiral arm. In other directions there is a wide tail to the spectrum showing the existence of inter-arm hydrogen which is emitting at the frequency expected on the Oort rotation model. The observed temperature is then read off the spectrum at the frequency expected at the tangent point in the line of sight. In the inter-arm regions the effect of the adjacent spiral arm at a slightly higher frequency is removed before the temperature can be derived. This temperature distribution with distance from the galactic centre shows maxima at $R = 4·0$, $5·8$ and $7·8$ kiloparsecs. The temperatures are then converted to optical depths assuming a kinetic temperature of 125° K. for the gas. Since the line of sight is at a tangent to the observed distribution, the depth from which this emission comes is difficult to obtain since it is a function of the beam-width, the band-width of the receiver and the velocity distribution (η) of the clouds. As a first approximation the depth of emission is taken to be the same throughout the distribution and its value is then obtained by equating the observed number of hydrogen atoms derived from spectra in the line of sight to the galactic centre with the number obtained using the above distribution of optical depth with distance.* The number of hydrogen atoms between the source and the sun was computed from the absorption spectrum of the source to be $1·9 \times 10^{21}/$ cm.², compared with $6·5 \times 10^{21}/$cm.² in line of sight to the centre of the Galaxy. The result places the source at a distance of $2·7$ kiloparsecs from the sun. The errors involved in this estimate are difficult to assess. Although the close agreement with the previous method is probably fortuitous, the results suggest that the source is much nearer than 8 kiloparsecs and is probably of the order of 3 kiloparsecs from the sun.

* Account is taken of the greater values of η observed near the galactic centre (see B.A.N. no. 452).

(c) *Discussion of the Sagittarius source.* Further information is available to clarify the nature of this source. Its diameter at 21 cm. can be obtained from a comparison between the intensity observed with the present equipment using a beam-width of 1°6 and the intensity observed by Hagen, Lilley and McClain [6] using a beam-width of 0°9. From an examination of the Jodrell Bank records it was evident that the size of the source was less than the beam-width because no broadening of the beam pattern was observed. The reduced ratio of intensity of the Sagittarius to the Cassiopeia source on the N.R.L. survey could be attributed to the Sagittarius source having a diameter between 0°8 and 1°0. This result is similar to that obtained at the lower frequencies [13], and corresponds to a region about 50 parsecs across at 3 kiloparsecs.

The spectrum already published [14] has now been extended to 3·15 cm. by Haddock and McCullough [15] confirming that it is due to emission from an optically thin body.

The above evidence led Davies and Williams [14] to suggest that the Sagittarius source may be associated with a group of 38 O- and B-type stars and emission nebulae observed by Hiltner [16] and Sharpless [17] to lie in this direction at a distance of 3 kiloparsecs. Such an identification has since been tentatively proposed by Haddock and McCullough.

REFERENCES

[1] Ryle, M. *Rep. Prog. Phys.* **13**, 218, 1950.
[2] Mills, B. Y. and Thomas, A. B. *Aust. J. Sci. Res.* A, **4**, 158, 1951.
[3] Williams, D. R. W. and Davies, R. D. *Nature*, **173**, 1182, 1954.
[4] van de Hulst, H. C., Muller, C. A. and Oort, J. H. *B.A.N.* **12**, 117, no. 452, 1954.
[5] Baade, W. and Minkowski, R. *Ap. J.* **119**, 210, 1953.
[6] Hagen, J. P., Lilley, A. E. and McClain, E. F. *N.R.L. Report*, 4448; *Ap. J.* **122**, 361, 1955.
[7] Oort, J. H. *Gas Dynamics of Cosmic Clouds* (H. C. van de Hulst and J. M. Burgers, editors), North-Holland Publ. Co., p. 20, 1955.
[8] Greenstein, J. L. and Minkowski, R. *Ap. J.* **118**, 1, 1953.
[9] Beals, C. S. and Oke, J. B. *M.N.R.A.S.* **113**, 530, 1953.
[10] Oort, J. H. *Gas Dynamics of Cosmic Clouds* (H. C. van de Hulst and J. M. Burgers, editors), North-Holland Publ. Co., p. 20, 1955.
[11] Wilson, O. C., Merrill, P. W. *Ap. J.* **86**, 44, 1937.
[12] Kwee, K. K., Muller, C. A. and Westerhout, G. *B.A.N.* **12**, 211, No. 458, 1954.
[13] Mills, B. Y. *Aust. J. Sci. Res.* A, **5**, 266, 1952.
[14] Davies, R. D. and Williams, D. R. W. *Nature*, **175**, 1079, 1955.
[15] Haddock, F. T. and McCullough, T. P. *A.J.* **60**, 161, 1955.
[16] Hiltner, W. A. *Ap. J.* **120**, 41, 1954.
[17] Sharpless, S. *Ap. J.* **118**, 362, 1953.

21-CM. ABSORPTION EFFECTS

J. P. HAGEN, A. E. LILLEY AND E. F. McCLAIN

Naval Research Laboratory, Washington, D.C., U.S.A.

In the discussion which follows, there are presented three areas of the Naval Research Laboratory's study of absorption effects in the spectra of discrete sources produced by interstellar hydrogen. The first topic presented is the theory of the two mechanisms which are operative in producing the observed absorption effects. The second area of the discussion presents the observations and their interpretations for sources which lie in directions where the effects of galactic rotation are an important aid in the interpretation of the observed absorption. When galactic rotation effects are small, special techniques must be developed to supplement the absorption effect in order to determine distances to the discrete source. An example of the latter technique is contained in the third topic presented here, in a study of the source Sagittarius A. All the data discussed here were obtained with the Naval Research Laboratory's 21-cm. radiometer in conjunction with the 50-ft. antenna. A signal band width of 5 kc. (1 km./sec.) was employed throughout.

I. THEORY OF THE ABSORPTION EFFECT

We closely follow here the discussion first presented by Lilley at the Princeton Meeting of the American Astronomical Society, 1955 [1]. The basic problem of interpretation of the observational data is contained in Fig. 2, Cassiopeia A, which shows that the 21-cm. profile for the exact direction of a discrete source is radically different compared to any adjacent profile. We are immediately led to suppose that, in the absence of the discrete source, the 21-cm. profile for the source direction would be equivalent to the profiles which characterize the adjacent part of the sky. We shall call this profile the 'expected profile' and the profile actually obtained in the direction of the discrete source the 'observed profile'.

If the radio source lies in the centre of the antenna reception beam, the observed absorption of the discrete source radiation by interstellar H I is effective only over the small solid angle, Ω_S, subtended by the discrete source. However, all the H I gas in the antenna beam of larger solid angle Ω_B contributes to the H I emission. Thus the difference between the expected and observed profiles is due only to that gas lying within the solid angle subtended by the discrete source.

Consider a discrete source embedded in the interstellar medium and viewed by an antenna (Fig. 1). Three spatial volumes are obvious in the figure. The projection of the configuration is shown schematically at the right. Call $\Delta T_1(\nu)$, $\Delta T_2(\nu)$ and $\Delta T_3(\nu)$ the 21-cm. profiles which would result from observations of these regions individually in the absence of the

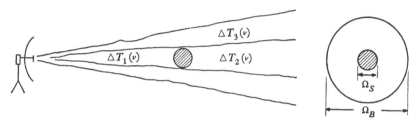

Fig. 1. Schematic representation of the three zones of H I which may be considered separately in the development of the absorption effect. Zones 1, 2, and 3, are those in which $\Delta T_1(\nu)$, $\Delta T_2(\nu)$, and $\Delta T_3(\nu)$ are found in the diagram. The projected configuration, showing the discrete source in the antenna beam, is shown at the right.

others, and in the absence of the source. Zones 1 and 2 have H I opacities given by $\tau_1(\nu)$ and $\tau_2(\nu)$. The apparent antenna temperature of the discrete source is T_A, and its continuous opacity is given by τ_S.

Then the expected profile is given by

$$\overline{\Delta T}(\nu) = \Delta T_1(\nu) + \Delta T_2(\nu)\, e^{-\tau_1(\nu)} + \Delta T_3(\nu) \qquad (1)$$

and the observed profile by

$$\Delta T'(\nu) = \Delta T_1(\nu) + \Delta T_2(\nu)\, e^{-\tau_S} e^{-\tau_1(\nu)} + \Delta T_3(\nu) + T_A e^{-\tau_1(\nu)} - T_A, \qquad (2)$$

where the last T_A in (2) is subtracted instrumentally.

Since we may write $\Delta T_2(\nu)$ as

$$T_K \left(1 - e^{-\tau_2(\nu)}\right) \frac{\Omega_S}{\Omega_B}$$

the difference between the expected and observed profiles then becomes

$$\overline{\Delta T}(\nu) - \Delta T'(\nu) = T_A(1 - e^{-\tau_1(\nu)}) + T_K(1 - e^{-\tau_2(\nu)})\,(1 - e^{-\tau_S})\, e^{-\tau_1(\nu)} \frac{\Omega_S}{\Omega_B}. \qquad (3)$$

Note that if the source is extra-galactic $(\tau_2(\nu) = 0)$ or if the continuous opacity of the source is small ($e^{-\tau s} \sim 1$) or if the angular size of the source is small, $\dfrac{\Omega_S}{\Omega_B} \ll 1$, then the last term in (3) may be neglected. One may then solve for the opacity $\tau_1(\nu)$ with the aid of observational and derived data as,

$$\tau_1(\nu) \simeq -\ln\left(1 - \frac{\overline{\Delta T}(\nu) - \Delta T'(\nu)}{T_A}\right). \tag{4}$$

2. OBSERVATIONAL RESULTS FOR CASSIOPEIA A, CYGNUS A AND TAURUS A

The most striking case of the effect is the results obtained for the Cassiopeia source (Hagen, Lilley and McClain, 1955) [2]. The basic data are shown in Fig. 2. In Fig. 2, the profile for the comparison position is characteristic

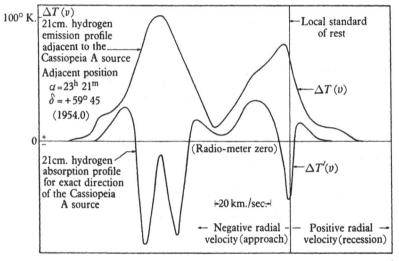

Fig. 2. Profile-type absorption measurement. Two 21-cm. hydrogen-line profiles are shown, one for the direction of the Cassiopeia A source and one for a comparison position approximately one degree away.

of the Cassiopeia region and is very nearly equivalent to the 'expected profile' which is obtained through a careful average of adjacent profiles at galactic latitudes identical with Cassiopeia A. The features for the profile obtained when the source was in the beam clearly show the effect of the H I gas within the small solid angle subtended by the radio source. The absorption features are interpreted as the effect of separate interstellar H I condensations which lie within the solid angle Ω_S. The distances of the H I clouds must be evaluated before a minimum distance to the radio source may be determined. A simple galactic rotation analysis places one cloud in the immediate vicinity of the sun, the local arm, and two in the second spiral arm, the Perseus arm, at a distance of about 3 kiloparsecs. On this basis, the minimum distance to the radio source is 3 kiloparsecs.

There remains the doubtful possibility that the three H I clouds are all actually nearby, and simply have radial velocities which are of the proper magnitude to fall in the radial velocity range of the second arm.

Absorption studies have been made in a similar fashion for Taurus A and Cygnus A. The Cygnus source is several degrees out of the galactic plane and the expected profile shows a strong emission from the near arm but a weaker intensity from the far arm. An easily detectable absorption exists in the near arm, but any effect in the far arm is too small for the 50-ft. antenna and present radiometer to detect. The measurements then place the source beyond the first arm and do not deny the possibility that the source is an extra-galactic object.

There is a clear case of absorption for the Taurus A radio source, which is located in the anti-centre region where galactic rotation analysis is almost impossible. The absorbing hydrogen cannot therefore be easily placed in terms of galactic distances. The profile for Taurus A is very nearly centred on the local standard of rest. The absorption effect covers the peak of the profile and extends to the positive radial velocity wing. The negative wing of the profile exhibits no absorption, indicating that the gas responsible for the negative wing of the profile lies beyond the radio source.

3. THE SPECIAL CASE OF SAGITTARIUS A

Figs. 3 and 4 summarize the work to date of McClain (1955)[3] in the direction of the galactic centre. Fig. 3 shows the expected profile $\overline{\Delta T}(\nu)$, the observed profile $\Delta T'(\nu)$ and the derived optical depth $\tau_1(\nu)$, at the position $l = 327°8$ and $b = -1°4$. It is seen that gas in the wings of the profile is not effective in absorbing continuum radiation from the radio source and hence may be interpreted as being beyond the source. This figure does not yield information about the absolute distance of various features in the profile. An alternative explanation that the non-absorbing gas is in the edge of the antenna beam is unlikely for two reasons. First, studies of the angular distribution of gas in the wings indicate that while the gas is of limited angular extent, the angular size is well in excess of the beam-width and is centred very close to the source. Secondly, if only a portion of the beam were filled with a small gas cloud the emission intensity would suffer by the ratio of the solid angle of the beam to the solid angle of the gas

cloud. The emission temperature of the observed profile in the range
-15 km./sec. to -20 km./sec. where absorption is known to be absent is
of the order of $50°$ K. This temperature is characteristic of the region at
this velocity and would seem to preclude a partially filled beam.

An independent distance scale has been obtained by assuming a constant
thickness for the hydrogen gas and by relating the angular extent of the gas
to its radial velocity.

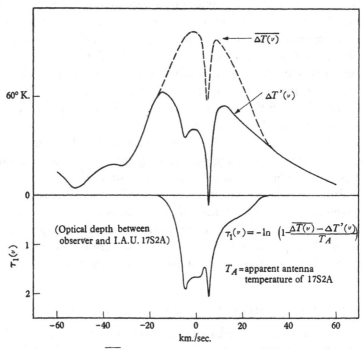

Fig. 3. The expected profile $\overline{\Delta T}(\nu)$ and the observed profile $\Delta T'(\nu)$ for the direction of 17S2A
are shown at the top. The computed value of the optical depth is shown at the bottom on a
common velocity scale.

If one assumes that the Galaxy is a flat disk of thickness t an observer
inside the disk can relate the angular width ϕ of the disk to the distance d
at which the angle was measured by the relation

$$\tan \frac{\phi}{2} = \frac{t}{2d}.$$

The assumption that the Galaxy is of uniform thickness is obviously not
rigorous but a more complicated model would hardly be justified. The
angular extent of the gas was measured by two different methods. First,
a number of profiles were taken at positions along a line normal to the

84

plane at $l = 327°8$ (Lund pole). The positions were symmetrically spaced about $b = -1°4$. The central position $l = 327°8$ and $b = -1°4$ is the measured position of 17S2A at 21 cm. Profiles were taken at $±0°5$, $±1°0$, $±1°5$, $±2°5$ and $±4°$ with respect to $b = -1°4$. Temperatures at corresponding radial velocities on the profiles were then plotted as a function of angle. Profiles taken in this manner do not yield a satisfactory measure of angular extent at velocities lower than $±10$ km./sec. This is a result of the large Ophiuchus dark complex (Heeschen and Lilley, 1954) [4] which makes an angle of about 20° with the galactic plane. For this reason

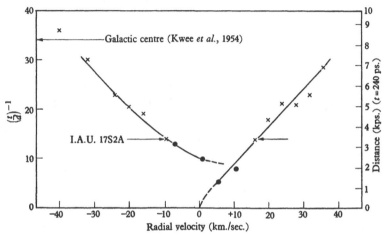

Fig. 4. Diagram showing the relation between distance and radial velocity obtained from the measurements reported herein. The left-hand ordinate is the dimensionless quantity $(t/d)^{-1}$. This quantity is the cotangent of the full half-intensity width. The right-hand ordinate is a distance scale derived by assuming a uniform thickness of 240 parsecs for the Galaxy.

declination sweeps (which avoid the complex) corrected to equivalent widths normal to the plane were used as the measure of angular extent at the lower radial velocities. The angular extent of the gas was taken as the measured one-half intensity point.

The data obtained in these measurements are illustrated in Fig. 4. The left-hand ordinate is the dimensionless quantity $\left(\dfrac{t}{d}\right)^{-1}$ (cotangent of the full one-half intensity width). The right-hand ordinate is the distance scale obtained by assuming a thickness of 240 parsecs. The abscissa is the radial velocity with respect to the local standard of rest. The crosses are the values taken from the hydrogen profiles and the dots are those obtained by means of angular sweeps. Both the positive and negative wings are seen to reach values commensurate with the value of $1°7$ at 8·2 kiloparsecs given by

Kwee, Muller, and Westerhout (1954) [5] for the galactic centre. Of primary interest, however, is the region within 4 kiloparsecs of the sun. It will be noted that there is an indication of a double-valued distance at about +5 km./sec. This is a necessary condition if one invokes the self-absorption hypothesis of Heeschen (1955) [6]. An attempt at a direct measurement of the angular extent of the more distant gas at +5 km./sec. leads to ambiguities and its existence can only be inferred by the notch in the profiles surrounding the centre.

Up to this point no account has been taken of differential galactic rotation. The effect of differential rotation when an antenna beam of finite size is pointed at the centre will be to increase the apparent velocity dispersion of the observed profile. Quantitative checks of this effect using the Dutch rotational velocities reveal that the observed dispersion at intermediate distances is several times that expected from pure rotation. For example, at a distance of 5 kiloparsecs from the sun the dispersion expected from a one degree beam would be about 2 km./sec. Fig. 4 reveals, however, that the observed velocities present at 5 kiloparsecs are of the order of 25 km./sec. Therefore, a consideration of the Dutch rotational velocities in conjunction with the measurements reported here leads to the conclusion that the random velocity η probably increases rather rapidly beyond 2 kiloparsecs from the sun, and Fig. 4 may be considered a measure of this dispersion.

In applying the distance scale to radio source 17S2A one is forced to consider the random motions present at the distance of the source. These random motions have the effect of making it difficult to associate a particular value of radial velocity with the onset of absorption. In Fig. 3 it is seen that $\tau_1(\nu)$, and hence absorption, occurs from -16 km./sec. to $+32$ km./sec. If one assumes that the correct value of $\tau_1(\nu)$ lies between 0·1 and 0·5 of the maximum value, the limits placed on the associated radial velocities are -7 km./sec. to -12 km./sec. for the negative wing and $+10$ km./sec. to $+24$ km./sec. for the positive wing. These velocities place the distance of the source between 3 and 4 kiloparsecs for the negative branch of Fig. 6 and between 2 and 5 kiloparsecs for the positive branch. Since the positive and negative branches have different slopes it is of interest to determine at what value of $\tau_1(\nu)$ and hence at what velocity the positive and negative branches yield the same distance. The value of $\tau_1(\nu)$ at which this occurs is 0·5 (25 % of the maximum value) and the distance indicated by both branches of the distance scale is 3·4 kiloparsecs. This value is indicated by arrows in Fig. 4.

A consideration of the measurements reported herein leads to the conclusion that the radio source responsible for the intense emission peak at $l = 327°8$ and $b = -1°4$ is probably not associated with the galactic centre but rather is superimposed on the broad central galactic background. A consideration of the uncertainties of the measurement places it between extreme limits of 2 and 6 kiloparsecs with a probable distance between 3 and 4 kiloparsecs. It is interesting to note that Haddock and McCullough (1955) [7] have suggested that this source is thermal and possibly associated with an H II region in the group of OB stars studied by Hiltner (1954) [8]. The mean distance given by Hiltner for this group is 3 kiloparsecs. It should be noted that the apparent position of a complex source might vary with the wave-length at which the observation is made and with the antenna beam size if the sources are not exactly concentric. Such variation has been noted for this source.

REFERENCES

[1] Lilley, A. E. (Abstract) *A.J.* **60**, 167, 1955.
[2] Hagen, J. P., Lilley, A. E. and McClain, E. F. *N.R.L. Report* 4448, 18 October 1954; *Ap. J.* **122**, 361, 1955.
[3] McClain, E. F. *Ap. J.* **122**, 376, 1955.
[4] Heeschen, D. S. and Lilley, A. E. *Proc. Nat. Acad. Sci.* **40**, 1095, 1954.
[5] Kwee, K. K., Muller, C. A. and Westerhout, G. *B.A.N.* **12**, 211, no. 458, 1954.
[6] Heeschen, D. S. *Ap. J.* **121**, 569, 1955.
[7] Haddock, F. T. and McCullough, T. P. (Abstract) *A.J.* **60**, 161, 1955.
[8] Hiltner, W. A. *Ap. J.* **120**, 41, 1954.

Discussion

Lilley: We have a distance discrepancy to Cassiopeia A of approximately a factor of six between the radio analysis of the absorption profile and the optical determination which has just been mentioned by Minkowski. The radio distance depends directly on the assumption that the absorption features at -38 and -48 km./sec. are produced by H I gas located in the second spiral arm. If the smaller optical distance is correct we must differently interpret the radio absorption profile and account for the negative velocity features. The possibility of nearby interstellar clouds has been mentioned. A second possibility also exists—the cool absorbing gas may be dynamically associated with the radio source itself. This situation could conceivably satisfy the velocity and geometrical projection requirements of the absorbing H I gas.

Greenstein: The conclusion that there is an interaction between the Cas source and the recoiling H I clouds in the vicinity can hardly be avoided. Could the Cas source have a cold H I envelope? The sharpness of the absorption components in the Cas A source is very unusual and striking.

COMMENTS ON McCLAIN'S OBSERVATIONS OF IAU 17S2A

F. K. EDMONDSON

Goethe Link Observatory, Indiana University, Bloomington, Indiana, U.S.A.

Reference to Fig. 1, p. 20 of this symposium will show that a 1° beam directed toward the galactic centre receives a range of velocities that is not negligible. This is due to galactic rotation, which causes material in one edge of the beam to approach and in the other edge to recede.

Assuming a Gaussian antenna pattern and using the rotational velocities from *B.A.N.* no. 458, we can compute σ_A, the 'antenna dispersion'. If the beam is filled, this dispersion is what would be observed even if random motions were zero. Table 1 gives σ_A as a function of distance from the sun toward the galactic centre for the N.R.L. 50-ft. antenna. Identical values are obtained from circular orbits ($\phi = 0°$) and spiral orbits ($\phi = 4°$)

Table 1

r (kpc.)	σ_A (km./sec.)	r (kpc.)	σ_A (km./sec.)
0–3	$< \pm 1 \cdot 0$	8·2	$\pm 22 \cdot 8$
4	1·3	9	7·8
5	2·0	10	3·7
6	3·1	11	2·3
7	5·4	12	1·6
8	18·2	13+	$< 1 \cdot 0$

If we define the velocity range of a Gaussian profile as corresponding to an intensity equal to 5% of central intensity, then the range equals $\pm 2 \cdot 5\sigma$. Assuming that the random motions have a dispersion of ± 10 km./sec., Table 2 gives the expected velocity range at different distances

Table 2

r (kpc.)	$2 \cdot 5 \sqrt{10^2 + \sigma_A^2}$ (km./sec.)
0–6	$\pm 25 \cdot 0$ to $\pm 26 \cdot 3$
7	$\pm 28 \cdot 5$
8	$\pm 52 \cdot 0$
8·2	$\pm 62 \cdot 3$
9	$\pm 31 \cdot 8$
10+	$\pm 26 \cdot 8$ to $\pm 25 \cdot 0$

toward the galactic centre which should be observed with the N.R.L. 50-ft. antenna.

The general features of McClain's profile of I.A.U. 17S2A can be explained in terms of the figures in Table 2. The relatively small velocity range of the observed absorption is consistent with McClain's estimate of distance. The unabsorbed wings extending out to ± 60 km./sec. are probably due to material at the distance of the galactic centre being picked up in the edges of the beam. If this explanation is correct, a larger antenna should reduce the observed intensity of the unabsorbed wings, while making no change in the velocity range of the absorption.

DETECTION OF THE SPECTRAL LINE OF DEUTERIUM FROM THE CENTRE OF THE GALAXY ON THE WAVE-LENGTH OF 91·6 CM.

G. G. GETMANZEV, K. S. STANKEVITCH AND V. S. TROITZKY
Gorky State University, U.S.S.R.

As was shown by I. S. Shklovsky, a radio spectral line from interstellar deuterium on $\lambda_D = 91\cdot6$ cm. ($f_D = 327\cdot38424$ Mc./s.) may be expected [1]. According to Shklovsky the difference between the effective temperature T_{eff} in this line and outside it must be of the order of $1/500\ T_{\text{eff}}$ in the direction of the galactic centre, if the concentration of deuterium equals 10^{-3} that of hydrogen. In this case, contrary to the emission by hydrogen, an absorption line is expected. As T_{eff} equals about $300°$ the depth should be of the order of $0\cdot6°$.

For measurements of this spectral line an aerial consisting of a 4-metre paraboloid with a half-wave dipole located in its focus was used. The time constant of the arrangement is about 90 sec. The noise factor $N \sim 10$ was measured by means of a standard generator. The width of the high-frequency band was about 15 kc./s., and the fluctuation threshold of the sensitivity of the arrangement T_{min}, expressed in the aerial temperature, is about $4°$.

In the process of the observations the frequency of the second heterodyne was varied smoothly in an interval near f_D. These observations embrace the period of time from October 1954 till June 1955 inclusive. About fifty-three records were obtained during this period. As the expected effect is less than the value of T_{min}, the ordinates for a great number of individual records are averaged. The resulting mean curve is a typical record of a radio absorption line. The width of the line contour determined from the mean curve is of the order of 30 kc./s. The effective temperature, corresponding to the 'depth of the line', determined from the records, equals $2°5$ (probable limits $1°5$ to $4°5$). This value corresponds to a concentration of the deuterium of about $1/300$ atom./cm.³ in the direction towards the centre of the Galaxy.

Monochromatic radiation near f_D was absent when the aerial was directed to the pole of the Galaxy. This result might be interpreted in the sense that the concentration of deuterium in the direction to the galactic poles is much less than in the direction to the galactic centre, which seems to be quite natural.

REFERENCE

[1] Shklovsky, I. S. *A.J. U.S.S.R.* **29**, 144, 1952.

*Discussion**

Pawsey: Corresponding observations were attempted in Sydney by G. J. Stanley and R. Price in 1954. They used an 80-ft. diameter paraboloid directed near to the zenith (declinations around $-34°$) and examined several regions near the galactic centre. Their results were negative but since their estimated uncertainty in intensity was a degree or so the results were not inconsistent with those reported by Getmanzev and his colleagues. (Note dated 24 October 1955.)

Hey: We are also attempting to detect the deuterium line at the Radar Research Establishment (Malvern, England). We have suffered from a certain amount of interference but the results indicate no absorption line of more than $1/2°$ K. aerial temperature in a direction near the galactic centre. Our aerial has a beam-width of approximately $10°$, and owing to certain limitations due to the position of the site and the aerial beam-width we are making observations with a fixed aerial direction such that the galactic equator crosses the axis of the beam about $5°$ from the galactic centre. More observations are needed before we can be quite definite. (Note dated 8 November 1955.)

---

* These data were not given in detail at the symposium but were submitted in later correspondence, as noted. [*Editor.*]

MICROWAVE AND RADIO-FREQUENCY RESONANCE LINES OF INTEREST TO RADIO ASTRONOMY

C. H. TOWNES

Physical Laboratory, Ecole Normale Supérieure, Paris

A review of microwave and radio-frequency spectral lines which might possibly be detected by the techniques of radio astronomy is attempted here. Brief discussions of this type have already been given by several authors [1, 2, 3, 4]. However, the present treatment is somewhat more complete than previously published material, and has the advantage of more recent information about certain transition frequencies. It includes a general discussion of types of spectra which might be found, expected intensities, and some characteristics and known frequencies of the lines which may be of interest in radio astronomy.

Transitions which lie in the microwave or radio-frequency region can be expected to come from atomic or molecular hyperfine structure, from atomic or molecular fine structure, and from molecular rotational frequencies. It is of course possible that some odd circumstance or chance makes two electronic levels which are not fine structure of the same level fall so close together that a transition between them corresponds to a microwave frequency. However, there seem to be no known 'accidents' of this type that are of interest to radio astronomy. It is also possible that some molecular vibration be hindered by just the precise amount required to reduce its frequency to the microwave or radio-frequency region. This occurs in the well-known case of ammonia, but there seems to be no possibility of another case that would be of importance to radio astronomy. The only other varieties of sharply resonant microwave or radio-frequency transitions which have so far been found in the laboratory are transitions between Zeeman or Stark components of one or neighbouring energy levels. However, such transitions fall in the interesting frequency regions only when the magnetic or electric fields are rather large, and to be sharp and easily detectable they require a homogeneity of the field that is not likely to occur over any resolvable region of an astronomical object. Hence we shall be concerned only with hyperfine structure, fine structure, molecular rotational frequencies and the inversion spectrum of ammonia.

I. INTENSITIES OF LINES

The intensity of a microwave or radio-frequency transition may be represented to a reasonable approximation by an absorption coefficient given by [5]*

$$\gamma = \frac{8\pi^2 n \, |\mu|^2}{3ckT} \frac{\nu^2 \Delta\nu}{(\nu-\nu_0)^2+(\Delta\nu)^2},\tag{1}$$

where γ is the absorption coefficient per unit length,

n is the density of atoms or molecules in the lower of the two levels between which the transition occurs,

$|\mu|^2$ is the square of the dipole moment matrix element (averaged over the magnetic quantum number M),

c, k, T are, as usual, the velocity of light, Boltzmann's constant, and the absolute temperature respectively,

ν is the frequency being considered or measured,

ν_0 is the resonant or central frequency of the transition,

$\Delta\nu$ is the half-width of the resonance at half-intensity.

Expression (1) applies strictly only when $h\nu \ll kT$ and when a Lorentz-type line shape occurs. However, it is sufficiently accurate for our purpose in any case which will be encountered below.

At the peak intensity where $\nu=\nu_0$, (1) becomes

$$\gamma = \frac{8\pi^2 n \, |\mu|^2 \nu^2}{3ckT\Delta\nu}.\tag{2}$$

If the line width is primarily due to collisions with other atoms or molecules, $\Delta\nu$ is proportional to the total pressure and is of the order 10 Mc./s. at a pressure of 1 mm. Hg, or when the total density N of molecules is approximately $3\cdot5 \times 10^{16}$. Hence, inserting values, (2) becomes

$$\gamma = \frac{2 \times 10^{-19} n\nu'^2 \, |\mu'|^2}{NT},\tag{3}$$

where γ is in cm.$^{-1}$,

ν' is in Mc./s.

μ' is in Debye units (10^{-18} e.s.u.) or 10^{-18} e.m.u.

---

* Background for, and details connected with, much of the following material can be found in reference [5].

Except for the atmospheres of planets or stars, the line broadening in astronomical objects usually comes primarily from Doppler effects rather than from collisions (or pressure broadening). In this case, $\Delta\nu$ depends on the distribution of velocities in the medium being observed. Assuming that velocities of the gases observed vary more or less randomly by about $\pm$ 10 km./sec., $\Delta\nu$ may be taken as $\dfrac{\nu}{3 \times 10^4}$, so that (2) becomes

$$\gamma \approx \frac{6 \times 10^{-15}n \mid \mu' \mid^2}{T\lambda} \approx \frac{5 \times 10^{-19}n \mid \mu'' \mid^2}{T\lambda}, \qquad (4)$$

where γ is in cm.$^{-1}$ if λ is in cm.,

n is the number of atoms or molecules in the lower state of the transition per c.c.,

μ' is the dipole matrix element in Debye (10^{-18} e.s.u.) or 10^{-18} e.m.u.,

μ'' is the dipole matrix element in Bohr magnetons ($0\cdot922 \times 10^{-20}$ e.m.u.).

Consider now the case where a radio telescope is directed towards a uniform gas which completely fills the field of view and behind which there is a uniform background of radiation with some effective temperature T_B as judged by the amount of radio-frequency radiation at the frequency of interest. If the effective temperature of the gas (determined by the relative population of the upper and lower states of the transition considered) is T, the change in apparent temperature due to the gas resonance is

$$\Delta T = (T - T_B)\,(1 - e^{-\gamma L}), \qquad (5)$$

where L is the length of the column of gas. For detection of this change, ΔT must be about as large as $1°$, in which case we may expect that $\gamma L \ll 1$, and (6) can be written

$$\Delta T \approx (T - T_B)\,\gamma L. \qquad (6)$$

Frequently $T_B \ll T$, and (6) combined with (4) gives

$$\Delta T \approx \frac{6 \times 10^{-15}n \mid \mu' \mid^2 L}{\lambda} \approx \frac{5 \times 10^{-19}n \mid \mu'' \mid^2 L}{\lambda}. \qquad (7)$$

Thus for detection, the column of gas must contain about $2 \times 10^{14}\lambda$ molecules of the right type per cm.^{2} if the matrix element is due to an electric dipole ($\mid \mu' \mid \approx 1$), or $2 \times 10^{18}\lambda$ if it is due to a magnetic dipole ($\mid \mu'' \mid \approx 1$).

94

It is of course the number of atoms or molecules having a resonance at the desired frequency that must be considered. Molecules of a given species may in some cases be distributed among a wide variety of levels, so that only a small fraction of them effectively respond to a given radio frequency. However, at low temperatures, molecules or atoms will in many of the cases considered below be concentrated in only a very few levels, so that the total number of a given species is not much greater than the number in a particular state involved in the radio-frequency transition.

If one considers interstellar gas in our own or a similar Galaxy, then the largest dimension is about $L \approx 10^{23}$ cm., and the average molecular density must be as large as $n > 2 \times 10^{-9} \lambda$ or

$$n > 2 \times 10^{-5} \lambda \qquad (8)$$

respectively for electric or magnetic dipole moments.

As has been suggested [6], it may be possible to use sufficiently large antennas so that some of the bright continuous sources give effective background temperatures T_B several powers of 10 greater than the gas temperature, and thus considerably reduce the molecular (or atomic) density indicated by (8) as necessary for detection.

Normally, transitions between different levels of fine or hyperfine structure involve no electric dipole moment matrix element, but rather a magnetic dipole matrix element of the order of one Bohr magneton ($|\mu''| \approx 1$) or in some cases only a nuclear magneton ($|\mu''| \approx 10^{-3}$). Molecular rotational transitions, on the other hand, normally involve electric dipole matrix elements, and are of the order of 1 Debye ($|\mu'| \approx 1$). Hence fine or hyperfine structure transitions are often much weaker than those between rotational levels, and require a much larger density n for detection. There are, however, some important exceptions to this generalization, which occur when so-called fine structure is not due to the effects of magnetic moments, or when the fine or hyperfine structure is comparable in magnitude to the separations between levels unsplit by the effects of electron spin or nuclear moments. These include the following cases where there is an electric dipole matrix element between fine or hyperfine structure levels:

(1) Transitions between ns and np levels of hydrogen-like atoms. In principle, other very close coincidences between two electronic levels might occur in more complex atoms, but no other cases are known that are of concern here.

(2) Transitions between Λ-doublets, as in the OH microwave spectrum.

(3) Transitions between hyperfine levels of a symmetric top molecule

95

that involve K-degeneracy. No cases of this type are likely to be of importance to radio astronomy.

(4) The so-called 'fine' structure of ammonia, the inversion spectrum.

Tables 1, 2, 3 and 4 give information about most of the microwave and radio-frequency transitions which are at present known and which may have sufficient intensities to warrant consideration for the purposes of radio astronomy.*

No transitions are listed with frequencies greater than 300,000 Mc./s. ($\lambda = 1$ mm.), since sensitive radio-detection can probably not soon be achieved at these high frequencies. In many cases lines are listed for completeness or comparison which are probably too weak for detection in any astronomical object. Since all microwave and radio-frequency lines are rather weak, it seems improbable that any will be found in astronomical objects before their frequencies are fairly accurately known from laboratory or very precise theoretical work. Because of this great importance of precise frequencies, there is a rather generous listing of transitions whose frequencies are now known, even though in some cases it seems unlikely that they are sufficiently intense for detection by radio telescopes.

2. HYPERFINE TRANSITIONS

Atomic hyperfine structure occurs only for nuclei with spin greater than zero (except for isotope shifts, which are sometimes called hyperfine structure, but are of no interest here). Such nuclei are relatively rare, the commonest being H^1, H^2 and N^{14}. Atoms of the latter two are less abundant than those of H^1 by a factor of several thousand, according to the usual estimates [7]. All other nuclei with non-zero spin are less abundant even than H^2 and N^{14} by a factor of about 100, and hence in most cases their hyperfine transitions can probably not be detected by radio astronomical techniques, except under some very special circumstances. Table 1 lists characteristics of these hyperfine lines, including all atomic hyperfine transitions with known frequencies from atoms of abundance not less than about 10^{-6} that of H^1.

In addition to H^1 and H^2, the h.f.s. of various stages of ionization of N^{14} and of He^3 II seem to be interesting cases for radio astronomy. Unfortunately, the frequency for N I only has so far been measured experimentally, and this frequency is so low that its detection seems quite difficult. The h.f.s. frequencies for N VII and He^3 II can be calculated with some

* Most of the information in these tables and references to original sources may be found in references [5] and [8].

Table 1. *Some hyperfine structure transitions of interest to radio astronomy*

The approximate abundances of each nuclear species relative to H^1 for our Galaxy are given. These are not necessarily the relative abundance in interstellar gases, where H^1 has a density of 1 atom./cm.^{3}. The matrix element for each transition is due to a magnetic dipole, and is approximately one Bohr magneton ($|\mu''|=1$). The quantity F is the vector sum of electronic angular momentum J and nuclear spin I.

Nucleus	Abundance relative to H^1	Transition		Frequency (Mc./s.)	Comments
H^1	1	H I $^2S_{1/2}$,	$F=0-1$	1420·4051	Already detected in interstellar gases
H^2	2×10^{-4}	H I $^2S_{1/2}$,	$F=1/2-3/2$	327·3842	
He^3	10^{-5}	He II $^2S_{1/2}$,	$F=1-0$	8659·3*	Calculated frequency without allowance for finite nuclear size
He^3	10^{-5}	He I $^3S_{3/2}-{}^3S_{1/2}$		6739·71	Metastabile state
N^{14}	5×10^{-4}	N I $2^3p\,^4S_{3/2}$,	$F=3/2-5/2$	26·12	
			$F=1/2-3/2$	15·67	
		N II $2p^2\,^3P_1$,	$F=1-2$	~ 500	Approx. 60 cm.$^{-1}$ above ground state
			$F=0-1$		
		N III $2p\,^2P_{1/2}$,	$F=1/2-3/2$	~ 800	
		N V $2s\,^2S_{1/2}$,	$F=1/2-3/2$	~ 3000	
		N VII $1S\,^2S_{1/2}$,	$F=1/2-3/2$	53·060	Calculated frequency without allowance for finite nuclear size
Na^{23}	1×10^{-6}	Na I $3s^2\,^2S_{1/2}$,	$F=1-2$	1771·61	
Al^{27}	3×10^{6}	Al I $3p\,^2P_{1/2}$,	$F=2-3$	1450	
P^{31}	4×10^{-7}	P I $3^3p\,^4S_{3/2}$,	$F=2-3$	112	
Cl^{35}	4×10^{-7}	Cl I $3p^5\,^2P_{3/2}$,	$F=3-2$	670·018	
			$F=2-1$	355·244	
			$F=1-0$	150·145	

* *Note added in proof:* Novick (private communication) has now calculated this transition frequency for the He³ II ground state as 8665·66±0·18 Mc./s. from recent measurements on the He³ II 2s state [R. Novick and E. Commins, *Phys. Rev.* **103**, 1897, 1956]. The calculation includes small electrodynamic corrections evaluated by Mittelman (to be published).

accuracy, and their precise experimental determination would be of considerable interest to physics as well as to astronomy because of a slight modification of the frequencies due to the finite nuclear size (hyperfine structure anomaly). Frequency changes of this type are, however, not very large, and particularly in the case of He³ the resulting uncertainty should not be large enough appreciably to complicate a search for this line. Assuming abundances similar to those on earth, He³ is considerably less abundant than H². However, the fact that its resonant wave-length is so much shorter than that for H² makes it approximately as favourable (cf. equation (8)), assuming that a large fraction of He atoms are ionized.

3. FINE STRUCTURE

It is only the lightest atoms for which the fine structure in the ground or lower states is sufficiently small to give transitions in the microwave region. In fact, the most interesting cases are limited to H, D, He II and perhaps N I which are listed in Table 2. Although small fine-structure spacings do occur in excited states of heavier atoms, these states are so short-lived that their microwave spectra would not be detectable in astronomical objects. The $n = 2$ state of hydrogen-like atoms has some characteristics which may make it detectable by radio telescopes, even though the abundance of this excited state may be very small. In particular, the transitions between $2s$ and $2p$ states involve a rather large electric dipole moment ($|\mu'| = 6\cdot2$ for H). Furthermore, the frequency is accurately known and the $2s$ level is metastable (with a lifetime of 8 sec. in the case of hydrogen). Hence an appreciable number of atoms may accumulate in it.

Table 2. *Atomic fine structure transitions of interest to radio astronomy*

The relative abundances given are those for atoms in the ground states. All cases listed are metastable excited states.

Atom	Abundance relative to H	Transition	Frequency (Mc./s.)
H	I	H I $^2S_{1/2}$–$^2P_{1,2}$	1057·8
		H I $^2P_{3/2}$–$^2S_{1/2}$	9912·6
He	10^{-1}	He II $^2S_{1/2}$–$^2P_{1/2}$	14·020
		$^2P_{3/2}$–$^2S_{1/2}$	161·510
N	5×10^{-4}	N I $2p^3\ ^3D_{5/2}$–$^3D_{3/2}$	240·000

Since the $2p$ levels are not metastable, they will contain much fewer atoms than do the $2s$ levels. Hence these transitions will be strengthened by approximately the factor $\dfrac{kT}{k\nu_0}$, where ν_0 is the transition frequency. However, the instability of the $2p$ levels greatly widens the resonance transition so that $\Delta\nu \sim 50$ Mc./s. (for H). This decreases the peak intensity by a factor of about 1000 over the cases where Doppler broadening predominates. This factor of 1000 is not all loss, since the receiver band-width should be increased to about 100 Mc./s. for detection of this line, and the minimum detectable temperature change should thus be decreased to perhaps $0\cdot1°$ C. The net result is that for the $^2S_{1/2}$–$^2P_{3/2}$ or the $^2S_{1/2}$–$^2P_{1/2}$ transition, the minimum detectable density of atoms in interstellar space is of the order of 10^{-11}/cm.³, or the total number required in the line of sight is about 10^{12}/cm.² assuming $T_B < T$ (cf. equation 6). An accurate estimate of the density of H atoms in the $2s$ state is difficult. However, the numbers required do not seem impossibly large. A more detailed discussion has been given by J. P. Wild [10]. J. E. Blamont (private communication) has considered radiation of this type from planetary nebulae.

4. MOLECULAR ROTATIONAL SPECTRA

In addition to fine and hyperfine structure, molecules have rotational levels which are often sufficiently close to produce microwave transitions. Microwave rotational transitions which may be of interest to radio astronomy are listed in Table 3.

Table 3

Some microwave transitions between molecular rotational levels of interest to radio astronomy. The inversion spectrum of NH_3 is also listed.

Molecule	Electronic state	Transition	Frequency (Mc./s.)	Matrix element in Debye units	Comments
CaH	$^2\Sigma$	$J=3/2-1/2$	254,080	Electric dipole unknown	
		$J=1/2-1/2$	252,650		
CO	$^1\Sigma$	$J=1-0$	115,270·6	0·10	
CO$^+$	$^2\Sigma$	$J=3/2-1/2$ } $1/2-1/2$ }	117,980	Electric dipole unknown	
CS	$^1\Sigma$	$J=1-0$	48,991·0	2·0	
NO	$^2\Pi_{1/2}$	$J=3/1-1/2$	150,176·3	0·07	Also other nearby lines due to h.f.s.
H_2O	$^1\Sigma$	$6_{1,6}-5_{2,3}$	22,235·22	0·16	
N_2O	$^1\Sigma$	$J=1-0$	25,123·28	0·17	Other lines at multiples of the frequency given. Very small h.f.s. present
HCN	$^1\Sigma$	$J=1-0, F=1-1$	88,600·1	1·72	Also l-doublet transitions may occur
		$F=2-1$	88,601·5	2·22	
		$F=0-1$	88,603·6	0·99	
CH_2					Structure and spectrum unknown, but may produce some microwave lines
NH_2					Structure and spectrum not well known, but probably produces some microwave lines
NH_3	$^1\Sigma$	Inversion, $J=1$, $K=1$	23,694·48	1·0	Also other inversion transitions at nearby frequencies
O_3	$^1\Sigma$	$1_{11}-2_{02}$	42,832·62	0·17	Also other rotational transitions
		$1_{11}-0_{00}$	118,364·3	0·53	

Probably only the diatomic molecules of Table 3 exist in detectable abundance in interstellar space; CaH and CN seem to be the only ones yet found there by optical spectroscopy. Of the common hydrides, CaH has the lowest rotational frequency which, even so, is near the border of the range of present microwave techniques. The other molecules listed may possibly be found in planetary atmospheres, in comets, or in some other regions of gaseous concentrations. A suitable telescope with resolution near to one minute of arc, for example, could very probably detect NH_3 lines in the atmosphere of Jupiter. Omission from the table of certain common molecules such as N_2 and CO_2 may strike the reader. They lack electric dipole moments and hence have no rotational spectra.

5. MOLECULAR HYPERFINE STRUCTURE

The only molecules of sufficient abundance and with sufficiently large hyperfine structure to be interesting to radio astronomy seem to be the hydrides which are in $^2\Sigma$ or $^3\Sigma$ states. Hydrides in electronic $^2\Pi$ states have considerably smaller h.f.s. (cf. OH, Reference [5]). Of the $^2\Sigma$ and $^3\Sigma$ hydrides, H_2^+, NH, OH^+ and CaH are probably the most abundant. They should all give hyperfine structure of the same order of magnitude as that found in the $1S$ state of atomic hydrogen. Unfortunately, these hyperfine transitions involve only magnetic dipole matrix elements as in atomic H, and this coupled with the lower abundance of the molecules and uncertainty in the precise frequencies of transition will make their detection quite difficult.

The most interesting of the above hydrides appears to be H_2^+* because it may be more abundant than the others, and also because its hyperfine frequencies can be obtained with some accuracy from calculations, even though there seems to be little immediate hope of their measurement in the laboratory. The ground rotational state of H_2^+ is ionized parahydrogen, and hence has no hyperfine structure. Orthohydrogen does have hyperfine structure and exists in the states $J = 1/2$ and $J = 3/2$ which are about 60 cm.$^{-1}$ above the ground state. In regions of low temperature it will of course be considerably less abundant than parahydrogen.

In orthohydrogen, the electron spin-nuclear spin coupling constant is of the order of one or two thousand megacycles/sec., whereas the electron spin-rotational motion coupling should be about a factor of ten less. The electron and nuclear spins would then be coupled together fairly strongly, and their resultant added vectorially to the end-over-end rotation, giving coupling case $(b_{\beta S})$ (cf. Reference [5], sec. 8–2).

---

* The possible value of this molecule to radio astronomy was suggested some time ago by B. Burke.

One may expect the lowest rotational state of orthohydrogen with either $J = 1/2$ or $J = 3/2$ to have hyperfine structure transitions with frequencies somewhere near the value for the hyperfine transition already detected in atomic hydrogen. A much more exact theoretical estimate of the hyperfine frequencies in H_2^+ can be made by some numerical work, since the electronic wave-functions can be obtained with precision [9]. However, a precise estimate will also involve evaluation of the fine-structure doubling (in this case much smaller than the hyperfine structure), which gives a deviation from pure coupling case ($b_{\beta S}$).

6. MOLECULAR FINE STRUCTURE

Molecular fine structure occurs with widely varying characteristics. The most interesting type of molecular fine structure from the point of view of radio astronomy is Λ-type doubling, since transitions between Λ doublets can involve electric dipole moments and hence are enormously more intense than most of the other fine-structure transitions, which involve only magnetic dipole moments. The most abundant molecules with this type of spectra are CH, OH and SiH.* Data on these and some other molecules are given in Table 4. Only in the case of OH are the Λ-doublet frequencies accurately known. This molecule appears, in fact, to be susceptible to detection by radio telescopes. Its average abundance in interstellar space needs to be about 10^{-6}/cm.^{3} for detection against a normal background, or considerably less for detection by absorption of radiation from a high-intensity source. Furthermore, the frequency of the strongest transition lies near the region where considerable apparatus has already been developed for study of the hydrogen lines.

Fine structure splittings between $^2\Pi_{\frac{1}{2}}$ and $^2\Pi_{\frac{3}{2}}$ levels correspond roughly to atomic fine structure, and are too large to fall in the microwave region for the lowest states of any molecule which would be abundant enough to be of interest here.

There are a variety of molecules in $^2\Sigma$ or $^3\Sigma$ states that involve spin (ρ-type) doubling or tripling with separations corresponding to microwave or radio frequencies. These transitions are weak because they have only magnetic dipole matrix elements. However, some, such as those of O_2, can certainly be observed by absorption in the earth's atmosphere, and perhaps might be found in comets or other planetary atmospheres if telescopes of sufficiently high resolution are available. The more favourable cases of these types of fine structure transitions are also listed in Table 4.

Table 4. *Some molecular fine-structure transitions of interest to radio astronomy*

Transitions between Λ-doublets [electric dipole, $|\mu'| \sim 1$]

Molecule	Electronic state	Transition	Frequency (Mc./s.)	Comments
CH	$^2\Pi_{1/2}$ $^2\Pi_{3/2}$	$J = 1/2$ $J = 3/2$	~ 1000	Also transitions at higher frequencies $\sim 1000 N(N+1)$ in excited rotational states
OH	$^2\Pi_{3/2}$	$J = 3/2, F = 2-2$ $F = 1-1$	$1665 \cdot 0$ $1667 \cdot 0$	Also, weak hyperfine satellites and higher frequencies in excited rotational states
AlH	$^1\Pi$	$N = 1$	380	Also higher frequencies in excited rotational states
SiH	$^2\Pi_{1/2}$	$J = 1/2$	2400	Also higher frequencies in excited rotational states
SH	$^2\Pi_{3/2}$	$J = 3/2$	114	Also higher frequencies in excited rotational states

Transitions between spin fine-structure levels [magnetic dipole, $|\mu''| \sim 1$]

Molecule	Electronic state	Transition	Frequency (Mc./s.)	Comments
H_2^+	$^2\Sigma$	$N = 1$	~ 100	Approx. 60 cm.$^{-1}$ above ground state
NH	$^3\Sigma$	$N = 1, J = 0-1$ $J = 1-2$	26,000 13,000	
OH$^+$	$^3\Sigma$	$N = 1, J = 0-1$ $J = 1-2$	133,000 63,000	
MgH	$^2\Sigma$	$N = 1$	1000	
CaH	$^2\Sigma$	$N = 1$	1500	
CN	$^2\Sigma$	$N = 1$	360	
CO$^+$	$^2\Sigma$	$N = 1$	450	
O_2	$^3\Sigma$	$N = 1, J = 0-1$ $J = 1-2$ $N = 3, J = 2-3$ $J = 3-4$	$118,750 \cdot 5$ $56,265 \cdot 6$ $62,487 \cdot 4$ $58,446 \cdot 2$	Also known frequencies for other rotational levels

7. GENERAL CONCLUSIONS

There is a limited number of radio frequency resonances for which immediate search with radio telescopes appears justified. Of these, the Λ-doubling of OH, the hyperfine structure of H^2 and perhaps of He^3 II are outstanding. A number of additional resonances can probably be found if their frequencies can be first accurately determined in the laboratory. These include the Λ-doubling of CH and the hyperfine structure of ionized N^{14}.

The construction of radio telescopes with very high resolving power should considerably increase the possibility of finding line spectra of astronomical interest. Such telescopes should yield a considerable increase

* References [1] and [2] list also CH$^+$ as producing Λ-doubling. This is apparently due to the incorrect assumption that the CH$^+$ molecule occurs in a $^2\Pi$ state. Its ground state is $^1\Sigma$.

in background temperature when directed towards one of the intense continuous sources and hence increased sensitivity for detection of absorption lines. They should also allow examination of the regions of high concentrations of gaseous matter, such as occur in planetary atmospheres, in comets, or in small nebulae.

Additional absorption or emission lines due to gases in the earth's atmosphere are probably immediately detectable; the most obvious of these cases is the rich microwave spectrum of O_2.

REFERENCES

[1] Shklovsky, I. S. *Astr. Zhur.* **26**, 10, 1949.
[2] Shklovsky, I. S. *Astr. Zhur.* **29**, 144, 1952.
[3] Purcell, E. M. *Proc. Amer. Acad. Sci.* **82**, 1347, 1953.
[4] Townes, C. H. *J. Geophys. Res.* **59**, 191, 1954.
[5] Townes, C. H. and A. L. Schawlow, *Microwave Spectroscopy*, McGraw-Hill Book Co., N.Y. (1955).
[6] Lilley, A. E. *Ap. J.* **122**, 197, 1955.
[7] Brown, H. *Rev. Mod. Phys.* **21**, 625, 1949.
[8] Herzberg, G. *Spectra of Diatomic Molecules*, Van Nostrand Co., N.Y. (1950).
[9] See, for example, Wilson, A. H. *Proc. Roy. Soc.* A, **118**, 617, 1928.
[10] Wild, J. P. *Ap. J.* **115**, 206, 1952.

PART II

POINT SOURCES: INDIVIDUAL STUDY
AND PHYSICAL THEORY

OPTICAL INVESTIGATIONS OF
RADIO SOURCES

INTRODUCTORY LECTURE BY

R. MINKOWSKI

Mount Wilson and Palomar Observatories, Pasadena, California, U.S.A.

Loose agreement of a radio position of low accuracy with that of some object listed in the NGC is not sufficient to provide the identification of a radio source. Even satisfactory coincidence of a precise position with that of an astronomical object requires supporting evidence. Agreement of the size of the source with that of the visible object, at least in order of magnitude, is an important argument in favour of an identification; exact agreement of sizes can be expected only where radio and optical emission are physically connected. The radio spectrum, the optical spectrum, and the physical characteristics of the visual object also have to be taken into account. Observations of the radio spectrum should be particularly useful to support the identification of sources with H II regions which can be recognized from their thermal emission even if they are obscured and optically inaccessible. If all data are available, satisfactory agreement exists between optical and radio observations. The best example of this kind at the moment is perhaps NGC 2237, the Rosette nebula, reported as a source by Ko and Krauss (1955) [1] and also observed by Mills, Little and Sheridan (1956 [11]; see also paper 18).

The number of positions observed with adequate accuracy has increased considerably during the last year. But for very few sources has it been possible to make identifications or even to find objects worthy of the detailed optical investigation necessary to support a possible identification. Since the optical investigations usually require the use of a large telescope, almost regularly the 200-inch telescope, their cost in observing time is high. Work of this type appears justified only for sources with positions whose uncertainty is of the order of a minute of arc. High positional accuracy is therefore of basic importance.

The reasons why a high percentage of the sources cannot be seen or identified will depend upon the various types of sources. Extended sources should be easy to identify. But among twenty-five extended sources for which positions were communicated by Ryle in advance of publication there is only one which is perhaps not too far removed from some recently discovered peculiar filaments to assume that this source is due to a filamentary nebulosity; most of this nebulosity could be obscured and the few visible filaments could be near its edge [12]. Of these twenty-five sources ten are in more or less heavily obscured areas at low galactic latitudes. That they cannot be seen is not particularly surprising. Some may be H II regions which should be established by investigation of the radio spectra.

The remaining fifteen sources in unobscured areas at high latitudes are a group of considerable interest. Objects of this kind have also been found by several other observers. These objects cannot be H II regions, nor can they be nearby extra-galactic systems of any known type. Since they are invisible, radio observations will have to find an answer to the puzzle. Some of the sources may represent blends of several objects, but it does not appear statistically reasonable to assume this explanation for more than a small fraction. The question thus arises whether they are really extended, in which case gas clouds with non-thermal radio emission unaccompanied by optical emission are at this moment the only plausible explanation, or whether the extent is not intrinsic, but due to some scattering effect, e.g. in interstellar clouds. This question is also raised by the excess of the radio dimensions of some extra-galactic objects over their optical size.

The situation is quite different for small sources which may be optically inconspicuous objects, such as peculiar galaxies and possibly stars. Cygnus A is near the limit where a peculiar object of this type can still be recognized by simple *inspection* with the 200-inch telescope. Another source of this type is Hercules A which has quite recently been found to be a double galaxy with high-excitation forbidden line emission at a distance of $1 \cdot 4 \times 10^8$ parsecs, about $1 \cdot 5$ times the distance of Cygnus A. More distant objects of this type might provide many identifications, but positions of the accuracy achieved by F. G. Smith and considerable work with the 200-inch telescope are required. The limit at which objects of the Cygnus A type can still be recognized should be at a distance of 1 to 2×10^9 parsecs.

Much less conspicuous than the peculiarities of colliding galaxies is that of M 87. This galaxy would be very difficult to recognize as containing a peculiarity if seen from another direction or at a much greater distance. The nature of the peculiarity is still not understood. But M 87 demonstrates clearly that even relatively nearby galaxies might have peculiarities which are very difficult to observe, if at all. It is therefore impossible to deny that an apparently normal distant galaxy might be the proper identification for a radio source. However, an identification requiring strongly enhanced radio emission from a normal-appearing galaxy seems acceptable only if based on a position of the highest accuracy.

To identify stars as sources requires impossible positional accuracy. At the moment, the best hope to identify stars as radio sources is offered by the observation of variability of certain sources by Slee and by Kraus. Intrinsic rapid variability is conclusive proof for the stellar character of an object. However, radio variability does not necessarily imply observable optical variability.

I. SINGLE GALAXIES

Even now a considerable variety of objects have been identified as sources which offer diverse observational problems. Normal galaxies are safely established as sources. One main problem at the moment is the relation between optical and radio magnitude. The present observations are not sufficient to decide how much of the scattering around a constant difference between the two magnitudes is intrinsic and how much of it is due to differences between the various types of galaxies. Results by Mills (1955 [13]; see also paper 18) favour the point of view that such differences exist, but the material has to be substantially increased before a final conclusion seems permissible. The two main difficulties met by such an investigation arise from optical absorption effects and from weak peculiarities. As an example of the first difficulty one may consider NGC 891 which is seen edge-on and has a very heavy absorption lane. Some part of the excess radio emission which follows from the observation of this nebula by R. Hanbury Brown is obviously due to the reduction of optical brightness by the absorption lane. It is, however, hardly possible to explain in this way a magnitude discordance of about 2 mag. But the absorption lane may also hide a peculiarity; it is not possible to be certain that NGC 891 is a normal galaxy. Obviously, objects of this type are not suitable samples. The second difficulty may be illustrated by NGC 1068 which Mills finds about 1 mag. brighter than the average Sb galaxy. Actually, NGC 1068 is long known to be one of the otherwise normal appearing galaxies whose small and bright nuclei emit a spectrum similar in its composition to that of a planetary nebula but with lines of great width corresponding to velocities of several thousand km./sec. This phenomenon is as yet unexplained. If the peculiar state of the nuclear gas involves radio emission with a volume emissivity equal to that in the Crab nebula, the excess radiation would be more than fully explained. In any case, NGC 1068 is not a good sample of a normal nebula.

It has already been mentioned that the peculiarity of a galaxy may evade optical observations. A good example is NGC 1316 which also serves well to demonstrate the complexity of the identification problem. The survey position (Mills, 1952) [2] and the precise position (Mills, 1952) [3] of the source in question were communicated in 1952 in advance of publication. The proximity of NGC 1316 to the source was noted, but the suggested identification did not seem acceptable to the Australian observers who considered the positional agreement in declination as inadequate. Moreover, Bolton and Mills were unable to agree on the exact right ascension and the intensity measured on different interferometers and on this suggested an extended source. Shklovsky (1953) [4], knowing only the survey position by Mills but not the later precise position, suggested again the identification with NGC 1316 and tried to support it by claiming similarity of NGC 1316 to NGC 5128. This similarity was also suggested by de Vaucouleurs (1953) [5]. However, there can be no doubt that the impression of similarity was due to an over-interpretation of a small-scale photograph. Actually, NGC 1316 (see Baade and Minkowski, 1954) [6] is an early-type galaxy with some absorption patches which suggest the beginning of spiral formation, possibly of a barred spiral. NGC 5128 (see Baade and Minkowski, 1954) [7] seems to be a combination of an early-type system with the major axis in position angle about 45° and a late system containing much gas and dust in position angle about 135°. Moreover, NGC 5128 has a G-type absorption spectrum with emission lines strong enough to have been seen by Hubble on objective prism exposures. Its spectrum shows considerable internal motions which support the point of view that two galaxies are in interaction. On the other hand, NGC 1316 has a G-type spectrum without emission lines; not even [O II], $\lambda 3727$ is observable. The similarity of the two systems cannot be maintained. This does not imply, however, that NGC 1316 has to be a normal galaxy. Actually, a recent position by Mills with the cross shows the presence of a source in excellent positional agreement with NGC 1316. This seems to admit no other conclusion than that NGC 1316, probably as part of a blend giving the appearance of an extended source, is actually a galaxy with fairly strong enhanced radio emission but without obvious strong optical peculiarity other than the appearance of a few absorption patches in an early-type galaxy.

2. COLLIDING GALAXIES

The identification of a source due to a collision between galaxies presents relatively little difficulty. Large nearby systems of this kind can easily be recognized from their appearance. Actual collisions seem to lead to conspicuous emission of forbidden lines of high excitation which can still be observed in objects too distant and thus too small to reveal their nature from their appearance. While it seems not yet possible to predict the strength of the radio emission from the appearance of a given object, the sequence of conditions seems basically clear. Close pairs of galaxies, such as NGC 4575/4576, without visible signs of interaction do not show enhanced radio emission. Close pairs with large tidal interactions, such as NGC 4038/4039, may give somewhat enhanced emission. Unpublished observations by Mills suggest that this object may be an emitter of somewhat more than normal intensity. Spectroscopic observations of NGC 4038/4039 show that the radial-velocity difference between the systems is small. The interaction would thus be expected to be weak, but at the same time of long duration favouring strong distortion. A strongly distorted galaxy such as NGC 2623 might be of somewhat similar kind. This system coincides closely with a source observed by Ryle. Spectroscopic observations show little gas, revealed by faint emission of [O II], $\lambda 3727$, but a spectrum of type A, a relatively rare type for galaxies. It might be that the other system is now on the far side and hidden. The interaction might be near its end and could have removed most of the gas from NGC 2623, while still-surviving early-type stars of population I could explain the early spectral type.

Strong interaction is represented by systems like Cygnus A, Hercules A, and NGC 1275. The most powerful source of this type, Cygnus A, is unfortunately too distant, thus too small and faint to permit a detailed optical investigation. The radio interferometric results suggest that the radio emission arises in outlying parts of this system which are optically unobservable. The full dimensions of the system indicated by the radio results are large, but tidal filaments in some systems extend much farther. There is thus no obvious contradiction to astronomical facts.

NGC 1275 is the only known sample of colliding galaxies which is suitable for a detailed optical investigation. That this object and not the Perseus cluster of galaxies is to be identified with the small source in its position is now safely established. The system consists of a tightly wound spiral of early-type and a strongly distorted late-type spiral. Inspection of blue and red exposures shows some unusual features regarding the colours and distribution of emission patches connected with the distorted spiral arms of the late-type system. A detailed spectroscopic investigation leads to the surprising results shown in Fig. 1. Spectrograms were obtained with the positions of the slit indicated on the left half of the figure. To the right are shown the sections of these spectrograms which contain the [O II] lines. In the northern part of the object, two sets of lines appear, separated by about 3000 km./sec., which show unmistakably the presence of two separate gas masses. Comparison of the structural details on the direct photographs with the spectrograms shows that the velocities of the early spiral and of the late-type system are about $+5200$ km./sec. and $+8200$ km./sec. respectively. Since the absorption patches of the late-type systems are obviously in front, the northern part of the late-type system moves toward the early spiral. Going from the north towards the nucleus of the early spiral one finds this same condition, but a fundamental change occurs between the nucleus and the nearest position to the north of it which has been investigated until now. Superposed on the nucleus, and extending beyond it into the object, appears a spectrum containing one set of high excitation emission lines of very great width, slightly asymmetrical towards the violet, which were first observed by Humason and later investigated in some detail by Seyfert (1943) [8]. This same spectrum appears also to the south of the nucleus. It should be noted that this is an important difference from the nebulae with nuclei containing emission lines in which the high excitation spectrum is confined to the nucleus.

The interpretation of these results is obvious. The two galaxies, which are seen nearly face on, are in collision. They are inclined towards each other in such a way that originally the southern part of the late system was closer to the early spiral. As the galaxies moved towards each other, the collision started in the south and progressed to the north, where interaction between the separate gas masses of the two galaxies is now going on. Farther to the south, probably from some line slightly to the north of the nucleus on, the collision is over, leaving the combined gas mass formed by the collision in a highly heated and excited condition. The velocity of the combined gas differs little from that of the early spiral; this requires that the mass of the gas in this system was considerably higher than that in the late-type galaxy. This result does not appear improbable. The early spiral is a system of very high luminosity having an absolute magnitude $M_{pg} \approx -19$ and may be expected to contain a correspondingly large mass of gas. More detailed discussions of the conditions in this system have to wait for additional observations.

If one uses a crude model in which the gas masses are considered as

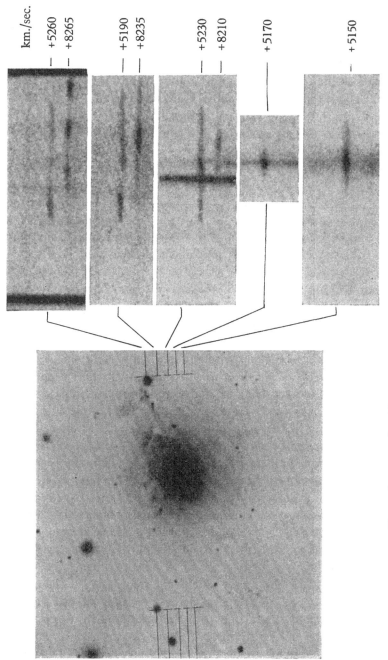

km./sec.

+5260
+8265

+5190
+8235

+5230
+8210

+5170

+5150

Fig. 1. Spectroscopic evidence that NGC 1275 consists of two galaxies in collision.

plane parallel layers of 500 parsecs thickness, one finds that the observed conditions require an angle of 15° to 20° between the galactic planes of the two systems. On this basis, the total duration of the collision is of the order 10^6 years. The intensity of the radio emission during this period should first increase, reach one or possibly several maxima depending on the distributions of gas in the two galaxies, and finally decline. If the present power represents about the average, the total energy emitted during the collision must be of the order 10^{47} erg, still a small fraction of the total kinetic energy of 10^{49} to 10^{50} erg available in a collision between galaxies with a relative velocity of 3000 km./sec.

3. GALACTIC NON-THERMAL SOURCES

Except H II regions which are thermal sources, only six galactic non-thermal sources have been definitely identified. They comprise apparently different objects of which it is not known whether and how they are related [14].

Remnants of supernovae of type I are represented by the Crab nebula. A source found by Hanbury Brown near the position of Tycho's nova may not be connected with this object; the Cambridge position of this source is too far from Tycho's position to admit this identification [15]. The investigation of the Crab nebula has brought most important progress in the discovery of polarization in the central region which emits a continuous spectrum. This discovery by Vashakidze and Dombrovsky is now being followed up by Oort and Walraven with most interesting results which leave no doubt that the optical continuous spectrum and the radio emission have a common origin, being produced by radiation of relativistic electrons in a magnetic field, as was suggested by Shklovsky (1953) [9].

Three sources have been identified with IC 443, the big loop in Cygnus and a very faint nebulosity in Auriga which might be similar in character to the first two nebulae [16]. The first two are definitely similar. They are slowly expanding with relatively high internal motions; the velocities are of the order 50 to 100 km./sec. The large size of these objects, which have diameters larger than a degree, involves slow progress for a detailed investigation such as is now under way for the Cygnus loop. The nature of these objects is obscure. Oort has suggested that they may be supernova shells slowed down by interaction with the interstellar medium. At this time, no data exist which support or disprove the suggestion.

The Cassiopeia source has been identified with a remarkable nebulosity with unique characteristics [10]. Puppis A seems to be connected with a nebulosity of similar kind. The investigation of the Cassiopeia nebulosity has been continued with interesting results the most important of which are measures of proper motions of the diffuse condensations by Baade which establish beyond doubt that the nebulosity is a rapidly expanding object. The sharp bits, however, do not show any observable trace of motion.

The nebulosity consists of two different types of filaments which are so drastically different in every way that it is hard to avoid the conclusion that two masses of gas are involved in some way. Sharp broken bits, ranging down to condensations of almost stellar appearance, can be photographed only in the red. Diffuse condensations are seen both in the blue and the red. As can be expected from this, the spectra are radically different. The sharp bits show H_α, the neighbouring [N II] lines with intensities comparable or slightly larger than H_α and very faintly the [O I] auroral lines. The diffuse condensations show lines of [O I], [O II], [O III], [S II], [Ne III], but no trace of H_α or [N II]. The red lines of [S II] are the most intense lines in some of the diffuse filaments. The interpretation of the line intensities meets difficulties which are most probably due to the fact that the ionization is not in equilibrium. From estimated values of the surface brightness and intensity ratios of lines belonging to the same ion, such as [O III], $\lambda 4363$ and $\lambda 4959/5007$ and [S II], $\lambda 4067/78$ and $\lambda 6713/31$, one may conclude that the sharp bits have relatively low electron temperature, possibly below 10,000° K., and moderately high electron density; if, as seems likely, the excitation is collisional, the ionization is incomplete and the total density may be very high, of the order 10^6 atom./cm.$^{-3}$. The diffuse filaments have high electron temperature, 20,000° K. or higher, and moderate electron density of the order 10^3 cm.$^{-3}$. To understand the relative line intensities it is necessary to assume a colour excess due to interstellar reddening of the order of 1 mag., corresponding to a total absorption $A_{ph} \approx 4$ mag., for the brightest northern region of the nebulosity. If the faint diffuse filaments near the centre are intrinsically as bright as those in the northern arch, the total absorption here may be about 6 mag. That the interstellar absorption in the field is large and spotty is obvious from its appearance. From star counts on the 48-inch Schmidt plates Greenstein finds that the total absorption exceeds $2^{m\cdot}5$ by an undetermined amount; this supports the conclusions drawn from line intensities.

The radial velocities observed until now are plotted in Fig. 2 as a function of the distance of the condensations from the centre of expansion. The velocities of the broken bits cover a range more than an order of magnitude smaller than that for the diffuse condensations. The broken bits nearest to the centre show somewhat higher velocities than those farther away. This could represent a small systematic expansion, but the scatter is too large and the number of measured points too small to permit any definite conclusion. The diffuse condensations show velocities which seem to

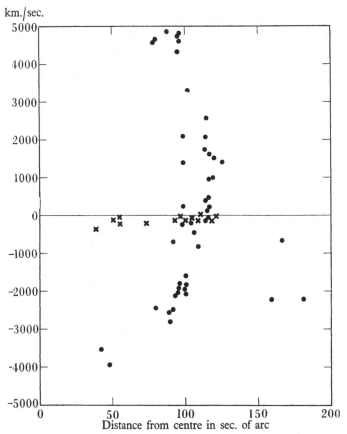

Fig. 2. Radial velocities in the Cassiopeia source as function of distances from the centre.

scatter without any regular arrangement. The highest velocities are not nearest to the centre, the smallest velocities are at an intermediate distance and the most distant condensations have still very high velocities. If the object were a complete expanding shell of some thickness, all points should lie between two ellipses; their axes in the velocity direction would represent the minimum and maximum velocities while the axes in the distance direction would give inner and outer diameters. Thus, the points with low

116

velocities at about 110″ from the centre may represent the inner diameter of the shell, but the outer diameter and the maximum and minimum velocities cannot be found since the limiting ellipses cannot be determined. It is clear, however, that the maximum velocity is not very much larger than 5000 km./sec.

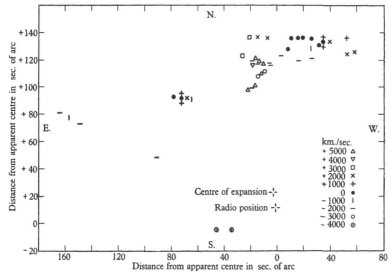

Fig. 3. Distribution of radial velocities in the Cassiopeia source.

What has happened becomes clear when the distribution of observed velocities over the area of the object is inspected. As can be seen in Fig. 3, the velocities are not scattered at random over the nebula, but velocities in certain ranges are grouped in certain regions. For instance, only velocities between +1500 and +2500 km./sec. occur in a region centred about 130″ N. and 45″ W. from the apparent centre of the nebulosity. Five such groups can be recognized in the region of the bright northern arch of the nebulosity. Their mean velocities and approximate co-ordinates in Fig. 2 are:

$$+4585 \text{ km./sec.}, \quad \alpha+15'', \quad \delta+110''$$
$$+2430 \qquad\qquad +15'' \quad +135''$$
$$+1305 \qquad\qquad -45'' \quad +130''$$
$$+\ \ 60 \qquad\qquad\ -20'' \quad +135''$$
$$-2130 \qquad\qquad\ -\ 5'' \quad +120''$$

Obviously, each group represents a cloud of condensations moving in a certain direction. It is not impossible, but not very probable that the original process has resulted in the ejection of a few batches of material in some directions. It seems much more likely that the gas was originally ejected in all directions with more or less spherical symmetry and that interstellar material in the region in which the explosion occurred has stopped most of the ejected shell. On this assumption the two gas masses whose presence is so strongly suggested by the existence of two distinctly different types of condensations are an expanding shell and interstellar gas.

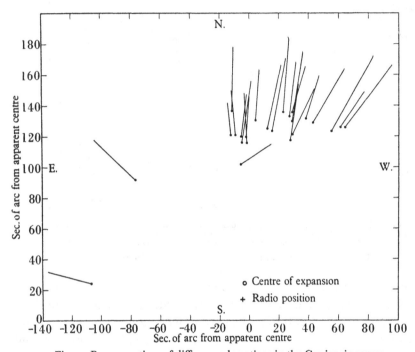

Fig. 4. Proper motions of diffuse condensations in the Cassiopeia source.

The fast-moving condensations which are now still visible must have moved through parts of the interstellar gas with lower than average density where they have lost little of their initial velocity. The sharp bits represent the effect of stopping of ejected material; at this time the question has to be left open whether they are compressed parts of the original ejected material which have lost their outward motion almost completely or whether they represent phenomena such as shock zones in the interstellar material caused by the braking of parts of the moving gas.

As mentioned earlier, the proper motions of the diffuse condensations have been measured by Baade with an interval of three years. The results of these measurements are shown in Fig. 4, where the length of the lines extending outwards from the positions of the condensations gives the motion for 100 years in the direction of the lines. The systematic expansion is obvious. The position of the centre of expansion determined by a least squares solution agrees well with the position of the radio source; it is, as a matter of fact, just about as far north preceding of the radio position as the apparent centre of the visible nebulosity is south preceding. The error of measurement is about 2 % of the average motion. Obviously there are noticeable random components superposed on the regular expansion; these might have been produced if the moving condensations had been braked in varying degrees by the interaction with the interstellar gas. The existence of velocity groups is not directly visible in Fig. 4 for the very simple reason that not all velocity groups are represented. For instance, all of the condensations in the cloud closest to the centre with an average velocity of $+4585$ km./sec. have changed so much in three years that their motions, while clearly visible, cannot be measured accurately. Actually, only three velocity groups are represented: the groups with mean radial velocities -2132, $+59$, and $+1307$ km./sec. Radial velocities are not yet available for all points for which the motion could be measured

If the absolute values of the motions are plotted as a function of the distance from the centre (Fig. 5), one finds considerable scatter around the straight line through the origin which would correspond to a simple expansion. The increase of motion with distance is, however, obvious and the mean values for points with more and less than the average distance, plotted as large full circles in the figure, are quite close to the line. The velocity groups appear separated to some degree. The straight line

$$s = 3 \cdot 52 \times 10^{-2} \, r,$$

where r is in sec. of arc and s in sec. of arc per year, describes the mean expansion satisfactorily. The time elapsed since the start of the expansion is then 280 years. If the material observed today has been slowed by interaction with interstellar gas, the time actually elapsed would be shorter.

If velocity of expansion and motion are known for an expanding shell, the distance can be determined. For the Cassiopeia nebulosity, such a determination is not possible in the usual way which requires essentially that the major and minor axis of a velocity ellipse in a plot like that in Fig. 2 be known. Since this is not the case, arbitrary assumptions would have to be made to apply the usual computation. Almost any value of the distance can be obtained in this way. It is therefore necessary to attack the problem differently. It seems reasonable to assume that in each velocity group the random velocities, which appear clearly both in radial velocities and motions, are independent of the direction from which the object is viewed. The distance can then be determined from the dispersions in radial velocities and in motions. If the random velocities in a group have spheroidal distribution, one has

$$d = \frac{\overline{(v - \bar{v})}}{4\cdot73 \times \dfrac{\pi}{2}\,\overline{(s - \bar{s})}} \text{ parsecs.}$$

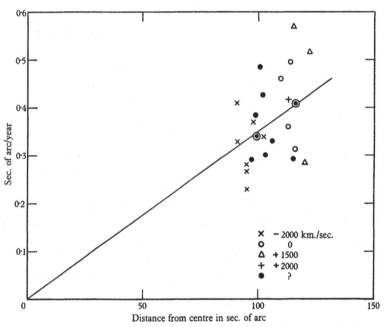

Fig. 5. Proper motions of diffuse condensations in the Cassiopeia source as function of distances from the centre.

Here, v is the velocity in km./sec. of a point of a group with the mean velocity $\bar{v}$, s the absolute motion in sec. of arc of a point of this group whose mean motion is $\bar{s}$. The values for the three groups for which data are available are given in Table 1. The three independent values agree satisfactorily. Thus, the mean value of the distance 540 parsecs seems to deserve some confidence [17]. From the apparent diameter of about 6', the linear diameter of the nebulosity is then about 1 parsec. To reach this diameter in 280 years, a velocity of about 4000 km./sec. is needed. This is the velocity of gas seen near the edge of the object. It should be noted that distance values of the same order are also obtained if one assumes that the first and third group move at an angle of 45° to the direction of vision. This assumption, however, is entirely arbitrary. It can certainly not be applied to the second group for which a distance of a few parsecs would follow; this group seems to represent a cloud near the inner edge of the shell.

Table 1. *Distance of the Cassiopeia source*

Km./sec.		"/year		Parsecs
$\bar{v}_r$	$(v-\bar{v})$	$\bar{s}$	$(s-\bar{s})$	d
−2132	274	0·318	±0·051	715
+ 59	311	0·407	0·074	575
+1307	248	0·446	0·098	340

A considerably larger distance has been derived from observations of 21-cm. absorption which show the presence of a very small interstellar cloud with a velocity of about −40 km./sec. in front of the object. If this velocity is interpreted as due to the effect of galactic rotation, one has to conclude that the object is at least in, if not behind, the second spiral arm, at a distance of more than 2000 parsecs. Such a distance would lead to an implausibly large linear size, about 5 parsecs, and velocity of expansion about 20,000 km./sec. As a matter of fact, there is no reason to believe that the velocity of a small cloud such as that seen in absorption represents the average velocity of the gas in any spiral arm; the concept of galactic rotation could at best be applied to that average velocity.

Finally the question as to the origin of the object has to be raised. Only one type of astronomical object is known at present in which expansion velocities of the order 5000 km./sec. have been observed: supernovae of type II. The average absolute photographic magnitude of this class of supernovae is near −14 mag. (new scale). If the photographic absorption near the centre of the object is 6 mag., the apparent brightness of such a supernova at 540 parsecs becomes −0^m·4. No nova like this has been recorded for about 280 years. An object of this brightness would certainly not go unnoticed today. However, no record of any nova exists in the period from the second half of the seventeenth century to the end of the eighteenth century. It seems not unreasonable to assume that this indicates a general lack of interest in novae; several ordinary novae of comparable brightness have occurred in the first half of the present century and it does not seem likely that the phenomenon was missing for more than a century. It is, of course, not impossible that the interstellar absorption is even heavier than assumed and that the nova was fainter. For the time being the assumption that the Cassiopeia source is the remnant of a supernova of type II seems to be the most plausible interpretation.

121

REFERENCES

[1] Ko, H. C. and Kraus, J. D. *Nature*, **176**, 221, 1955.
[2] Mills, B. Y. *Aust. J. Sci. Res.* A, **5**, 266, 1952.
[3] Mills, B. Y. *Aust. J. Sci. Res.* A, **5**, 456, 1952.
[4] Shklovsky, I. S. *A.J. U.S.S.R.* **30**, 30, 1953.
[5] de Vaucouleurs, G. *Observatory*, **73**, 252, 1953.
[6] Baade, W. and Minkowski, R. *Observatory*, **74**, 130, 1954.
[7] Baade, W. and Minkowski, R. *Ap. J.* **119**, 215, 1954.
[8] Seyfert, C. K. *Ap. J.* **97**, 195, 1943.
[9] Shklovsky, I. S. *Reports of the Academy, U.S.S.R.* **90**, 983, 1953.
[10] Baade, W. and Minkowski, R. *Ap. J.* **119**, 206, 1954.

REFERENCES AND NOTES ADDED DECEMBER 1956

[11] Mills, B. Y., Little, A. G., and Sheridan, K. V., *Aust. J. Phys.* **9**, 218, 1956.
[12] Photographs with the 48-inch Schmidt telescope and a plate-filter combination with a narrow passband for $H\alpha$ have shown that the filaments are indeed part of an extremely faint nebulosity which coincides in position and size with the radio source HB 21 (R. Hanbury Brown and C. Hazard, *M.N.R.A.S.* **113**, 123, 1953).
[13] Mills, B. Y., *Aust. J. Phys.* **9**, 368, 1955.
[14] The number of identified galactic non-thermal sources has now increased to nine.
[15] An extremely faint nebulosity which is the remnant of Tycho's nova has recently been found. The approximate centre of the nebulosity follows by about 30 sec. the accepted position of the nova. It now appears probable that this position is less reliable than was formerly assumed and that the source is to be identified with Tycho's nova. Since a radio source in the position of Kepler's nova of 1604 has been found by Mills, Little and Sheridan [11] the remnants of all three known galactic supernovae of type 1 are now identified radio sources.
[16] The nebulosity with which the source HB 21 has been identified [12] also belongs in this group.
[17] If changes of shape of the diffuse filaments with time make a noticeable contribution to the random components of the proper motions, the distance of 540 parsec is only a lower limit for the true distance.

CURRENT PROGRESS IN DEVELOPMENT AND RESULTS OBTAINED WITH THE 'MILLS CROSS' AT THE RADIOPHYSICS LABORATORY

J. L. PAWSEY

Division of Radiophysics, Commonwealth Scientific and Industrial Research Organization, Sydney, Australia

The study of continuous spectrum cosmic radio waves in the Radiophysics Laboratory is based mainly on two large 'Mills Crosses': one on 85 Mc./s. which has been in use for a year, and another on 20 Mc./s. which is under construction. The principle of this instrument has been described by Mills and Little (1953) [1]. It utilizes two long thin arrays arranged in the form of a cross and gives an effective 'pencil-beam' response which records both the background and discrete sources simultaneously. Essential data relating to these Crosses are given in Table 1.

Table 1

	85 Mc./s. Cross	20 Mc./s. Cross
Length and orientation of arms	N.-S. 1500 ft. E.-W. 1500 ft.	N.-S. 3600 ft. E.-W. 3400 ft.
Width of dipole arrays forming arms	2 dipoles	1 dipole
Angular resolution (between half-power points, near zenith)	50'	85'*
Sensitivity limit (ideal conditions)	10^{-26} w.m.$^{-2}$ (c./s.)$^{-1}$	—
Present status	Working since 1954	Under construction

* Design figure.

The instrument is used as a transit instrument with the declination adjusted by phase adjustments in the north-south arm. For maximum sensitivity the beam remains at a fixed declination and the output is recorded as the sky passes overhead. For survey work and plotting iso-photes the beam is switched successively (on the 85 Mc./s. Cross) between five declinations separated by about half a beam-width and the output on each declination recorded. This method utilizes one-fifth of the total time on each declination and the sensitivity is correspondingly reduced by a factor of about $\sqrt{5}$. During the past year the programme of observations on

85 Mc./s. was based on the study of optical objects while at the same time work continued on the improvement of the equipment. This is now ready for a comprehensive survey of the sky visible to the instrument. The beam deteriorates at angles far removed from the zenith due to foreshortening effects, and the approximate limits of coverage are $\pm 45°$ from the zenith or, say, from declination $+10°$ to the south pole.

The observations described in this paper include a study of normal galaxies which is the subject of a paper by B. Y. Mills (1955) [2], and studies of the radio emission from various classes of objects: colliding galaxies, novae, and emission nebulae, which are reasonably advanced

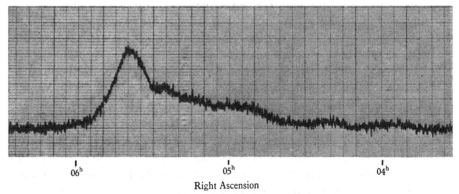

Right Ascension

Fig. 1. A typical record of an extended source. The Large Cloud of Magellan traversed at declination $-69° 26'$.

items selected from those being studied by Mills and his colleagues, A. G. Little, K. V. Sheridan and O. B. Slee. The statistics of the thousand discrete sources so far listed are discussed in a separate note (paper 40, Pawsey) and some observations of discrete sources which appear to show variations (paper 31, Bolton).

Sample records are shown in Figs. 1 and 2. Fig. 1 illustrates the effect of an extended object, the Large Cloud of Magellan recorded on declination $-69° 26'$. Fig. 2 illustrates discrete sources of various intensities. In the upper record NGC 55 is near the limit for detection and several records were necessary before its existence was accepted. Its flux density is $1·8 \times 10^{-26}$ w.m.$^{-2}$ (c./s.)$^{-1}$. The middle record shows an isolated source NGC 253 of flux density $1·1 \times 10^{-25}$ w.m.$^{-2}$ (c./s.)$^{-1}$ and the lowest one shows no detectable source in the position of NGC 4594. Over the greater part of the sky the detection limit is set by sensitivity but large areas exist which are confusion-limited.

The different types of objects studied will now be discussed in turn.

I. NORMAL GALAXIES

The study of the radio emission by normal galaxies is based on observations of our own galaxy, of the Clouds of Magellan which are resolved by the Mills Cross, and of samples of various types of external galaxies which are so distant that only the integrated emission is observed.

Sample records of traverses across the Milky Way shown in Fig. 3 strongly suggest the dual distribution put forward by Shklovsky (1952, 1953) [3], a disk-like one presumably associated with population I objects and another having a very extensive, roughly spherical, distribution unlike

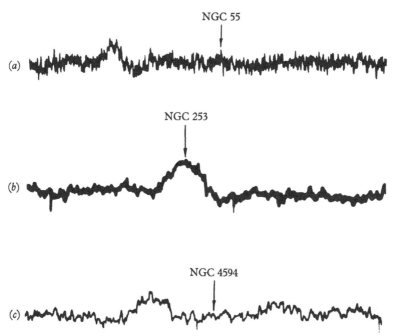

Fig. 2. Sample records showing discrete sources of small size. The sources noted are external galaxies. Flux densities NGC55 − 1·8 × 10⁻²⁶; NGC253 − 11 × 10⁻²⁶; NGC4594 (undetectable) < 1·3 × 10⁻²⁶ w.m.⁻² (c./s.)⁻¹.

any known stellar population. Both distributions have spectra which show that the predominant emission at low frequencies is non-thermal.

Isophotes of the Clouds of Magellan are shown in Fig. 4. These conform better with the distributions of interstellar atomic hydrogen as given by 21-cm. observations and of bright stars than with other optical objects studied and suggest that the radio emission is associated in the Clouds with population I.

Nine globular clusters were examined and only one, NGC6121, can possibly be identified with a radio source. This source, however, appears extended so that its identification with the star cluster itself is unlikely. The clusters examined included the bright ones 47 Tucanae and NGC 362 and,

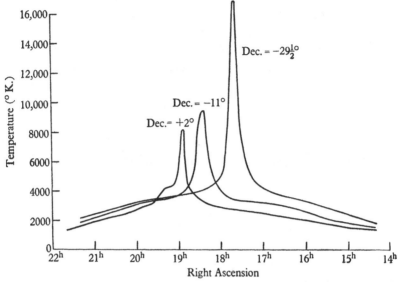

Fig. 3. Sample traverses at constant declinations through the Milky Way in the vicinity of the centre of the Galaxy (Dec. = $-29\frac{1}{2}°$). (Accuracy 10 or 20%.)

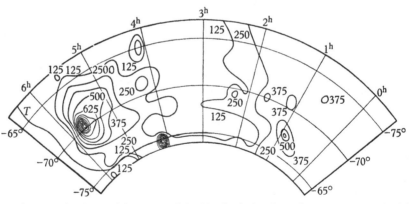

Fig. 4. 85 Mc./s. isophotes of the region of the Clouds of Magellan. Contour interval of brightness temperature 125° K.; all values specified as excess over a base level T estimated as 700° K.

in the former case if the ratio of radio to optical emission were the same as for late-type spirals, a signal 500 times the minimum detectable level would have been expected. Globular clusters with their type II population appear to be very poor radio emitters.

Attempts were made to observe thirteen bright southern galaxies of various types. These results, and those for the two bright globular clusters mentioned above, are listed in Table 2 together with photographic magnitudes. In order to have a convenient measure of the relative ratio of optical and radio emission the flux densities have been expressed on a magnitude scale defined by

$$m_{R(158)} = -53 \cdot 4 - 2 \cdot 5 \log_{10} S_{(85 \text{ Mc./s.})} + 0 \cdot 47$$

which is similar to that used by Hanbury Brown and Hazard (1952) [4] with the addition of the final term $0 \cdot 47$ which is an estimated correction to convert from 85 Mc./s. to the latter's 158 Mc./s. The final column, $m_R - m_p$, is a measure of the required ratio in magnitudes. In Table 4 $m_{R(85)}$ is defined similarly except for the omission of the term $0 \cdot 47$ converting from 85 to 158 Mc./s.

Table 2. *A comparison of the radio and photographic magnitudes of some southern galaxies and globular clusters*

Nebula	Type	Flux density at 85 Mc./s. (w.m.$^{-2}$ (c./s.)$^{-1}$ × 10^{-26})	$m_{R(158)}$	m_p	$m_R - m_p$
LMC	(M)	2000	3·8	0·5	3·3
SMC	(M)	300	5·9	2·0	3·9
NGC55	(M)	1·8	11·4	7·8	3·6
NGC253	Sc	11	9·4	7·6	1·8
NGC300	Sc	4·8	10·4	8·5	1·9
NGC5236(M83)	SBc	19	8·9	7·4	1·5
NGC4945	SBc	12	9·2	7·8	1·4
NGC6744	SBbc	3·0	10·7	9·1	1·6
NGC1068(M77)	Sb	19	8·9	9·6	−0·7
I5267*	Sb	2·5	11·1	10·8	0·3
NGC4594(M104)	Sab	<1·3	>11·8	8·9	>2·9
NGC1291	SBo	<1·5	>11·6	9·5	>2·1
NGC3115	E7	<1·2	>11·9	10·15	>1·7
47 Tucanae	Globular cluster	<2·6	>11·0	3·0	>8·0
NGC362	Globular cluster	<2·9	>10·8	6·0	>4·8

* Identification uncertain.

Having used Hanbury Brown and Hazard's scale it is possible to increase the sample by including six northern galaxies observed by them (1935) [5], namely, NGC5194–5, 224, 3031, 4258, 2841 and 891. The ratio $m_R - m_p$ for both series was plotted against galactic type and a definite trend became evident. Spiral galaxies of types Sb and Sc tend to have a higher ratio of radio to optical emission ($m_R - m_p$ smaller) than do either elliptical or magellanic types.

In discussing this evidence Mills has suggested that the two types of distribution, diskoidal or population I and spherical, evident in our galaxy, are present in greater or less extent in normal galaxies. In the magellanic types the population I would appear to predominate. The spheroidal types lack this component and presumably the other is ill-developed. In the Sb and Sc types the spherical system appears to predominate and to be the main source of the relatively strong emission. Mills discusses the probable role of relativistic electrons in the emission.

2. COLLIDING GALAXIES

Colliding galaxies have been suggested as radio sources because three strong radio sources can be identified with galaxies which have peculiarities suggesting that a collision may be in progress. An attempt has now been made to detect radiation from four other galaxies which have been suggested as being in collision. The results shown in Table 3 are negative except for one example, NGC 4038/39, where a slightly abnormal emission exists (about 2 magnitudes brighter than an average Sc galaxy). This abnormal emission is still very much less than that of the first three galaxies mentioned. If the four galaxies tested are really in collision it may be concluded that a collision does not always result in greatly enhanced radio emission.

Table 3. *Observations of galaxies suggested to be colliding*

(a) New observations

Galaxy and radio source	Flux density at 85 Mc./s. (w.m.$^{-2}$ (c./s.)$^{-1} \times 10^{-26}$)	$m_{R(158)}$	m_p	$m_R - m_p$
NGC 4038/39*	4·7	10·3	10·6	− 0·3
NGC 3256†	< 1·5	> 11·6	10·6	> 1·0
NGC 1487†	< 2·5	> 11·0	11·9	> − 0·9
NGC 520†	< 4·0	> 10·5	11·6	> − 1·1

(b) Previously known

Cygnus A	14,000	2·0	18·0	− 16·0
NGC 1275, Perseus A	130	7·0	12·0	− 5·0
NGC 5128, Centaurus A	5,000	2·8	6·0	− 3·2

* Suggested by R. Minkowski. † Suggested by G. de Vaucouleurs.

3. NOVAE

Supernovae and novae have been suggested as radio sources because of the identification of several strong sources as the remnants of supernovae. A selection of ten novae and two supernovae, one extra-galactic, has therefore been examined and the results are given in Table 4. Of these twelve objects the only certain identification is with Kepler's Star, the galactic supernova. Taken in conjunction with previously known identifications these results support the suggestion that supernovae give rise to intense radio sources, but give no support to the corresponding suggestion for novae. Quantitatively, the radio emission from the remnants of novae studied must be less than that from supernovae by at least a factor equal to the ratio of the light emitted at the maximum of each. Unfortunately the majority of novae occur near the galactic centre where the sensitivity of the equipment is much reduced.

Table 4.

Southern novae observations

Nova	Flux density at 85 Mc./s. $(\text{w.m.}^{-2}\,(\text{c./s.})^{-1} \times 10^{-26})$	$m_{R(85)}$	m_p at max.	$m_R - m_p$	Remarks
1604 Ophiuchi	38	7·5	−2	$9\frac{1}{2}$	Kepler's star, a supernova
1860 Scorpii	< 10	> 9	7	> 2	
1895 Carinae	< 10	> 9	8	> 1	
1895 Centauri	< 5	> 10	7	> 3	A supernova in NGC 5253
1898 Sagittarii	< 5	> 10	4·7	$> 5\frac{1}{2}$	
1899 Sagittarii	< 10	> 9	8·5	$> \frac{1}{2}$	
1899 Aquilae	< 10	> 9	7	> 2	
1910 Sagittarii	< 20	> 8·3	7·5	$> \frac{1}{2}$	
1917 Ophiuchi	< 10	> 9	6·5	$> 2\frac{1}{2}$	
1918 Aquilae	< 10	> 9	−0·7	$> 9\frac{1}{2}$	Wrongly suggested as a strong radio source by Bolton, Stanley and Slee
1925 Pictoris	< 5	> 9·7	1·1	$> 8\frac{1}{2}$	
1942 Puppis	< 5	> 9·7	0·4	$> 9\frac{1}{2}$	
		Northern novae			
1054 Tauri	1800	3·4	−6	$9\frac{1}{2}$	The Crab nebula, a supernova
1572 Cassiopeiae	170*	6·0*	−4	10*	Tycho Brahe's star, a supernova

* Estimated from Manchester 158 Mc./s. observations.

4. EMISSION NEBULAE

It is known from theoretical considerations that emission nebulae should emit thermally in the radio spectrum by the free-free transition mechanism, and the intensity of emission at each frequency can be estimated from optical data. Such objects should be observed in emission if the background brightness temperature were lower than the electron temperature in the nebula, in absorption if higher. A number have been observed in emission at high frequencies (order of 1000 Mc./s.) where the thermal emission is relatively high, but none has been reported on lower frequencies where it is relatively low.

Table 5. *Observations of emission nebulae*

| | Flux density at 85 Mc./s. | |
| | Observed | Estimated from optical data |
Nebula	(w.m.$^{-2}$ (c./s.)$^{-1}$ × 10^{-26})	(w.m.$^{-2}$ (c./s.)$^{-1}$ × 10^{-26})
η Carinae nebula	150	—
NGC 2237	130	160
30 Doradus	30	35*
Orion nebula (M 42)†	37	15*
I 2177	15‡	—
NGC 2264	8‡	—

* Estimated assuming the nebula is optically thick and electron temperature = 10,000° K.

† Also observed at N.R.L. at frequencies of 1400 and 3200 Mc./s. with flux density 450.

‡ The background in these neighbourhoods is complex so that the identifications are not yet certain. The question should be settled when detailed contours over the region become available.

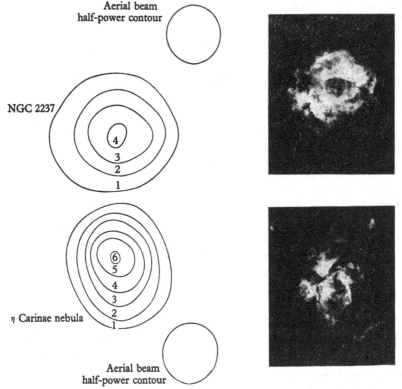

Fig. 5. A comparison of 85 Mc/s. isophotes with photographs for two emission nebulae. The aerial beam-width to half-power is shown in each case.

Our new equipment, however, has sufficient sensitivity and resolution to permit observation of a considerable number. Six bright nebulae situated in regions of relatively faint background have probably been observed in emission, five of these having not previously been reported observed at radio frequencies (see Table 5). In the cases where adequate optical data exist the observed radio emission has been computed and this value is also shown. The larger nebulae can be resolved and the radio isophotes are compared with photographs in Fig. 5. The evidence indicates that many emission nebulae emit thermally in the expected manner and do not contain intense non-thermal sources.

REFERENCES

[1] Mills, B. Y. and Little, A. G. *Aust. J. Phys.* **6**, 272, 1953.
[2] Mills, B. Y. *Aust. J. Phys.* **8**, 368, 1955.
[3] Shklovsky, I. S. *U.S.S.R. Acad. Sc. Ast. J.* **29**, 418, 1952.
    Shklovsky, I. S. *U.S.S.R. Acad. Sc. Ast. J.* **30**, 15, 1953.
[4] Hanbury Brown, R. and Hazard, C. *Phil. Mag.* **43**, 137, 1952.
[5] Hanbury Brown, R. and Hazard, C. *M.N.R.A.S.* **113**, 123, 1953.

Note added in proof.

Subsequent calibrations indicate that the flux densities in this paper are too low by a factor of approximately 2. This does not seriously affect physical conclusions.

RECENT RESULTS IN RADIO ASTRONOMY AT THE OHIO STATE UNIVERSITY

J. D. KRAUS, H. C. KO, R. T. NASH AND D. V. STOUTENBURG

Ohio State University, Columbus, Ohio, U.S.A.

Research in radio astronomy at the Ohio State University covers a wide range of astronomical and engineering phases. In this paper the principal topics currently under investigation are briefly discussed. Reference is made to more detailed papers or reports published within recent months or now in the press.

1. MAPPING PROGRAMME

Last year a map of the summer sky including the region of the galactic nucleus was published [1]. This map is shown in Fig. 1. The radio-isophotes have a temperature difference of about 20°. It was obtained by means of the 96 helix antenna. At a frequency of about 250 Mc./s. this antenna has a beam width of about 1° in Right Ascension and 8° in declination.

During the winter and spring of this year the same antenna was used to make a map of the radio radiation from the winter sky including the region of the galactic anti-centre. This map is shown in Fig. 2. It is both more accurate and more detailed than the one of the summer sky. The contour interval corresponds to a temperature difference of about 7° or about one-third the interval on the other map. Moreover, the map is corrected for the effects of individual intense localized and extended sources so as to show only the residual background radiation. This map has been published in preliminary form [2] and in more complete form [3].

2. INTENSE RADIO SOURCES IN THE REGION OF GALACTIC ANTI-CENTRE

In the map of the winter sky (Fig. 2) nineteen intense localized and extended sources are shown by small circles. The solid circles correspond to sources of less than 1° angular extent while the open circles correspond to sources of 1° angular extent or greater. The size of the open circle is equal to the source extent assuming a uniform source distribution.

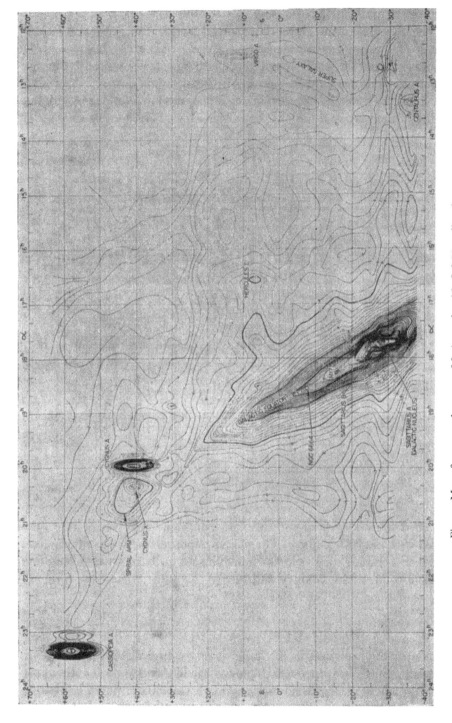

Fig. 1. Map of summer sky at 250 Mc./s. made with O.S.U. radio telescope.

133

The power flux of the sources is expressed by their radio magnitude with sources of the first magnitude having a power flux of at least 5×10^{-24} janskys (1 jansky = 1 watt per square metre per cycle per second). On this magnitude scale the Crab nebula is a first-magnitude source.

A list of the nineteen intense sources is presented in Table 1. A striking feature of the map and table is the existence of a considerable number of

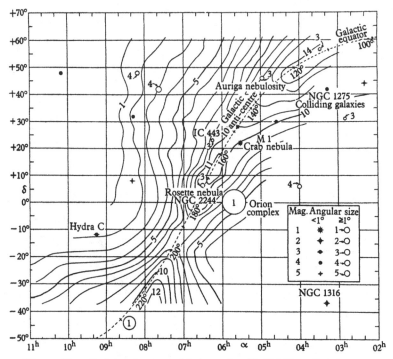

Fig. 2. Map of winter sky at 242 Mc./s. made with O.S.U. radio telescope.

extended sources lying close to the galactic plane and presumably associated with our own Galaxy. The existence of such galactic radio sources and their concentration near the galactic equator has also been reported by Bolton, Westfold, Stanley and Slee [4], by Mills [5], and by Hanbury Brown, Palmer, and Thompson [6].

3. RADIO EMISSION FROM H II REGIONS AND IN PARTICULAR FROM THE ROSETTE NEBULA

A number of the sources in the table correspond to H II emission regions, as, for example, the extended first magnitude sources in Orion and Puppis. Of particular interest is our source No. 12 which corresponds in position

Table 1

No.	R.A. (1950) h. m.	Dec. (1950) (°)	Flux density $(10^{-26} \text{ w.m.}^{-2} \text{ (c./s.)}^{-1})$	Other catalogue number	Remarks
1	02 19 ±2	+44±2	30	R02.01, BH4, BH5	NGC891?
2	02 48 ±2	+31±2	90	KKM AriB	1° extended source
3	03 16 ±1	+42±2	60	KKM PerB, R03.02, M03+4, BH6, BSS40	NGC1275
4	03 20 ±1	−37±2	200	M03−3, BSS10, S03−4	NGC1316
5	03 27·5±1·5	+56±2	80	BH7	
6	03 59 ±2	+ 6±2	70	KKM Tau E	1°5 extended source
7	04 35 ±1·5	+30±2	90	KKM Tau B, R04.01? M04+3, BWSS−B	
8	04 57 ±1	+46±1	90	KKM Aur A? M05+4 BH9, BSS76	Auriga nebulosity 1°7 extended source
9	05 31·5±0·5	+22±1	800	KKM Tau A, R05.01, M05+2, BSS2, HMS2	M1, Crab nebula
10	05 42 ±3	0±3	—	KKM Ori A, BWSS−D	Orion complex 9° extended source
11	06 15 ±1	+23±1	160	KKM Gem B, HMS22, HMH13, Baldwin and Dewhirst	IC443 1°2 extended source
12	06 30·5±0·5	+ 5±1	160	KKM Mon A	Rosette nebula, NGC2244 1°5 extended source
13	07 38 ±2	+42±2	45	R07.02	2° extended source
14	08 11 ±2	+48±2	60	R08.01, BH11, BSS84	1°5 extended source
15	08 18 ±2	+32±2	45	KKM Lyn A, R08.03	
16	08 20 ±2	+ 8±4	20	KKM Hya P	Fluctuating source
17	08 24 ±3	−44±4	—	KKM Pup A (Vel A), BWSS−F	H II region 5° extended source
18	09 15·5±1	−12±2	170	KKM Hya C, M09−1A, BSS 26, S09−1	
19	10 12 ±2	+48±2	60	New	

to the Rosette nebula. This source was also detected in our survey of 1953 [7] but the position was not accurate enough to permit an identification at that time.

Fig. 3 is a sample signature of the Rosette nebula taken with the O.S.U. radio telescope. Shown also in this illustration are photographs of the nebula in red (above) and blue (below) taken by Dr Minkowski of the Mount Wilson and Palomar Observatories.

A discussion of our radio observations of the Rosette nebula is given in *Nature* and in an *O.S.U. Radio Observatory Report* [8].

Our measurements indicate an apparent black-body temperature for the nebula of 200° K. Assuming an electron temperature of $10^{4°}$ K. in the

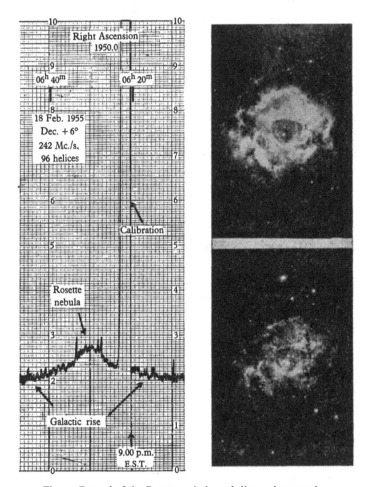

Fig. 3. Record of the Rosette nebula and direct photographs.

nebula, the average optical thickness at 242 Mc./s. over this region is then 0·02. According to the analysis of Greenstein and Minkowski [9] this will correspond to an emission measure of 3000. Assuming the nebula is spherical with a diameter of 37 parsecs at 1400 parsecs, this leads to an average electron density of 9 per cubic centimetre. This value is in good agreement with the recent value of 14 per cubic centimetre given by

136

Minkowski [10] as based on improved measures of the surface brightness by Kron and supports the hypothesis of thermal emission by free-free transitions from the Rosette nebula.

4. FLUCTUATING RADIO SOURCE

During the mapping survey of the winter sky a source was found that exhibits remarkable fluctuations. Since either scintillations (of ionospheric or tropospheric origin) or inherent variations in intensity of the source, or both, may be involved, the term 'fluctuation' is used to describe the effect whatever its cause.

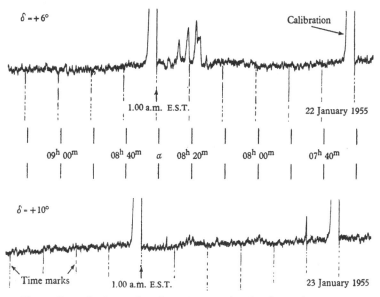

Fig. 4. Records of 22 and 23 January 1955, showing fluctuating source.

The first fluctuation record of the object was obtained 22 January 1955. A photograph of this record is shown by the upper trace in Fig. 4. It is to be noted that the peaks of the fluctuations (near $\alpha = 8^h\ 20^m$) outline an envelope having the shape of the antenna pattern. The source was detected again as a slight rise but without the marked fluctuation on 23 January. A photograph of this record is shown by the lower trace in Fig. 4. The large deflections on the record at 1.00 a.m. are calibrations impressed automatically once each hour. The marks extending below the traces at 10-minute intervals are time marks impressed automatically from WWV time signals at the beginning of each 600 c./s. tone interval.

Observations of the source were interrupted after 23 January by the routine sky survey but were resumed again in April and May. A total of fourteen good records of the source have been obtained. On four occasions the source showed marked fluctuations as on 22 January but on the other ten days the source was almost indistinguishable from the background (as on 23 January).

Without fluctuations the source intensity is about 1×10^{-25} janskys. With fluctuations the peak intensity rises by a factor of 5 or 10. The fluctuations have a period of 2 to 3 minutes.

Owing to the weakness of the source and its fluctuating nature it is difficult to determine its position accurately. Our present best position (1950·0) is:

$$\text{R.A.} \quad 08^h \quad 19^m \quad \pm 1^m$$
$$\text{Dec.} \quad +8° \quad \pm 3°$$

This source is close to the one reported in our list of 207 sources as Hydra P [11].

No fluctuation effect such as observed on this source had been noted by us in all of our previous observing at 250 or 242 Mc./s. Accordingly, the source was regarded with great interest since if the fluctuations were scintillations they might indicate a source of very small angular extent, perhaps a true radio star.

However, some factors suggest that the fluctuations may be intrinsic source variations. These are:

(1) Marked fluctuations have been observed near midnight and also near sunset. This result does not correlate with the usual diurnal variation observed in scintillations.

(2) Marked fluctuations have never been observed on two consecutive days or in fact at a more frequent interval than eleven days.

(3) It has not been possible to establish any correlation between the marked fluctuations and any solar or ionospheric phenomena.

If the fluctuations are intrinsic variations in the source an explanation is that the source is a star with variable radio emission having a period of hours or days and also a short-period variation of a few minutes. Or the fluctuations may be a true scintillation effect observed only when the source is at or near maximum intensity and ionospheric conditions are suitable. Whatever the explanation, the observations suggest that the source may be a true star or a new class of radio object. Attempts to identify the source with an optical object have not yet been successful. There are several 8th and 9th magnitude stars near the radio position.

A 12th magnitude star, KZP 1284, in Kukarkin's list of stars suspected to be variable, is also close to the radio position. During the summer the source has been in a poor position for observing but it is planned to resume observations later this year. A more complete discussion is given by Kraus, Ko, and Stoutenburg in *Nature* [12].

5. STUDIES OF THE GALACTIC NUCLEUS

Fig. 5 is a profile through the galactic nucleus obtained from a record taken at 242 Mc./s., with the O.S.U. radio telescope at a declination of −29° on 26 February 1955. The ordinate is inches of deflection of the recorder (proportional to power) and the abscissa is the hour angle. The peak signal-to-noise ratio on this profile is over 5000 to 1. Since the

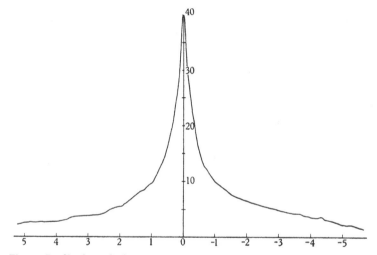

Fig. 5. Profile through the galactic nucleus taken with O.S.U. radio telescope.

antenna pattern is only about 5 minutes in time between half-power points the effect of antenna smoothing on this record is significant only near the very peak of the record.

Using the profile out to an angle of about 10° but not including the smoothed peak, the observed distribution is found to correspond to that which would be produced by a spherically symmetrical source having an intensity per unit volume that varies as the inverse 2·5 power of the radius. This value is in agreement with one deduced theoretically for the nucleus by Keller.

The radio position of the nucleus in galactic (Ohlsson) co-ordinates is: $l = 327°8$, $b = −1°4$ (Kraus and Ko [13]). This position is close to the centre of a highly obscured emission region which shows up as a dark band on the remarkable photograph published by Morgan, Strömgren, and Johnson [14]. It is apparent that their photographic technique enables one to observe emission regions near the galactic nucleus in spite of a tremendous amount of obscuring matter.

6. RADIO MODELS OF THE GALAXY

Based on our radio maps of the Milky Way and studies of the region of the galactic nucleus, a radio model for our Galaxy has been deduced. This model has been developed by Ko [15].

7. ANTENNA RESOLUTION AND SOURCE DISTRIBUTION STUDIES

We have given considerable attention to a number of problems related to the fundamental characteristics of radio-telescope antennas [16, 17]. These are:

(1) The effect of the source distribution on the observed antenna pattern and the inverse problem of determining, in so far as possible, the source distribution from the observed pattern.

(2) The resolution of radio-telescope antennas and the ultimate number of celestial sources that a radio telescope can resolve. According to Ko's criterion this number is numerically equal to the directivity of the antenna.

(3) The range of radio telescopes and other problems.

8. RESEARCH ON RECEIVERS

Research and development on the receiving equipment used at the Ohio State University is continuing. As a result of this work the noise figure of the present 242 Mc./s. receiver is 1·5 and the ultimate sensitivity for a single record and 30-sec. time constant is about 1×10^{-26} janskys, or less than one-tenth of a degree Kelvin.

9. DESIGN FOR A SUPER RADIO-TELESCOPE AND ITS TEST BY MEANS OF A SCALE MODEL

Progress in radio astronomy is to a large extent dependent on the development of larger radio telescopes. A design for a transit telescope which has been under development at Ohio State provides a maximum of effective aperture per dollar of cost [18]. The incoming celestial waves are deflected by a tiltable flat reflector into a standing paraboloid and thence to a horn at the prime focus. An antenna 2000 ft. across by 200 ft. high would have half-power beam-widths of about one-tenth degree in right ascension and 1° in declination at a wave-length of 1 metre. The ground is electrically part of the antenna acting as an image plane for the structure above it. One of the great economies of the design results from the fact that the ground itself is the principal supporting structure.

To test the design a scale model was constructed for operation at 1·2 cm. The model is 12 ft. across, which would correspond at 1 metre to a length of 1000 ft. The tests indicate excellent characteristics with side lobes 1 % or less of the main lobe power. The beam-widths between half-power points are one-quarter degree in right ascension by 1·9° in declination.

The performance of the scale model has been so satisfactory that the model itself is now being put into use as an astronomical instrument for solar and galactic observations at a wave-length of 1·2 cm. The 1·2 cm. receiver is nearing completion and it is expected that actual observations will begin within a few weeks.

Acknowledgment

Radio astronomy at the Ohio State University is supported by grants from the Development Fund and the fund for basic research of the Ohio State University and from the National Science Foundation.

REFERENCES

[1] *Sky and Telescope*, **14**, 22, November 1954.
[2] *Sky and Telescope*, **14**, 371, July 1955.
[3] Ko, H. C. and Kraus, J. D. *O.S.U. Radio Observatory Report*, no. 4, 22 June 1955.
[4] Bolton, J. G., Westfold, K. C., Stanley, G. J. and Slee, O. B. *Aust. J. Phys.* **7**, 76, 1954.
[5] Mills, B. Y. *Aust. J. Sci. Res.* A, **5**, 266, 1952.
[6] Hanbury Brown, R., Palmer, H. P. and Thompson, A. R. *Nature*, **173**, 945, 1954.
[7] Kraus, J. D., Ko, H. C. and Matt, S. *A.J.* **59**, 439, December 1954.
[8] Ko, H. C. and Kraus, J. D. *Nature*, **176**, 221, 30 July 1955; also *O.S.U. Radio Observatory Report*, no. 4.
[9] Greenstein, J. L. and Minkowski, R. *Ap. J.* **118**, 1, 1953.
[10] Minkowski, R. *I.A.U. Symposium*, no. 2, 1955.
[11] Kraus, J. D., Ko, H. C. and Matt, S. *A.J.* **59**, 439, December 1954.
[12] Kraus, J. D., Ko, H. C. and Stoutenburg, D. V. *Nature*, **176**, 304, 13 August 1955.
[13] Kraus, J. D. and Ko, H. C. *Ap. J.* **122**, 139, July 1955.
[14] Morgan, W. W., Strömgren, B. and Johnson, H. L. *Ap. J.* **121**, 611, May 1955.
[15] Ko, H. C. in his doctor's dissertation, Ohio State University, August 1955.
[16] Matt, S. and Kraus, J. D. *Proc. Institute of Radio Engineers*, **43**, 821, July 1955.
[17] Kraus, J. D. *Trans. Institute of Radio Engineers*, p. 445, Special 1956 issue by the Professional Group on Antennas and Propagation on Symposium at Ann Arbor, Mich., 24 June 1955.
[18] Kraus, J. D. *Scientific American*, **192**, 36, March 1955.

SPECTRA OF SOME RADIO SOURCES

J. P. HAGEN

Naval Research Laboratory, Washington, D.C., U.S.A.

At the tenth General Assembly of U.R.S.I. held in The Hague, Netherlands, in August 1954, the author presented a set of curves showing spectra
of five bright, non-thermal, discrete radio sources. The spectra were based

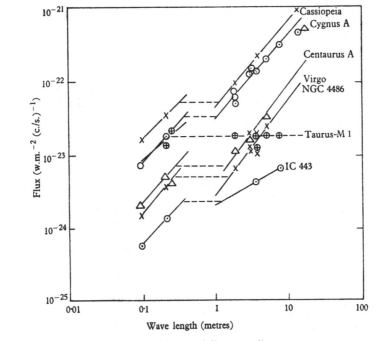

Fig. 1. Spectra of discrete radio sources.

on the recent measurements at centimetre wave-lengths made with the
50-ft. antenna at the Naval Research Laboratory and on earlier published
values for the flux at longer wave-lengths. The spectra are shown in Fig. 1.
It is seen that there is in each case a discontinuity in the region between
30 and 100 cm. wave-length. Unfortunately, there had been no measure-

ments made in this middle region. The slope of the two ends of the curve is nearly the same. There was considerable doubt at the time as to whether the spectra were in fact S-shaped as they would be if the two ends were joined in the simplest fashion. One possibility was that, due to the different techniques used in the centimetre and metre regions both in antennas and receivers, there might exist a calibration error of sufficient magnitude to cause the offset. For many reasons, this was hard to believe. In particular, measurements of solar flux at long and short wave-lengths yield a consistent picture.

To establish the validity of the curve, it was decided to measure at least one of the sources using the 50-ft. antenna at some metre wave-length using receiver calibration techniques at that wave-length similar to those used at centimetre wave-lengths. Mr C. Grebenkemper and Mr E. McClain assisted in making such a measurement at 155 cm. wave-length of Cygnus A and Cassiopeia A. The feed for the antenna consisted of a dipole with a reflector which gave a primary pattern of a type necessary to illuminate properly the $f/0.5$ reflector. The flux from Cygnus A can be estimated fairly accurately from the measured flux but removing the effect of nearby Cygnus X leaves some uncertainty. Cassiopeia A is, however, in a much clearer region and so it was decided to use its spectrum for the check. The measured values for the flux in each case were:

Cygnus A 35.4×10^{-24} w.m.$^{-2}$ (c./s.)$^{-1}$,
Cassiopeia A 62.4×10^{-24} w.m.$^{-2}$ (c./s.)$^{-1}$.

A new set of measurements at 3·2 cm. by Haddock and McCullough provides a further point in the centimetre region and gives greater weight to the slope of the lower end of the curve.

Fig. 2 shows the spectrum of Cassiopeia A using points at 3·2, 9·4, 21 and 155 cm. obtained at the Naval Research Laboratory with the 50-ft. antenna, points at 1·2, 1·9, 3·7 and 13·9 metres from earlier published information. The data are given in Table 1.

Table 1

Wave-length (m.)	Flux w.m.$^{-2}$ (c./s.)$^{-1} \times 10^{-24}$	Observer
0·0315	4	Haddock and McCullough
0·094	12	Haddock, Mayer and Sloanaker
0·21	26·6	Hagen, McClain and Hepburn
1·2	57	Kraus and Ko
1·55	62	Grebenkemper, McClain and Hagen
1·9	93	Brown and Hazard
3·68	220	Ryle, Smith and Elsmore
13·9	950	Hey and Hughes

It is seen that the new point at 1·55 metres is in good agreement with nearby measurements by Kraus and Ko and Brown and Hazard.

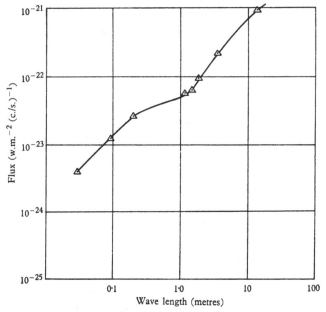

Fig. 2. Spectrum of the Cassiopeia source.

Since the techniques used at 1·55 metres were the same as those used in the centimetre region, one can now join the two sections of the curve with the resultant S-shaped spectrum. Referring to Fig. 1, it should thus be concluded that the spectra of the bright non-thermal sources, with the exception of Taurus A, are S-shaped.

THE LOW-FREQUENCY SPECTRUM OF CYGNUS A AND CASSIOPEIA A

R. J. LAMDEN AND A. C. B. LOVELL

Jodrell Bank Experimental Station, University of Manchester, England

The published measurements of the intensity of the radio sources cover a frequency range down to a lower limit of 22·6 Mc./s., at which measurements have been made on Cygnus and Cassiopeia by Hey and Hughes (1954) [1]. Information about the spectrum at still lower frequencies is difficult to obtain because of interference arising from ionospheric reflexion of distant radio transmitters. Some of this trouble can be alleviated by using a narrow pencil-beam radio telescope for reception and the present communication describes measurements made on frequencies of 16·5, 19·0, 22·6 and 30·0 Mc./s. using the 218 ft. transit radio telescope at Jodrell Bank.

Even with the relatively narrow beam of the transit telescope it was found that observations could be made only during the summer months when Cygnus and Cassiopeia were in transit during darkness. Under these conditions the 30 and 22·6 Mc./s. measurements were straightforward, but the difficulties were severe on 19 and 16·5 Mc./s., and only a small percentage of the total number of runs was sufficiently clear of interfering signals to be used in the analysis.

The direction of the beam of the telescope could be adjusted by altering the tilt of the mast carrying the primary feed so that, in principle, observations on Cassiopeia and Cygnus could be made during the same night. However, the beam was narrow enough to give a complete picture of the transit of both sources in this way only on 30 Mc./s. At the lower frequencies it was found advisable to observe the sources on separate nights.

I. RESULTS

The experiment was carried out between 1 May and 12 August 1955, and although the apparatus was run on most nights, only a few of the runs were suitable for analysis either because of the interference problem mentioned above or because the presence of scintillations prevented an accurate measurement of the deflexion. The details are given in Table 1 where the intensities refer to both planes of polarization.

Table 1

Date (1955)	Frequency (Mc./s.)	Intensities w.m.$^{-2}$ (c./s.)$^{-1} \times 10^{-23}$		Ratio Cygnus/ Cassiopeia
		Cassiopeia	Cygnus	
2 August to 12 August	30·0	45·0 ± 20 %	24·0 ± 20 %	0·53
1 May to 24 May	22·6	46·0 ± 20 %	24·4 ± 20 %	0·53
30 May to 11 June	19·0	27·2 ± 20 %	18·9 ± 20 %	0·69
26 June to 19 July	16·5	< 11 ± 20 %	14·5 ± 20 %	> 1·0

The main source of error in this experiment is a systematic one arising from uncertainty as to the actual collecting area of the transit telescope at these low frequencies. The value of 1000 m.² taken in the reduction is believed to be correct to within 20 %, and this is substantiated by the good agreement of the 30 Mc./s. measurement with the 38 Mc./s. measurements of Adgie [2] for which an accuracy of ± 10% is claimed. The chief cause for the random errors lies in the uncertainty of the extrapolation of the background radiation necessary to obtain the true deflexion of the source from the recording chart. This is particularly the case for Cygnus which lies on a steep slope of the background radiation. However, the occurrence of scintillation in some of the records provides a satisfactory index on which to base this extrapolation. No deflexion could be measured from Cassiopeia on 16·5 Mc./s. and the value given therefore represents an upper limit for the intensity. The ratio of intensities is of course known much more accurately than the absolute intensities since the systematic error due to the telescope is not involved.

2. DISCUSSION

The outstanding feature of these results is the abrupt fall in intensity of both sources at frequencies below 22·6 Mc./s., and the indication that the fall off in the intensity of Cassiopeia is more rapid than that of Cygnus. When the results are compared with those of other workers it is found that the 30 Mc./s. and 22·6 Mc./s. measurements follow naturally the slope of the curve at higher frequencies if the recent accurate results of Adgie at 38 Mc./s. are used. The 22·6 Mc./s. measurement differs markedly from that of Hey and Hughes (1954), their value being well outside the limits of error set in the present measurements. Hey and Hughes exclude the systematic error of their aerial gain in the assessment of errors and it is possible that this may be the source of the discrepancy.

These measurements were made at transit of the sources at latitude 53° N. and it is clear that absorption in the ionosphere cannot be the cause of this sudden fall in intensity. It seems most likely that the drop is due either to absorption in the interstellar medium or to some particular feature of the mechanism of generation of the radio energy in the sources.

REFERENCES

[1] Hey, J. S. and Hughes, V. A. *Nature*, **173**, 819, 1954.
[2] Adgie, R. Unpublished symposium communication.

10 2

PRELIMINARY OBSERVATIONS OF POINT SOURCES AT 12·5 AND 15·5 MC./S.*

H. W. WELLS

Department of Terrestrial Magnetism, Carnegie Institution of Washington, Washington, D.C., U.S.A.

At frequencies much below 30 Mc./s. radio astronomy is substantially affected by the earth's ionosphere. The principal effects of the ionosphere are to absorb signals from extra-terrestrial sources, and to propagate earthbound interfering signals over long distances. Successful observation at these relatively low frequencies requires (1) a clear observing channel, or (2) operation during the interval immediately preceding sunrise when the maximum usable frequency for oblique-incidence ionospheric propagation has fallen below the operating frequency. Since early attempts to locate a clear channel between 10 and 20 Mc./s. were fruitless, we accepted the condition in (2) realizing that ionospheric absorption would also be minimized, since the observing frequency, under these conditions, is more than three times greater than the ionospheric critical frequency at vertical incidence.

The useful time interval depends on the operating frequency and local ionospheric conditions. Our experience during the spring and summer of 1955 in the vicinity of Washington, D.C., revealed about 4 to 5 hours of potential observing time in the pre-sunrise period for 12·5 Mc./s. with a somewhat longer interval at 15·5 Mc./s.

However, this interval free from direct interference often included other unstable conditions which confused the observations. These may be described as (1) scintillations, (2) surges of short duration, and (3) bays lasting several hours. It has been established by Little, Lovell and Smith that scintillations are caused by irregularities in the terrestrial atmosphere. Undoubtedly the other effects are related to additional transient conditions in the outer atmosphere.

Other nights, however, were reasonably undisturbed and permitted the identification of several point sources. Although the 12·5 Mc./s. records seldom presented much of the classical interference patterns characteristic

* Presented by B. F. Burke.

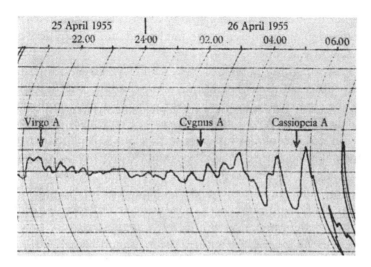

Fig. 1. Record showing interference pattern of Virgo A, Cygnus A, and Cassiopeia A at 12·5 Mc./s., obtained at Derwood, Maryland. (D.T.M.–C.I.W., 1955.)

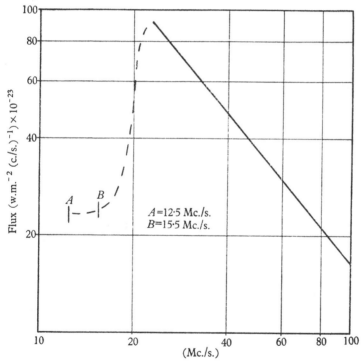

Fig. 2. Drop in flux of Cassiopeia A below 20 Mc./s. Derwood, Maryland. (D.T.M.–C.I.W., 1955.)

(*Note.* Subsequent developments reveal that values reported here are *too low* as a result of receiver saturation effects. Revised estimates are several times greater. However, ionospheric conditions resulting from the new sunspot cycle will prevent new measurements at these frequencies for 8–10 years[3].)

of the higher frequencies, the day-to-day repetition of certain features and their progression with sidereal time facilitated the recognition of Virgo, Cygnus A, and Cassiopeia A as in Fig. 1.

The relative magnitudes of Cassiopeia A and Cygnus A appear to be in the same ratio observed at higher frequencies; that is, roughly two to one. This may be confirmed later in the year when flux measurements will be made on the Cygnus A and other sources at frequencies below 20 Mc./s.

The estimated flux of Cassiopeia A at 12·5 Mc./s. is plotted at A, Fig. 2. The value of about 22×10^{-23} watts per square metre per cycle per second (w.m.$^{-2}$ (c./s.)$^{-1}$)* is obtained from a systematic procedure involving (1) selection and measurement of undisturbed interferometer traces, (2) regular daily calibrations, (3) measurement of loss in transmission lines, and (4) a determination of effective antenna aperture.

Antennas of the interferometer array were separated 946·5 ft. providing spacings of 12 and 15 wave-lengths at 12·5 and 15·5 Mc./s. respectively. Each antenna was a folded dipole with simple reflector. The effective antenna aperture was determined from observations at 27 Mc./s. with the same basic folded dipole-reflector antenna. Assuming the flux at 27 Mc./s. to be 80×10^{-23} w.m.$^{-2}$ (c./s.)$^{-1}$ after the Hey–Hughes report [1] the antenna aperture was found to be 1/2 square wave-length. Pending determination of the effective antenna aperture by other means, the internal accuracy is believed to be within $\pm 20\%$.

Ionospheric absorption is considered as negligible under these selected conditions because the vertical-incidence penetration frequencies have fallen well below one-third of the operating frequency [2].

In August 1955 the recording frequency was shifted to 15·5 Mc./s. and the antennas were scaled accordingly. The flux measures only slightly higher than at 12·5 Mc./s., being 24×10^{-23} w.m.$^{-2}$ (c./s.)$^{-1}$* as shown by B of Fig. 2. The same estimate of over-all accuracy applies. In Fig. 2 the solid line from 22 to 100 Mc./s. is due to observations compiled by Hey and Hughes. Compared to their results at 22 Mc./s. these values at 12·5 and 15·5 Mc./s. are down by a factor of four.

Subsequent measurements will be made at somewhat higher frequencies to 'fill-in the gap' followed by a return to the 12 to 15 Mc./s. region for observation of other point sources while certain basic assumptions are being verified.

REFERENCES

[1] Hey, J. S. and Hughes, V. A. *Nature*, **173**, 819–20, May 1954.
[2] Mitra, A. P. and Shain, C. A. *J. Atmos. Terr. Phys.* **4**, 204–18, 1953.
[3] Wells, H. W. *J. Geophys. Res.* **61**, 541–5, 1956.

* See 'Note' under Figure 2.

OBSERVATIONS OF DISCRETE SOURCES WITH THE 22 MC./S. MILLS CROSS

B. F. BURKE AND K. L. FRANKLIN

Department of Terrestrial Magnetism, Carnegie Institution of Washington,
Washington, D.C.

The 22·2 Mc./s. crossed array of the Carnegie Institution of Washington has been in use since 20 July 1954. This antenna system consists of two linear arrays 2047 ft. in length, each composed of sixty-six half-wave folded dipoles. The amplitude gains of the two arrays are, in effect, multiplied together by a phase-switching system similar to that used in phase-switching interferometers (Ryle, 1952) [1]. The design differs somewhat from the arrangement first used by Mills (Mills and Little, 1953) [2] in that the arrays are arranged in the form of a slightly flattened X. The resulting pencil beam is slightly elliptical in cross-section, measuring 1°6 by 2°4 at half-power points, and is directed by inserting lengths of line into the feeder system of each array, phasing the dipoles such that the maximum response is at the desired zenith angle.

Initial efforts were directed at improving the side-lobe response of the arrays, which causes particular difficulty in this system, since amplitude gains, not power gains, are involved. A conventional Tchebyscheff illumination was used which reduced all side responses to 30 db. below the main beam, a level which has proved tolerable. Of course, whenever a source passes through one of the fan-beams of either array, but not through the main beam, the actual rejection is only 15 db., and hence a strong source can cause confusion in certain positions. Since total power received is also recorded, it is easy to tell when a strong source is passing through one of the fan-beams, and consequently these troublesome side-lobe effects can be identified. In practice, only the Cygnus and Cassiopeia sources have caused difficulty, all other sources being too weak to cause a noticeable response.

A number of the principal radio sources have been observed for purposes of testing the instrument. We have preferred to assign relative intensities only, since these can be measured more accurately. Some corrections are needed, even for relative measurements, because the gains of the arrays are not the same for all phasings, while the polar diagrams of the individual elements are affected by the finite ground conductivity. An error of $\pm 10\%$ can probably be safely assigned to the results, except for the weakest sources, where the uncertainty is greater.

Table 1. *Relative source intensities at 22·2 Mc./s.*

Source	NGC 4486	M 1	IC 443	NGC 1275	Cyg A	Cas A
I.A.U. no.	12N1A	05N2A	06N2A	03N4A	19N4A	23N5A
Relative intensity	1·00	0·75	0·16	0·31	5·7	14·5

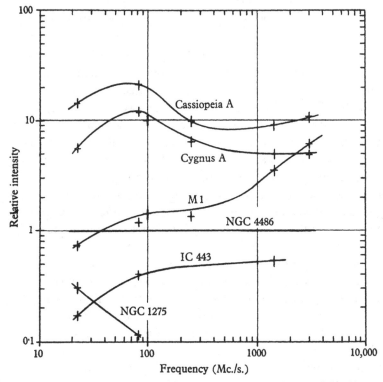

Fig. 1. Source intensities at various frequencies relative to NGC 4486.

The intensities of six sources are given in Table 1, where the Virgo source, NGC 4486, is taken as unity. These results are compared with observations at other frequencies in Fig. 1, where the relative intensities were computed from the values listed in the recent I.A.U. catalogue of reliably known discrete radio sources (Pawsey, 1955) [3].

It is interesting to note that no two sources have identical spectra, although the Cygnus and Cassiopeia sources are similar in qualitative behaviour. The intensity derived from the Perseus source, NGC 1275,

152

includes the integrated intensity received from the associated cluster of galaxies. While this contribution at higher frequencies represents only 25% of the total intensity (Baldwin and Elsmore, 1954) [4], the percentage may be higher at 22·2 Mc./s. Even with as much as half the intensity due to the cluster, however, this source appears to be increasing in intensity with increasing wave-length even more rapidly than the Virgo source. The other sources plotted in Fig. 1 all seem to be decreasing, relative to the Virgo source, as frequency decreases.

REFERENCES

[1] Ryle, M. *Proc. Roy. Soc.* A. **211**, 351, 1952.
[2] Mills, B. Y. and Little, A. G. *Aust. J. Phys.* **6**, 272, 1953.
[3] Pawsey, J. L. *Ap. J.* **121**, 1, 1955.
[4] Baldwin, J. E. and Elsmore, B. *Nature*, **173**, 818, 1954.

THE 400 MC./S. FLUX FROM CASSIOPEIA A

Ch. L. SEEGER

University Observatory, Leiden, Netherlands

The first trial of equipment designed for 400 Mc./s. absolute flux density measurement was completed just prior to this symposium. Using the midday sun, the gain of a 7·5 metre Würzburg (equipped with a $TE^{°}_{11}$ circular wave-guide feed) was found by comparison with an 'optimum' pyramidal horn of 20·2 square wave-lengths aperture. Cable losses and signal powers were measured against the available noise power from two co-axial resistors, one at ambient temperature and the other stabilized at the boiling-point of pure nitrogen.

Using the horn and the Würzburg in a 20 wave-length, total-power interferometer, the flux density from Cassiopeia A was found to be 56×10^{-24} w.m.$^{-2}$ (c./s.)$^{-1}$.* Because of the small amount of observational material, it is difficult to state a mean error for this determination. However, $\pm 15\%$ expresses the observer's personal impression of the reliability. Also, the brief experience with this equipment showed that with refined practice one might expect to reach an accuracy (for Cassiopeia A) of perhaps 2 or 3 %, i.e. to the uncertainty in the calculated effective area of the horn [1]. Details of this investigation have since been published [2].

REFERENCES

[1] Jakes, W. C. Jr. *Proc. I.R.E.* **39**, 160, 1951.
[2] Seeger, Ch. L. *B.A.N.* **13**, 100, no. 472, 1956.

* The preliminary value reported at the Symposium was 32×10^{-24}.

INTENSITIES OF THE DISCRETE SOURCES IN CASSIOPEIA, CYGNUS AND TAURUS AT λ 3·2 CM.

V. A. RAZIN AND V. M. PLETCHKOV

Gorky State University, U.S.S.R.

Measurements of the intensities of radio emission from the three most powerful discrete sources were carried out early in 1955 at the Gorky radio astronomical station 'Zimenky' (latitude 56° 9·5'). The arrangement used for these measurements is described elsewhere [1]. The main part of the aerial consists of a paraboloid, 4 metres in diameter, on an alt-azimuth mounting. The beam has an opening (between half-power points) equal to 32'. The effective area of the aerial was determined by comparison with the standard megaphone antenna for solar radio emission [2] and equals 10 m.². The efficiency of the aerial is determined according to the method of measurement of the proper radio emission of the aerial [3]. The reception device is of a modulation type. The fluctuation threshold of the sensitivity of the device for the used time constant of 20 sec. equals 0°6 C., which corresponds to a flux of non-polarized radiation of $1·65 \times 10^{-24}$ w.m.$^{-2}$ (c./s.)$^{-1}$ reaching the aerial.

The results of these measurements are summarized in the table, where every number represents the mean of a number of measurements.

Source	Intensity in units of w.m.$^{-2}$ (c./s.)$^{-1}$ $\times 10^{-24}$
Cassiopeia A	4·6
Taurus A	6
Cygnus A	6·6

The random errors in the mean values do not exceed $\pm 5\%$. Systematic errors may be in the range of $\pm 20\%$.

REFERENCES

[1] Troitzky, V. S., Rakhlin, V. L., Bobrick, V. T. and Starodubtzev, A. M. *Publications of the 5th Cosmogonical Conference*, Moscow, 1956, p. 37.
[2] Zelinskaja, M. P. and Troitzky, V. S. *Ibid.* p. 99.
[3] Troitzky, V. S. *J. exp. theor. phys. U.S.S.R.*

THE RADIO FREQUENCY SPECTRUM OF CASSIOPEIA A: A SYMPOSIUM SUMMARY

Ch. L. SEEGER

University Observatory, Leiden, Netherlands

Data presented at this symposium showed that there had been considerable progress during the previous year towards the solution of one of the basic problems in radio astronomy—the determination of the full radio frequency spectra of the discrete sources. The generally accepted plan is to

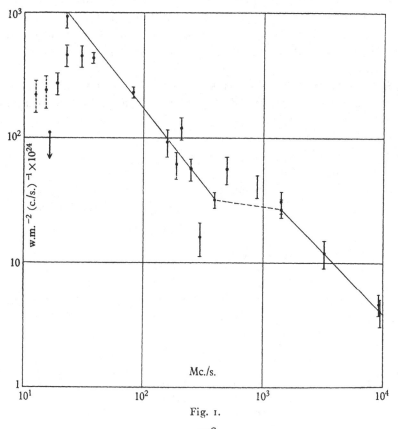

Fig. 1.

establish the spectra of a few 'reference sources'. At present, these reference objects must be chosen from among the most intense sources, Cassiopeia A, Cygnus A, Taurus A and Virgo A. The utility of both Cygnus A and Taurus A is impaired by their close proximity to other discrete sources of appreciable strength. If they are to be observed with small antennas, meticulous interferometry, employing both variable spacing and variable azimuth, will probably be required in order to obtain an accuracy of 10 %. Cassiopeia A and Virgo A appear to be sufficiently in the clear for present needs, though only the latter is visible in both hemispheres and it is about twenty times less bright than Cassiopeia A. However, Cassiopeia A has received most attention so far. The effects of radiometer non-linearity must be examined, particularly at the lower frequencies, when calibrating Virgo A in terms of Cassiopeia A.

Table 1. *Flux density vzs. frequency for Cassiopeia A* ($23N_5A$)

ν [Mc./s.]	S [w.m.$^{-2}$ (c./s.)$^{-1}$] $\times 10^{-24}$	O.P.I.R.* $\pm\%$	Observers and References
12·5	220	See paper	Wells, H. W. Paper 22
15·5	240	See paper	Wells, H. W. Paper 22
16·5	<110	20	Lamden, R. J. and Lovell, A. C. B. Paper 21
19	272	20	Lamden, R. J. and Lovell, A. C. B. Paper 21
22·6	460	20	Lamden, R. J. and Lovell, A. C. B. Paper 21
22·6	940	20	Hey, J. S. and Hughes, V. A. *Nature*, **173**, 4409, 1 May 1954
30	450	20	Lamden, R. J. and Lovell, A. C. B. Paper 21
38	435	10	Adgie, R. L. Unpublished symposium paper
81·5	232	10	Adgie, R. L. Unpublished symposium paper
158	93	25	Brown, R. Hanbury and Hazard, C. *Mon. Not. Roy. Astr. Soc.* **113**, 123, 1953
193·5	62	25	Grebenkemper, C., McClain, E. F. and Hagen, J. P. (in publication); see also paper 20
210	121	20	Adgie, R. L. Unpublished symposium paper
250	57	20	Kraus, J. D., Ko, H. C. and Matt, S. *A.J.* **59**, 11, December 1954
300	16	30	Razin, V. A. and Pletchkov, V. M., as communicated to symposium by Dr S. B. Pikelner
400	32	15	Seeger, C. L. Paper 24
500	56	25	Adgie, R. L. Unpublished symposium paper
900	33–50	—	Denisse, J. F. Private communication
1420	31	20	Westerhout, G. *B.A.N.* **12**, 309 (no. 462), 1956. This is 'pine-tree' calibration.
1420	26·6	+20–15	Hagen, J. P., McClain, E. F. and Hepburn, N. Paper 20
3200	12	25	Haddock, F. T., Mayer, C. H. and Sloanaker, R. M. Paper 20
9400	4·6	20	Razin, V. A. and Pletchkov, V. M. Paper 25
9500	4	25	Haddock, F. T. and McCullough, T. P. Paper 20

* Observer's personal impression of reliability.

157

Though the principle of accurate absolute flux density measurement appears to be simple and straightforward, present-day practice has turned out to be unexpectedly difficult. Even with Cassiopeia A, measurements by different observers sometimes disagree by a factor of two or more. Nevertheless, the data on Cassiopeia A seem to be sufficiently consistent, and the spectrum of such intrinsic importance, to warrant the above tabulation and the accompanying curve (Fig. 1). I have tried to check each entry with the individual observer, since some of the data are published here for the first time. In the event of any errors or omissions, I herewith tender my most sincere apologies.

INTENSITY DISTRIBUTION ACROSS THE CYGNUS AND CASSIOPEIA SOURCES

R. C. JENNISON

Jodrell Bank Experimental Station, University of Manchester, England

Measurements of the angular distribution of intensity across the intense discrete sources in Cassiopeia and Cygnus have previously been handicapped by lack of knowledge of the phase of the Fourier transform at very long aerial spacings. The technical difficulties of measuring the phase of the transform and also of calibrating the absolute amplitude have been solved by a new technique involving three stations. This method enables the phase to be measured relative to a frame of reference within the source and obviates the need for retaining the phase angles accurately constant on the removal of one of the aerial systems to a new site. The phase measurement is not limited to observations of the central fringe, and useful measurements may be made on all the fringes contained within the aerial polar diagrams.

The three stations are arranged in a straight line and the outputs are connected together to form three interferometer systems (Fig. 1). The principle evoked is that, under the illumination of a point source, the addition of the wave-forms of the fringes from the two inner systems AB, BC is always equal to the fringe pattern observed on the outer system AC, irrespective of any errors in the equipment prior to the final multiplier system. If the source is extended and contributes a phase component, the relative phase of the fringe pattern traced by the interferometer AC to the sum of patterns AB and AC will represent the relative phase of the transform over these spacings. For calibration purposes it is possible to place station B mid-way between A and C so that $AB = BC$; the relative phase of the fringes now simply refers to the single intermediate point. It is, however, possible to use the equipment when AB and BC are widely different spacings, provided that the phase of the transform over distance AB and distance BC has been, or will be, determined. Provision is incorporated for slowing down the fringe systems by means of separate continuously rotating resistance capacity quadrature phase shifters incorporated in a common local oscillator unit.

The three stations may be utilized to determine the absolute amplitude of the transform at any spacing provided that a total power recorder is incorporated in one channel.

The experimental procedure is to have the largest aerial system and total power recorder in a fixed station, whilst the two remaining stations are moved out successively to greater distances. The centre station may be

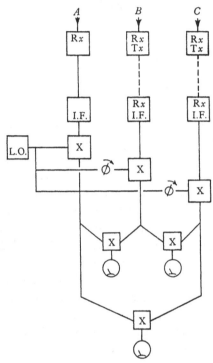

Fig. 1. Basic block diagram of the phase-sensitive interferometer. X=multiplier or mixer; Rx=receiver; Tx=transmitter. The dotted lines represent the coherent radio links from the outstations, B and C, and incorporate both local oscillator and signal channels.

transplanted to become the farthermost station in the next series of measurements so that only one station is actually moved to give a new reading. Both the phase and the amplitude of the transform are measured concurrently during each series of measurements.

PRELIMINARY RESULTS

The equipment has recently been applied to the measurement of the phase and amplitude of the Fourier transform of the intensity distribution across the intense sources Cygnus and Cassiopeia. The base-lines used have varied between 500 and 2000 λ in an east–west direction. The operating frequency is 127 Mc./s.

Preliminary results on Cassiopeia are inconclusive as the measurements so far have been carried out at spacings where the source is almost completely resolved, and further readings on short spacings are required in order to establish the general nature of this transform.

The results on Cygnus are shown in Fig. 2, in which the square of the amplitude, ρ^2, is used as an ordinate. The new measurements of ρ^2 are

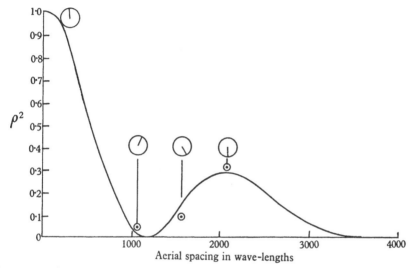

Fig. 2. The radio source in Cygnus, variation with base-line of the square of the amplitude of the Fourier transform (full curve) and phase angle, or argument of the Fourier integral (clock faces).

shown by the small circles whilst the phase is indicated by the position of a unit vector on a clock dial. The full curve is a reproduction of that obtained in a previous survey of the source using a post detector interferometer [1, 2]. The Fourier transform of the original curve when supplemented by the new phase measurements yields an intensity distribution which is generally in agreement with the earlier interpretation of measurements in which the phase was not recorded [1, 2]. The source consists of two prominent centres of emission but it now appears that the system as a whole is slightly asymmetrical.

REFERENCES

[1] Jennison, R. C. and Das Gupta, M. K. *Nature*, **172**, 996, 1953.
[2] Jennison, R. C. and Das Gupta, M. K. *Phil. Mag.* **1** (Ser. 8), 55, 1956.

THE ANGULAR DIAMETER OF DISCRETE RADIO SOURCES

H. P. PALMER

Jodrell Bank Experimental Station, University of Manchester, England

An interferometer of readily varied resolving power has been constructed at Jodrell Bank, and since 1953 it has been used to measure the angular diameters of all but the faintest of the discrete sources reported in the survey of Brown and Hazard [1].

I. EQUIPMENT

The equipment, which has been fully described by Brown, Palmer and Thompson [2], operates at a wave-length of 1·9 metres. The two aerials used are the 218′ transit radio telescope, which has a pencil beam of 2° at this wave-length, and a small broadside array with an aperture of 36 m.². The small aerial has been moved to several remote sites to permit observations at different aerial spacings. The two sections of the equipment were connected by a co-axial cable, or, at aerial spacings greater than 1 Km., by a radio link.

The interferometer was designed on the rotating lobe principle, which permits control of the period of the fringe pattern of the recorder. A long recorder time-constant can therefore be used even when there is a large spacing of the aerials. Without this facility the measurements of high resolving power could not have been carried out, and observations of faint sources, using these aerials, would be restricted to a resolving power of about $\frac{1}{2}$°.

Discrete sources of intensity greater than 30×10^{-26} w.m.$^{-2}$ (c./s.)$^{-1}$ can be identified on individual records. This also represented the limit of sensitivity, for successive records of this type cannot be averaged readily because the phase of the output signal varies in a random manner from one transit to the next.

2. THE DISCRETE SOURCES OBSERVED

This experiment was designed to obtain diameter measurements for some of the fainter sources revealed by the well-known survey of Brown and Hazard. They detected twenty-three discrete sources in this survey using the same transit radio telescope, with a field of view between declinations $+38°$ and $+68°$. Fifteen of these sources are sufficiently intense to be detected by the interferometer; the remainder, some of which were detected in the original survey only by averaging several records, were invisible.

The sources detected include the intense sources Cassiopeia A and Cygnus A (nos. 19 and 22), the diameters of which were measured by Brown, Jennison and Das Gupta [3, 4], while the source in Andromeda (no. 2, identified with M31) and also the source Cygnus X (no. 20) were reported to be extended sources.* (Diameter $> 2°$.)

Sources nos. 1 and 22 are too close to Cassiopeia A for interferometer measurements. The remaining eleven sources have been observed with the interferometer, and all but three of them have been resolved. These eleven sources fall clearly into the classes I and II suggested by Mills. The six class I sources are all relatively intense ($> 60 \times 10^{26}$); they all lie within $5°$ of the galactic plane, and were all found to have large angular diameters, greater than $1°$. The five class II sources were less intense, were more than $10°$ from the galactic plane, and were found to have angular diameters of a few minutes of arc.

3. THE CLASS I SOURCES

The measurements on the sources near the galactic plane have been reported previously [5]. All were resolved at aerial spacings $< \lambda 50$. One of these sources, in Auriga, has been tentatively identified by Minkowski with a faint nebulosity having a filamentary structure, which was photographed with the 48-inch Schmidt at Mount Palomar.

4. THE CLASS II SOURCES

The area of the field of view away from the galactic plane contained five faint sources which could be detected with this interferometer. They were all found to be of much smaller angular diameter, and are being studied in a second series of observations, using much larger aerial spacings,

* The reference numbers quoted here are those given in Brown and Hazard's catalogue[1].

11·2

extending, so far, to λ6720. The sensitivity of the equipment deteriorates slightly when a radio link is used, and these faint sources are then close to the limit of sensitivity. The results are given in Fig. 1 and in Table 1. The intensities in the table have been quoted from reference 1; the apparent surface temperature has been calculated from it, assuming uniform disks of the diameter quoted.

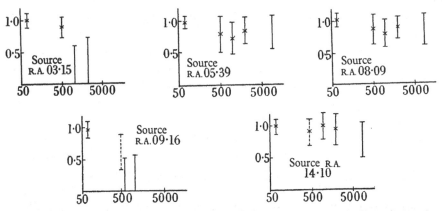

Fig. 1. Fringe visibility of five discrete sources as a function of aerial spacing in wave-lengths. ⊢×⊣ aerials spaced E.-W. |---×---| aerials spaced N.-S. ⊢—⊣ aerials spaced N.W.-S.E. (provisional results).

Table 1. *Preliminary results of observations with large aerial spacings*

No.	Source I.A.U. no.	R.A.	Galactic co-ordinates (lat.)	Galactic co-ordinates (long.)	Intensity w m.⁻² (c./s.)⁻¹ × 10⁻²⁶	Apparent temperature (° K.)	Angular diameter	Remarks
6	03N4A	03·15	−12·4	118·3	65	$2·2.10^5$	$2'·4 ± 0'·5$	NGC 1275
10	—	05·39	+11·4	129·4	50	$>5·5.10^6$	$<25''$	Unidentified
11	08N4A	08·09	+34·3	138·8	40	$>4·5.10^6$	$<25''$	Unidentified
12	09N4A	09·16	+45·8	142·7	30	$3·6.10^4$	$4'·0 ± 1'·0$	Unidentified
18	14N5A	14·10	+61·2	61·1	40	$>4·5.10^6$	$<25''$	Unidentified

Two of the sources of class II, at R.A. 03·15 (no. 6) and at R.A. 09·16 (no. 12) appear to have been resolved. The observations of R.A. 09·16 were difficult because it is the faintest source studied, and it may have been lost through the deterioration in sensitivity, rather than resolved.

The other three of the sources, at R.A. 05·39, 08·90, 14·10 (nos. 10, 11 and 18) were still detected at the largest aerial spacing. Because the observations are so close to noise it is only possible to say that the fringe visibility is greater than 50 % at an aerial spacing of λ6720. If these sources are assumed to be spherical their angular diameter must be less than 25″, and their apparent surface temperatures, at 158 Mc./s., of the order of 10^7°K. This is less than the apparent surface temperature of Cygnus A by only one order of magnitude. This experiment is being continued in a further attempt to resolve these three sources.

REFERENCES

[1] Hanbury Brown, R. and Hazard, C. *M.N.R.A.S.* **113**, 123, 1953.
[2] Hanbury Brown, R., Palmer, H. P. and Thompson, A. R. *Phil. Mag.* **46**, 857, 1955.
[3] Hanbury Brown, R., Jennison, R. C. and Das Gupta, M. K. *Nature*, **170**, 1061, 1952.
[4] Jennison, R. C. This Symposium Report, paper 27.
[5] Hanbury Brown, R., Palmer, H. P. and Thompson, A. R. *Nature*, **173**, 945, 1954.

THE DISCRETE SOURCE OF RADIO WAVES AT THE GALACTIC CENTRE

F. G. SMITH, P. A. O'BRIEN AND J. E. BALDWIN

Cavendish Laboratory, Cambridge, England

The discrete source of radio emission in Sagittarius is among the most intense in the whole sky, but its situation in the belt of emission from ionized hydrogen and other sources associated with the galactic plane makes it difficult to observe. The observations described in this paper were made at frequencies of 38, 81·5, 210, and 500 Mc./s.; at these low frequencies it is particularly difficult to obtain sufficient aerial resolving power to distinguish the discrete source from the background. Interferometer aerials were therefore used, and at 38 and 210 Mc./s. spacings up to $\lambda 60$ were used, sufficient to resolve the source completely. At 81·5 Mc./s. various sections of the large interferometer aerial were used.

These observations gave the following values of the flux density ($\pm 20 \%$) and of the diameter (to half-brightness) of the source:

Frequency (Mc./s.)	Flux density 10^{-22} w.m.$^{-2}$ (c./s.)$^{-1}$	Diameter (°)
38	1·0	2·0
81·5	1·0	—
210	1·0	1·1
500	0·73	—

The spectrum of the source may now be found by combining these results with observations at 1420 Mc./s. by Hagen, Lilley and McClain (1954) [1] and at 3200 Mc./s. by Haddock, Mayer and Sloanaker (1954) [2]. In calculating the flux density at these frequencies it has been assumed that the source is 1:1° in diameter throughout, corresponding to the measured diameter at 210 Mc./s. At the lower frequencies some absorption may be expected to occur in the galactic plane, and the optical depth at 38 Mc./s. would be expected to be about unity if the source is as far distant as the galactic centre. In the spectrum suggested in Fig. 1 the intensity at 38 Mc./s. has accordingly been increased by a factor of 2·5. Absorption of this order would also account for the larger diameter observed at this frequency.

The suggestion has been made by Davies and Williams (1955 [3]; see also paper 12) that in spite of the remarkable coincidence in position this source is not at the centre of the Galaxy, but considerably nearer. Their proposed distance of about 3 kiloparsecs is based on a measurement of the depth and width of the absorption line at 1420 Mc./s. They point out that an H II region exists at their suggested location.

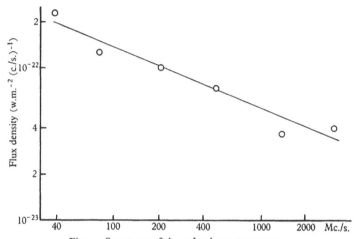

Fig. 1. Spectrum of the galactic centre source.

Our observations at low frequencies have led us to suggest a spectrum where flux density varies as $\lambda^{-0.4}$, which is closer to the spectra of the non-thermal discrete sources than to that of ionized hydrogen. The source observed at these frequencies cannot in any case be an ionized hydrogen region radiating thermally, since the brightness temperature exceeds $10^5\,°$K.—ten times greater than that expected and observed from other such regions.

The position of the source was measured on all four frequencies, and clear support was found for previous suggestions that the source lies very closely in the direction of the galactic centre. It seems that it might still be considered as being in fact at the centre, at which distance it would have a diameter of 150 parsecs.

REFERENCES

[1] Hagen, J. P., Lilley, A. E. and McClain, E. F. *N.R.L. Report*, no. 4448, 1954; *Ap. J.* **122**, 361, 1955.
[2] Haddock, F. T., Mayer, C. H. and Sloanaker, R. M. *Nature*, **174**, 176, 1954.
[3] Davies, R. D. and Williams, D. R. W. *Nature*, **175**, 1079, 1955.

Haddock: The source at the centre observed at 3 cm. with a beam-width of 9ʹ5 appears to have a small diameter (a fraction of a degree) and the surrounding region is complex. The 10 cm. result would have to be raised by a factor 8 if the diameter were 1°. I simply cannot accept such a large multiplication of the flux density.

Pawsey: Studies of the region around the galactic centre by members of the Radiophysics Laboratory indicate that this region is very complex. Observations taken under different circumstances yield different results (cf. McGee, Slee and Stanley (1955) [1]) and it seems most likely that the pattern is not only complex but changes radically with frequency. For example, while Sydney observations at 400 Mc./s. and Washington ones at a few thousand Mc./s. show a concentrated bright region in the suspected direction of the galactic centre, preliminary Sydney observations at 85 Mc./s. with the Mills Cross show isophotes with several adjacent maxima but a small depression in the actual direction of the high-frequency source.

In a complex region such as this, interferometer observations could be grossly misleading. There is a high probability that the measurements cited by Smith refer to the sum of contributions from a number of diverse objects with differing spectra differently compounded at each frequency.

Smith: Fine structure along the galactic plane should have been revealed also by the interferometer measurements. The point we wish to emphasize is that over the whole frequency range 38 to 500 Mc./s. the source is one discrete object, with an angular size of the order of 1°.

Bolton: However, Mills' fine structure is along a line perpendicular to the axis of your interferometer and therefore you cannot resolve it.

REFERENCE

[1] McGee, R. X., Slee, O. B. and Stanley, G. J. *Aust. J. Phys.* **8**, 347 (no. 3), 1955.

RADIO AND OPTICAL INTENSITY DISTRIBUTIONS IN THE CENTAURUS SOURCE (NGC5128)

G. DE VAUCOULEURS AND K. V. SHERIDAN

Yale-Columbia Southern Station, Mount Stromlo, Canberra, Australia, and Radiophysics Laboratory, Sydney, Australia

(1) A preliminary determination of the apparent intensity distribution in the Centaurus source has been made with the 1500-ft. Mills' Cross at 3·5 metres wave-length. At the declination of the source the beam is nearly circular and has a half-power width of $0°8$. Fig. 1 shows contours of apparent equal intensity above a smooth interpolated reference level. The inner 'point' source[1] which is not resolved by the present equipment appears to be embedded in an extended source strongly elongated in position angle $12°$. The secondary, isolated maxima and minima of intensity on the major axis of the extended source at distances greater than $1°$ from the central maximum are probably unreal and may represent side lobes in the diffraction image of the point source.

(2) The optical intensity distribution in NGC5128 was determined from a combination of photographic and photo-electric measurements with the 30-inch Reynolds reflector at Mount Stromlo. Fig. 2 shows the isophotes of the inner globular condensation determined directly on short- and long-exposure photographs in blue light. The unit of B is approximately 21 mag./sq. sec. corresponding to an integrated (total) magnitude of 6·1 (*pg*) for NGC5128.* The intensity distribution in the outer elliptical 'corona' was derived from photo-electric scans along the east–west direction through the nucleus. The photo-electric results agree well with those obtained from tracings of the long-exposure plate along the same direction. In the outer corona the isophotes are very nearly elliptical, with a major axis in position angle $30°$ and a nearly constant ratio of axes $b/a = 0·70$. Beyond $\bar{r} = 10'$, where $\log B = -0·85$, the slope of the curve of $\log B$ plotted against $\bar{r} = \sqrt{a \times b}$ is about $-4·3$ per degree. Along the east–west

---

* The results in Fig. 2 are in fair agreement with those of Evans[2].

direction the nebula has been traced out to about $0°5$ from the nucleus in both directions.

For comparison with the radio data the optical absorption in the dark lane was neglected and the isophotes treated as concentric circles, the hidden central maximum being taken as $\log B = 1·0$.

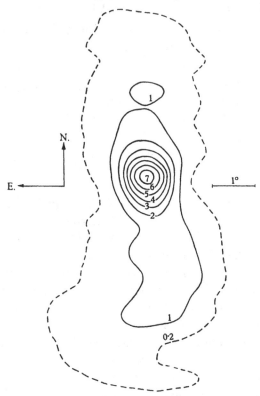

Fig. 1. Apparent radio isophotes of the Centaurus source at 3·5 metres determined with the 1500-ft. Mills Cross aerial at Sydney.

(3) The optical distributions along both the minor and the major axes were smoothed with the aerial beam pattern which can be approximated by a gaussian distribution for $r < 1°$ as drawn in Fig. 3 (small circles). The mean luminosity distribution in the nebula is reproduced on the same scale in the lower left corner of the figure, with the minor (m) and major (M) axes shown separately on a scale enlarged ten times in ordinates beyond $r = 5'$.

The apparent radio and optical distributions as 'seen' by the radio telescope are shown, reduced to a common maximum, for the nebula and the radio source.

(a) *The minor axes* of both the nebula and the source are barely resolved by the aerial and they appear to agree with one another within the uncertainty of the data. The hypothesis that the laws of radio and optical emission along the minor axes of the source and the nebula are similar is supported by earlier interferometer observations of the east–west distribu-

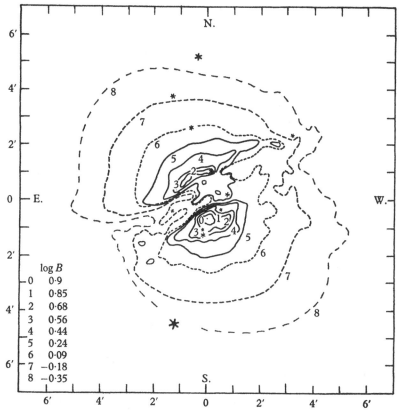

Fig. 2. Optical isophotes of the core of the nebula NGC 5128 in blue light determined on short- and long-exposure photographs taken with the 30-inch Reynolds reflector on Mount Stromlo.

tion across the source by Mills [1]. This is shown by a comparison of the Fourier transform of the amplitude spectrum of the east–west radio interferometer data with the east–west intensity distribution through the nucleus obtained by direct photo-electric scanning at the Cassegrain focus of the Reynolds reflector [3]. The two sets of data are, however, not directly comparable and there is some uncertainty in the interpretation of the radio observations.

(b) *The major axes* of the nebula and of the source are both well resolved by the aerial, but the major axis of the source appears much more extended than that of the nebula. The divergence is particularly marked in the outer parts where the radiation comes mainly from the 'corona'; in the inner parts the apparent distribution is dominated by the diffraction

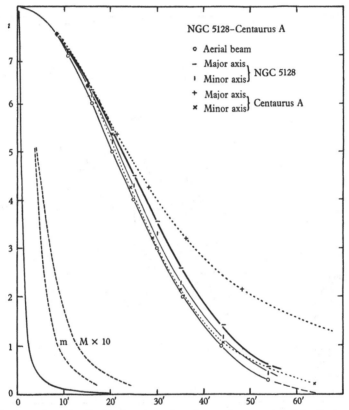

Fig. 3. Apparent intensity distribution along the minor and major axes of the Centaurus source and of the nebula NGC 5128 after smoothing with the 1500-ft. aerial beam pattern.

pattern of the 'point' source where any real discrepancy would tend to vanish. It should be noted that only 2° of the extended source, out of a total length of over 6°, are used here, i.e. a region of fairly high intensity where side lobe effects and uncertainties in the interpolated background should be unimportant.

Dr R. Minkowski has raised, in private discussions before and after the meeting, the question of the reliability of the interpolated background and of the possibility of reconciling the radio and optical distributions by a suitable modification of the background level. The answer to this suggestion appears to be that such an *ad hoc* manipulation of the data would result in producing an extended source of maximum brightness temperature about 2000° K. concentric with Centaurus A and of shape and elongation very much similar to the corona in Fig. 1. Such a coincidence seems at least unlikely, especially when it is recalled that a similar coincidence has been invoked in the case of Fornax A (NGC 1316) which appears to be another example of extra-galactic associations between a point source and an extended source. Fornax A is, however, barely resolved with the present cross-aerial and is too weak for reliable interferometer analysis.

We are indebted to Mr B. Y. Mills for the use of the large cross-aerial and to Dr S. C. B. Gascoigne for that of the Lick photometer on the Reynolds reflector.

REFERENCES

[1] Mills, B. Y. *Aust. J. Phys.* **6**, 452, 1953.
[2] Evans, D. S. *M.N.R.A.S.* **109**, 94, 1949.
[3] Vaucouleurs, G. de, *Occasional Notes R.A.S.* no. 18, 130, 1955.

APPARENT INTENSITY VARIATIONS OF THE RADIO SOURCE HYDRA A*

J. G. BOLTON AND O. B. SLEE

Radiophysics Laboratory, Commonwealth Scientific and Industrial Research Organisation, Sydney, Australia

During the course of a survey for radio sources using a sea interferometer changes in the apparent intensity of the source Hydra A were observed. The intensities on two consecutive days differed by more than 30 % although the intensities of sources appearing earlier and later on the same records showed no significant changes. The position of this source given by Mills (1952)[2] is R.A. 09^h 15^m 46^s, dec. $-11°$ 55' (epoch 1950) and its I.A.U. number 09S1A.

In April 1954 the survey was completed and a systematic study of this source using the 110 Mc./s. sea interferometer was undertaken by O. B. Slee. The source was observed at its rising above the eastern sea horizon and control observations were made of at least one of the sources Taurus A, Virgo A, Centaurus A and Fornax A within a few hours. In this way the intensity measurements of the Hydra source were largely freed from the effects of calibration errors. The daily measured intensities of Hydra A and a comparison source are shown in Fig. 1 *a* for the period April to December 1954. Here the comparison source is a composite, obtained by bringing the intensity measurements of the Taurus, Virgo, Centaurus and Fornax sources to a common scale; this procedure was necessary as normally only one comparison source was observed at any particular time of the year.

From December 1954 to July 1955 the Hydra source, together with the three comparison sources in Taurus, Virgo and Centaurus, was recorded almost daily during meridian transit, using the east–west arm of the 85 Mc./s. 'cross'-aerial (Mills, Little and Sheridan, in preparation [3]). This arm of the cross has a fan-shaped response pattern, 0°6 east–west and 50° north–south, between half-power points. The measured intensities of the Hydra source and the comparison source in Virgo for the period December to July 1955 are shown in Fig. 1 *b*.

* A full account is published in the *Aust. J. Phys.* [1].

From both the sea interferometer and the cross observations it can be seen that the variations in the intensity of the Hydra source are far greater than those of the comparison sources. This is also illustrated in the histograms of Fig. 2, where it is shown that essentially the same results were obtained on both instruments.

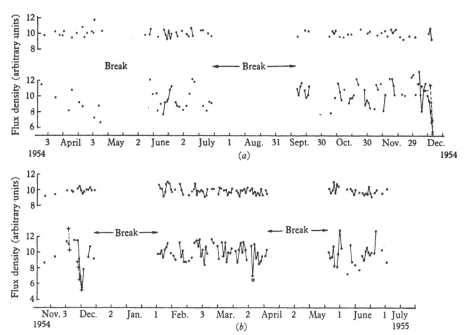

Fig. 1. Plots of the daily observed intensities of Hydra A and the comparison sources. All sources have been brought to a common arbitrary intensity scale of mean value 10 units. 1 a and 1 b refer to the rising and transit observations respectively. The upper diagram in each figure shows the results for the comparison source, and the lower for Hydra A. The open circle of 1 b refers to an interferometer measurement. The crosses plotted in December represent values transferred from the other series of measurements during the overlap of observations.

Two possible sources of error were investigated. In both series of observations the suspected variable was compared with sources which were of higher signal-to-noise ratio at the receiver input except for Fornax A in the sea interferometer observations and Taurus A in the transit series. It might be expected that the lower signal-to-noise ratio would produce a large scatter in the Hydra A intensities. The transit observations were therefore subjected to a statistical analysis in which it was found that standard deviations of the three comparison sources were very similar despite large differences in the signal-to-receiver noise fluctuation ratios. From this it was concluded that the noise fluctuation level was not high enough seriously to affect the accuracy of the intensity measurements on any of the source. A further possible source of error in the sea interferometer observations resulted from the difficulty of estimating intensities in the presence of scintillations. However, no correlation was found between the estimated intensities and the scintillation index.

From Fig. 1 it can be seen that the correlation between intensity variations recorded on both instruments during the overlap period in December 1954 is very good. Good correlation was also obtained between 'cross'

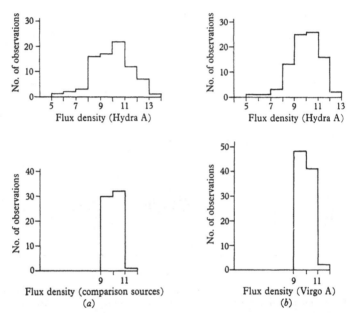

Fig. 2. Histograms showing the distributions of intensities for Hydra A and the comparison sources. (a) and (b) refer to the rising and transit observations respectively.

observations on occasions during April, May and June 1955 and observations made by Carter (unpublished work) using an interferometer operating at 100 Mc./s. with a baseline of either 90 or 1000 wave-lengths. With this instrument three very significant reductions amounting to as much as 70 % of the mean intensity have been observed, one of which was also observed with the 'cross'.

Taking all the observations together, a fairly strong case has been made out for the variability of the Hydra source. The variations do not appear to have any marked periodicity but are random from day to day. No marked changes have occurred during the period of observation with single instruments, and in some cases there is agreement between observations with two instruments 6 hr. apart. The average 'fading time' appears to be between 6 and 24 hr.

The origin of the observed intensity variations may be in the source or in the intervening medium. An identification of the source may help to decide which is the more likely possibility. The radio position obtained by Mills from the 85 Mc./s. 'cross' (unpublished) is R.A. $09^{\text{h}} 15^{\text{m}} 40^{\text{s}} \pm 4$, dec. $-11° 52'.5 \pm 2'$ (1950) which is practically coincident with a faint galaxy which has been photographed by Baade, and its spectrum obtained by Minkowski with the 200-inch telescope (personal communication). Although the galaxy has a double nucleus there are no spectrum abnormalities to suggest a possible collision as in the case of Cygnus. Unpublished measurements by Carter suggest a size for the source of about $1'.5$ between half-brightness points in the east–west direction which is consistent with the optical dimensions of the galaxy. If the identification is correct it is unlikely that the variations of such short period originate in such a large object. The physical dimensions of the source would be unlikely to exceed the distance a light signal could be propagated across it in a time of the order of the fading time. If the identification with the galaxy is neglected and the above argument is taken in conjunction with the angular size of $1'.5$ the source must be within a few parsecs of the sun.

If the variations are impressed between the source and the earth, these could occur in ionized regions in interstellar space or in the ionosphere. As the variations are not seen on a source of comparable angular size (Cygnus) the latter seems unlikely. If the former is the case one possible mechanism for the fading would be the rotation of the plane of polarization due to the Faraday effect in an ionized region with a magnetic field. However, to produce variations in intensity of the observed extent the source would have to be at least 50 % plane polarized; this should be easy to detect.

During the course of the transit observations a second possible variable source was found. Owing to the low resolving power of the aerial in the north–south direction, no reliable estimate of the declination of this object can be given; its R.A. is approximately $10^{\text{h}} 05^{\text{m}} \pm 5^{\text{m}}$. The apparent variations in the intensity of this source are much larger than that of the Hydra source; they take the form of occasional jumps to as much as three times the normal intensity for a period of a day or two. However, in the absence of a reliable position and confirming evidence from other instruments, the reality of this phenomenon must be questioned.

REFERENCES

[1] Slee, O. B. *Aust. J. Phys.* **8**, 487, 1955.
[2] Mills, B. Y. *Aust. J. Sci. Res.* A, **5**, 266, 1952.
[3] Mills, B. Y., Little, A. G. and Sheridan, K. V. (in preparation). 'A high resolution radio telescope for use at a wave-length of 3·5 metres.'

Discussion

Smith: Can part of the variation be due to scintillation?

Bolton: The scintillation index was found to be the same for the Hydra and Virgo sources and no correlation was found between scintillations and variability of Hydra.

THEORETICAL PROBLEMS OF DISCRETE RADIO SOURCES

J. L. GREENSTEIN

Mount Wilson and Palomar Observatories, California, U.S.A.

The range of theoretical problems connected with the discrete sources is very large. It is convenient to distinguish between the normal and abnormal radio-frequency emitters that have quite different energy outputs per unit volume. Since the estimates by Minkowski and Greenstein [1] there has been considerable progress in the identification of these objects and in the provision of quantitative data about them. Table 1 includes revised estimates of the total luminosity, L, the emitting volume, V, and the luminosity per cubic parsec, J. Here L has been obtained by multiplying the observed power at 100 Mc./s. by an effective band-width of 500 Mc./s. and using the newly estimated distances. J is expressed in units of the total energy output of the sun ($= 3.82 \times 10^{26}$ watt) per cubic parsec. The figure for the Galaxy and M31 is considerably higher than that given by Baldwin in this symposium, since I have not revised the estimate of the emitting volume from the original paper by Minkowski and Greenstein. If most of the emission comes from a larger galactic halo the specific luminosity, J, for the Galaxy and M31 should be considerably reduced. For certain of the extra-galactic sources it is not certain whether the volume of the whole

Table 1. *Observed radio power of discrete sources*

Object	r (psc.)	$\log L$ (watts)	$\log V$ (psc.^{3})	J ($L_\odot$ /psc.^{3})	Remarks
Galaxy	—	30·38	11·67	1.3×10^{-8}	
M31	5×10^5	30·08	11·38	1.2×10^{-8}	
NGC 1068	6×10^6	31·63	5·54	0·31	Nucleus
NGC 5128	10^6	32·00	$\begin{cases} 8\cdot11 \\ 12\cdot42 \end{cases}$	$\begin{matrix} 1.9 \times 10^{-3} \\ 9.6 \times 10^{-8} \end{matrix}$	Present collision region Whole nebula
M87	6×10^6	33·38	$\begin{cases} 6\cdot34 \\ 11\cdot55 \end{cases}$	$\begin{matrix} 2\cdot7 \\ 1.7 \times 10^{-5} \end{matrix}$	Jet Whole nebula
NGC 1275	3×10^7	33·88	$\begin{cases} 8\cdot00 \\ 8\cdot47 \end{cases}$	$\begin{matrix} 0\cdot19 \\ 0\cdot06 \end{matrix}$	Present collision region Present and past collision regions
Cyg A	9×10^7	36·80	$\begin{cases} 9\cdot78 \\ 11\cdot67 \end{cases}$	$\begin{matrix} 2\cdot6 \\ 0\cdot033 \end{matrix}$	Nucleus Whole nebula
Crab	10^3	26·00	0·21	0·16	
Cas A	540	26·48	−0·59	2·9	

nebula, or only of the parts now in collision (or peculiar in nature, like the jet in M 87) should be used in computing the specific emissivity, J. In Table 1, two values are then given.

The interesting feature, for abnormal radio-frequency emitters, is that the emission per cubic parsec is often of the order of or greater than one solar unit. Since in our own Galaxy the total optical emission in the galactic plane is very much less than the luminosity of the sun, it is clear that the radio-frequency emission process must be fundamentally of very short life, and must tap other energy than that of the stars.

There are two main subjects in which some definite theory is now possible, (1) the *thermal emission* from ionized regions, and (2) the possible sources of relativistic electrons that are needed if the abnormal radio emission is to be explained by means of the *synchrotron mechanism*. Before discussing some aspects of these two problems, I would like to state that (3) *plasma-type theories* should not be overlooked completely. Some recent theoretical work by Field (Princeton) and by Gould (California Institute of Technology) is suggestive in providing mechanisms for the escape of the plasma radiation, i.e. for the selective conversion of thermal energy into radio-frequency energy, with moderately high efficiency. Gould's mechanism involves the conversion of the longitudinal plasma oscillations into transverse waves by the normal thermal density fluctuations. And it need not be excluded that magnetized shock waves may provide high enough compression to yield plasma radiation at correct frequencies in discrete sources as well as in the sun.

I. THERMAL EMISSION FROM H II REGIONS

The possibility of detecting free-free emission of galactic emission nebulae was predicted by Reber and Greenstein in 1947 [2]; for the galactic plane, along spiral arms, it was predicted in 1949 [3]. With the detection by Haddock, Mayer and Sloanaker [4] of many diffuse nebulae at 3000 and 9500 Mc./s., it is apparent that a powerful new method of studying the ionized hydrogen has become available. I will first list some of the advantages of the radio-frequency measurements:

(1) Freedom from absorption by dust.
(2) Integration over the variable gas density.
(3) Detection at very large distances.

Detailed spectroscopy is of course not possible; methods for distance determination may yet be developed; brightness distributions may be determined interferometrically.

The effect of absorption by dust in altering the apparent shape, size and brightness of the H_α emission regions is quite large. An approximate ratio of the absorption at H_α to the photo-electric colour excess on the E_1 scale can be obtained from the $1/\lambda$ law, which gives the ratio as 6. Using Whitford's [5] value of the infra-red absorption, the lower limit to the ratio is 3·8 (assuming absorption at $1·1\mu$ as zero); an approximate value may also be derived from Morgan, Harris and Johnson [6] as 4·9. I will adopt 4·3. In that case the apparent emission measure

$$EM = EM' \times 10^{-1·72E_1},$$

where EM' is the true value of $\int n_i n_e \, dl$. As usual, EM and EM' will be expressed in parsec $\times$ cm.$^{-6}$. I have made a rough estimate of the colour excess of the exciting stars or objects in the region of some H II radio sources. The colour excesses range from $0^{m}·11$ for NGC 7000 to $0^{m}·40$ for M 17, NGC 6604. Consequently, nearby H II regions can be dimmed by factors of five, and these behind denser dark clouds by much larger amounts.

An inspection of 48-inch Palomar Schmidt plates suggests that the apparent size and shape of emission nebulae is often largely controlled by overlying dark lanes. For example, the Orion nebula suffers heavy obscuration just north of the Trapezium; objects like M 16 and especially M 8 and M 20 are located in extremely dense absorbing regions and show sharp dark markings superposed. Behind the central absorption lanes of the Milky Way even the brightest emission nebulae will disappear. I have made rough estimates of the required correction to the apparent size of some of the radio sources, for extinction in the dark lanes; in several cases the factor is greater than two. Thus the H_α surface brightness times the apparent area may be reduced by factors of the order of ten.

The variation of gas density (and possibly of excitation) produces large effects on the apparent surface brightness in H_α. For example, NGC 6604 has a diameter about 50' at low surface brightness, and has a core of strong H_α emission about 5' in diameter. Without isophotes it is difficult to predict the total H_α magnitude, and therefore the radio frequency emission. Fundamentally, it is even difficult to *define* a galactic nebula. From the rough moduli of the stars involved, from their closeness in the sky, similarity of surface brightness, and involvement with dark lanes, it is not certain that M 8 and M 20 are two distinct objects, although separated by 90', which is about the long dimension of M 8. Sharpless [7] suggests that NGC 6604, M 16 and M 17 may be a physically connected group. Radio isophotes at intermediate declinations would be decisive.

Some of the features of the radio emission may be best understood if the optical nebulae are brighter and relatively absorption-free condensations in larger units. The radio isophotes integrated over the larger areas then will give valuable information on the mean gas density in these giant H II complexes.

The observations of H II regions by Haddock, Mayer and Sloanaker [8] permit derivation of the true EM' from the observed power received. From the theory of free-free emission in an optically thin nebula (all are thin at these high frequencies), the power P_f at frequency f is approximately

$$P_f \approx 5 \times 10^{-23} EM' \, \theta^2 T_e^{-1/2} \text{ w.m.}^{-2} \text{ (c./s.)}^{-1},$$

where θ is the apparent size in radians. An approximate constant value of g, the logarithmic term in the absorption coefficient, has been used and $T_e \approx 10^{4} \,^{\circ}\text{K}$. The size must be corrected from the Schmidt plate measurements. Table 2 gives the resultant values of EM', i.e. essentially the predicted H_α surface brightness; EM is the value after reduction for interstellar absorption. The areas are given in square millimetres on the Schmidt prints, scale $1'\cdot 2/\text{mm}$.; F is the factor by which areas are multiplied to allow for the obscuration.

Table 2. *Emission measures deduced from radio power*

Object	$10^{25}P_f$	Area	Area $\times F$	EM'	EM	Remarks
M 20	11	100	150	$1\cdot 2 \times 10^5$	4×10^4	Part of M8?
M 8	20	1500	2000	$1\cdot 6 \times 10^4$	8×10^3	
NGC 6604	10	1800	2500	7×10^3	10^3	Total area
		25	25	7×10^5	10^5	Bright core only
M 16	16	1600	1600	$1\cdot 7 \times 10^4$	6×10^3	Total area
		300	600	$4\cdot 5 \times 10^4$	2×10^4	Bright core only
M 17	68	1200	1200	9×10^4	3×10^4	Total area
		300	750	$1\cdot 5 \times 10^5$	5×10^4	Bright core only
IC 4701	7	1000	1000	$1\cdot 2 \times 10^4$	7×10^3	Part of M17?
NGC 7000	—	—	—	4×10^3	2×10^3	From T_A
NGC 1976	45	1600	2500	3×10^4	10^4	Total area
IC 443	6	1200	1200	8×10^3	3×10^3	Filamentary

Several of these EM' are so large that the nebulae would be optically thick at about 200 Mc./s., i.e. show energy decreasing with frequency. Interesting problems are raised by the EM in Table 2. The very large range of brightness in H_α results for certain nebulae in two determinations, dependent on which size is assumed for the nebula. Without radio and optical isophotes we can go no further. The values observed optically for low-surface-brightness nebulae run down to 500 (Sharpless-Osterbrock [9]); the weakest radio emissions here studied are still far brighter than the faint H II regions detected optically. Strömgren's values of EM for some of the fainter diffuse nebulae go up to 7000 (for IC 405). The brightest areas in NGC 6604, M 17 and M 20 run up to 40,000. It is known (although detailed optical measurements are lacking) that the nebulae in Sagittarius are about the brightest, with the exception of NGC 1976. Thus, the present sensitivity and resolution are not quite sufficient for detailed studies of the distribution of ionized hydrogen in the Galaxy. Whether M 20 and M 17 are really as high in EM as derived from radio data depends critically on the areas, and higher angular resolution would decide. Although optical data are lacking, we may conclude that the identifications of bright H II regions seem correct, and the radio emission is reasonably accordant with the optical.

It is interesting to note that IC 443, which may be a superthermal emitter at low frequencies, has a quite reasonable EM, so that its 300 Mc./s. radiation may be thermal. Observations of the details of its high-frequency spectral energy would be valuable.

One of the most remarkable entries in Table 1 is the predicted EM for NGC 1976, the Orion nebula. In spite of its rather high P_f, its large total area results in $EM \approx 10^4$, averaged over the nebula. Strömgren [10] found $EM = 8 \times 10^6$ from the surface brightness at the Trapezium; Greenstein [11], averaging over the inner part of the nebula from the spectra, gave $EM \approx 10^5$, and H. Johnson [12] gave 6×10^4. There is undoubtedly an enormous gradient in H_α surface brightness which accounts for the disparity in these results, and the radio value of $EM = 10^4$ includes the outer extension. Osterbrock [13] has developed a new method using the ratio of $\lambda\lambda 3726$–3728 of (O II), which is pressure-sensitive, to determine the electron density in NGC 1976. At the Trapezium he gives $n_e \approx 4 \times 10^4$; at 4' distance $n_e \approx 10^3$; at 16', $n_e \approx 300$. This determination of $n_e(r)$ permits a numerical integration to give the effective mean EM averaged over the nebula. Assume it to be spherically symmetrical, of radius R. From a definition of the type:

$$\overline{EM} = \frac{N^2 \times \text{volume}}{\text{area}}$$

valid for a sphere of constant density N, we can formally define

$$\overline{EM} = \frac{4}{R^2} \int_0^R N^2(r) \; r^2 dr.$$

Even with Osterbrock's data the near-singularity of very high density at the centre of Orion makes integration difficult, the innermost 1' contributing nearly one-third the total. However, I find $\overline{EM} \approx 9 \times 10^5$, higher than previous spectroscopic averages and about thirty times the EM' deduced from the radio observations. It is probable that the outermost parts of the nebula were not included in the N.R.L. measurements (my area is about four times their beam-width). In conclusion, if there is any discrepancy between radio and optical emission, it is in the sense that the radio emission is too weak, so that we can exclude superthermal mechanisms at these high frequencies. Other nebulae with the high EM of NGC 1976 or M 17 could be detected at very great distances. M 17 is approximately 1000 parsecs from the sun; at 17,000 parsecs on the opposite side of the galactic centre it would have $P_f \approx 0\cdot2 \times 10^{-25}$ w.m.$^{-2}$ (c./s.)$^{-1}$, which does not seem beyond the reach of high-frequency technique.

One statistical comment may be of interest. If we assume that radio and optical emission are in proportion, the ratio of radio power to observable H_α surface brightness increases because of dust absorption. Consider a uniform, random distribution of H II regions and dust. Let the emissivity in H_α be ϵ_α, in radio frequencies ϵ_R, and let k_α be the absorption by dust. Then the surface brightness will be in the ratio:

$$ I_R/I_{H\alpha} = \frac{\int_0^{l_1} \epsilon_R \, dl}{\int_0^{l_1} \epsilon_\alpha e^{-k_\alpha l} dl}, $$

$$ I_R/I_{H\alpha} = \frac{\epsilon_R l_1}{\dfrac{\epsilon_\alpha}{k_\alpha}(1 - e^{-k_\alpha l_1})} = \frac{\epsilon_R \tau_1(\alpha)}{\epsilon_\alpha(1 - e^{-\tau_1(\alpha)})}. $$

Thus, as $\tau_1(\alpha)$, the optical absorption increases, the observed radio power compared with the H_α EM becomes of the order of $\tau_1(\alpha)$ times its value for an unobscured H II region. Gas along a spiral arm will probably become invisible optically if $\tau_1(\alpha) > 3$, i.e. if more distant than 3000 parsecs. Thus, along the direction of Cygnus, or through inner arms in Sagittarius, large numbers of invisible H II regions will be detectable at radio frequencies.

2. ENERGY CONSIDERATIONS IN GAS COLLISIONS

The strong radio-frequency emission of colliding gas masses suggests that some fundamental process becomes operative in almost any collision at sufficiently high velocities. Let us briefly consider the various energy contents of a gramme of hydrogen. The thermal energy is

$$ \frac{kT}{m_H} \approx U_{\text{th.}}, $$

the kinetic energy at collision velocity v is

$$\frac{v^2}{2} \approx U_{\text{kin.}},$$

and the nuclear energy (for conversion into helium) is

$$0 \cdot 007 c^2 \approx U_{\text{nuc.}} (H).$$

In general, the last expression gives too high a yield because the direct proton-proton interaction is very slow; in consequence the reactions of hydrogen with light nuclei are the only effective ones, and therefore it is only the deuterium reaction that should be included. With x_D the abundance of deuterium, which is about $1/7000$,

$$0 \cdot 007 x_D c^2 = U'_{\text{nuc.}}.$$

Let us compare these energies when collision velocities of 1000 km./sec. are considered:

$$U_{\text{th.}} \approx 8 \cdot 3 \times 10^7 \ T \ \text{ergs/gm.},$$
$$U_{\text{kin.}} \approx 5 \times 10^{15} \ \text{ergs/gm.},$$
$$U'_{\text{nuc.}} \approx 9 \times 10^{14} \ \text{ergs/gm.}$$

Ordinary gas temperatures are so low that $U_{\text{th.}} \ll U_{\text{kin.}}$, unless the shock-wave heating is included, when $U_{\text{th.}} \approx U_{\text{kin.}}$; even in the latter case the heating comes from the translational energy. Surprisingly, the easily realizable nuclear energy is somewhat less than the kinetic energy. In consequence, even if nuclear energy sources can be tapped, they will not greatly increase the available energy, and we must not expect direct effects of nuclear processes in the production of radio power. I believe, however, that high-velocity gas collisions could provide relativistic electrons for further acceleration by a Fermi or a betatron process in such objects as the Cassiopeia A source, or the Crab nebula. I compute below effects to be expected if gas clouds with a mass near one solar mass collide at velocities of 1000 km./sec. or higher. These computations could not apply to the Crab nebula if the mass is as low as Prof. Oort conjectures, they might apply to the Cassiopeia A source, to other supernova envelopes colliding with interstellar gas, or to the Cygnus A source with proper scaling factors.

3. NUCLEAR REACTIONS IN COLLIDING GAS MASSES

The very high observed velocities within different filaments of the Cassiopeia A source and the turbulent velocities in the outer filaments of the Taurus A source suggest the possibility of nuclear reactions. The hydrodynamics of shock fronts, with probable magnetic effects, are too complex for detailed treatment, but some rough considerations can be applied. Collisions may be internal, involve heating to very high temperatures and subsequent expansion into an appreciable volume where thermonuclear reactions occur, or nuclear collisions may occur only on the fronts themselves at the gas-collision velocities. The composition of the gas may be very abnormal (H deficient probably in Taurus A) or nearly normal, as in Cassiopeia A. The abundance of the most interesting nucleus, deuterium, might be expected to be low in an exhausted star, but on the other hand, D is produced abundantly in high energy and heavy element reactions. The most profitable situations for a first investigation are the (d, p) and the (d, d) thermonuclear reactions, assuming normal abundance.

A collision at relative velocity v results in a heating to

$$T \approx \frac{v^2}{2R} \frac{\gamma - 1}{4},$$

or 10^7 °K. for 1000 km./sec. relative velocity. (The compression is only by a factor of four.) At relative velocities of 5000 km./sec. such as are found within a diffuse filament of Cassiopeia A, $T \approx 250$ million degrees. Straightforward nuclear collisions at velocity 5000 km./sec. correspond to 0·13 MeV. for a proton (or $1·5 \times 10^9$ °K.). The latter energies correspond to relatively high Gamow barrier-penetration, and lead to appreciable reaction probabilities in spite of the very low densities.

The carbon-cycle reactions, however, are too slow even at these energies. $C^{12}(p, \gamma)$ N^{13} has a cross-section of 3×10^{-33} cm.^{2} at 150 KeV., while the total cross-section for $D^2(d, p)$ T^3 and $D^2(d, n)$ He^3 is 3×10^{-26}. Thus, unless $X_D^2/X_C X_H$ is less than 10^{-7}, the carbon reaction can be neglected; the probable value of the abundance ratios is about 10^{-3}. (However, if the temperatures were ever appreciably higher, the carbon reactions would be important.)

The thermonuclear rates, with T in million degrees, are given for the D–D reaction as:

$$p_{DD} = 4 \times 10^{10} \rho x_D \, T^{-\frac{2}{3}} \, e^{-42·6/T^{\frac{1}{3}}},$$

$$p_{DD} \approx 10^6 \rho x_D, \quad \text{(250 million degrees)}$$

per particle per second, while for the p–D reaction

$$p_{pD} = 7 \times 10^4 \rho x_H \, T^{-\frac{2}{3}} \, e^{-37·2/T^{\frac{1}{3}}},$$

$$p_{pD} \approx 5 \rho x_H, \quad \text{(250 million degrees)}.$$

The p–D reaction is slower by about 30, but represents a lower limit if D is exhausted in the source material, and still present in interstellar gas. The energy output is obtained from Q, the yield per reaction and the mass M or volume V of the reacting colliding clouds:

$$L = \frac{pQx_D M}{m_H} = pQx_D V n_H.$$

The number of reactions is L/Q. The values of p are small, so there is no exhaustion of the nuclei.

$$L_{DD} = 8 \times 10^{33} n_H x_D^2 M/M_\odot,$$

which for one solar mass, a volume of 6×10^{54} cm.^{3} (i.e. $n_H = 200$) gives $L_{DD} \approx 3 \times 10^{28}$ ergs/sec., and the number of reactions as 8×10^{33} per sec. The p–D reaction gives

$$L_{pD} = 5 \times 10^{28} n_H x_D x_H M/M_\odot,$$

which results in $L_{pD} \approx 1 \cdot 4 \times 10^{27}$ ergs/sec., and the number of reactions as 3×10^{32} per sec. Thus if we can conceive of as much as one solar mass in collision at 5000 km./sec. we will get something like 10^{34} reactions per sec. This number is not quite large enough for Oort's interpretation of the Crab nebula, as we shall see. The temperature coefficients of the reactions are small; $p_{DD} \propto T^{1\cdot6}$, and $p_{pD} \approx T^{1\cdot3}$, so that the collision velocities are not critical.

A larger yield is obtained if we do not allow the collisions to degrade into thermal motions, but imagine the same mass of gas impinging on a stationary target, and suffering reactions, a somewhat unrealistic picture, of course. Then $L/Q = \sigma v V x_D^2 n_H^2$ for the DD reaction. The cross-section at 5000 km./sec., i.e. 130 KeV., is found in the laboratory to be

$$\sigma = 3 \times 10^{-26} \text{ cm.}^2.$$

Then L/Q is 7×10^{34} sec.$^{-1}$ for one solar mass, which seems to be about as large a yield as can be obtained reasonably. It refers probably more to Cassiopeia A than to Taurus A.

These reactions yield different products. The D–D yields ultimately, per reaction, in addition to kinetic energy, one-half of an 0·02 MeV. electron (tritium decay), one-half of a 0·8 MeV. electron (from neutron decay), and one-half of a 1·5 MeV. proton (which may produce some fast electrons). Each p–D reaction yields a 5·5 MeV. quantum. While the reactions do not seem important as a direct energy source they provide two important sources of relativistic electrons. Consider the 5·5 MeV. γ-ray; with a cross-section of about 0·10 of the Thomson scattering cross-section it can produce by Compton scattering an electron with average energy of the order of 3 MeV. (and rising to 5 MeV.); the cross-section, $\sigma = 6 \times 10^{-26}$ cm.^{2}. Then if $n_H = 200$, thickness $l = 10^{18}$ cm., the probability of absorption is $1 \cdot 2 \times 10^{-5}$, and the p–D reaction produces 10^{-5} relativistic electrons per reaction. This is actually smaller than a nuclear process suggested privately by T. Lauritsen, called internal-pair-production, which is observed to occur in light nuclei. At about 5 MeV. a dipole γ-ray produces a positron and negaton within the nucleus, with probability about 2×10^{-3}; thus 4×10^{-3} are produced per reaction; the life of the positron is long at these densities. In résumé, from one solar mass we can obtain up to 3×10^{32} relativistic particles (≈ 3 MeV.) per sec. Since at 250 million degrees, kT is only $0 \cdot 02$ MeV., *the nuclear processes give a step-up in energy of over a hundred,* and may provide the fast electrons for a Fermi acceleration process. Their number is smaller than Oort has stated, in private communication, to be necessary for the probable rate of loss by synchrotron radiation (up to 10^{36} electrons sec.$^{-1}$). If regions of much higher density could exist, even if of smaller total mass, the output could be appreciably raised. No probable situation, at the present low gas density, could make the reactions significant as an energy source, but in the past the density would have varied inversely as the cube of the time.

In solar flares, for example, collision velocities are comparable, but densities are perhaps 10^{10} times higher. If a volume of 10^{27} cm.^{3} is adopted, with $v = 5 \times 10^8$ cm./sec., $10^{-25} n_H^2$ cm.$^{-2}$ sec.$^{-1}$, γ-rays would reach the top of the earth's atmosphere. If they originate at or near chromospheric densities such a flux of γ-rays might be detectable in balloon flights.

4. ABSORPTION OF FAST NUCLEI

If the cross-section for absorption of γ-rays is small, it is probable that the gas is transparent for other energetic particles, although the mean-free path at thermal velocities may be small. With $n_H = 200$, $l = 10^{18}$ the probability of collision is $2 \times 10^{20} \sigma$, small, since for energetic particles $\sigma \approx 6 \times 10^{-26} A^{\frac{2}{3}}$ cm.^{2}, where A is the atomic weight. Then the collision of an accelerated particle with a stationary proton occurs in time

$$\tau \approx 2 \times 10^7 \text{ years}/A^{\frac{2}{3}} n_H^2.$$

Since even for $A = 100$, τ exceeds the age of the sources, the heaviest elements are substantially unaffected, and there is no important loss to the accelerating mechanism.

One possibility is that accelerated protons could destroy heavy stationary nuclei, and produce light elements by spallation. Cosmic ray

emulsions show that a spallation reaction in a heavy nucleus usually frag-
ments it largely into isotopes of H and He. About 0·3 of the mass appears
as D. For the collision rate we need the number of relativistic protons,
N_H. Assume that for every electron accelerated there is also a proton; then
in 10^3 years there will be 3×10^{46} protons, a density of $N_H \approx 5 \times 10^{-9}$ cm.$^{-3}$.
The lifetime of a heavy nucleus is then 3×10^{14} years, so that the fraction
3×10^{-12} is destroyed, producing 3×10^{-11} D nuclei per original heavy
particle. If the heavy particles formed 1/6000 of the original mass, the
abundance of deuterium created becomes about 10^{-14} that of hydrogen—
much too low to be significant. There has been no approach to a 'steady
state' such as is imagined for galactic cosmic rays.

5. ABSORPTION OF LOW-ENERGY ELECTRONS

The nuclear reactions give relativistic electrons, with energies about 100
times the thermal. In a Maxwellian velocity distribution a completely
negligible fraction have such high energies, but an additional important
factor is the great stopping power of matter for low-energy electrons. A
complete theory of the energy loss in an ionized gas is not available from
low to relativistic electron energies. I use Bethe's expressions for the energy
loss per unit length in a neutral gas, $Z = 1$ (hydrogen):

$$-\frac{dE}{dx} = \frac{4\pi e^4 N}{mv^2} \log mv^2/I; \quad \text{(non-relativistic)}$$

$$-\frac{dE}{dx} = \frac{2\pi e^4}{mc^2} \log E^3/2mc^2 I^2; \quad \text{(relativistic)}$$

where N is the space density and I the mean excitation or ionization energy.
For a neutral gas the logarithmic term is usually approximated by
$I \approx 14$ volts; for an ionized gas the approximation may be used that I
corresponds to the potential energy at the mean distance to the nearest
neighbour; for a plasma more complex expressions could be used.
Fortunately, even large errors in the logarithmic term produce little
effect. The loss of energy per unit length is obtained from

$$\log E/E_0 = -\beta X,$$

as approximately:

$$\beta = \frac{2\pi e^4 N}{E^2} \log 2E/I,$$

or

$$\beta = \frac{2\pi e^4 N}{mc^2 E} \log E^3/2mc^2 I^2.$$

As for the cut-off energy, I, if it can be taken as $\gtrsim e^2 N^{\frac{1}{3}}$, which is about 10^{-6} electron-volts it increases β by about 2·5 over the value derived for $I = 14$ volts. Values of β are:

$E =$	10^3	10^4	10^5	10^6	10^7	eV.
$\beta/N = \sigma =$	6×10^{-18}	8×10^{-20}	7×10^{-22}	3×10^{-23}	3×10^{-24}	cm.^{2}

Electrons of a few kilovolts, or less thermal energy, have ranges of about $10^{19}/N$ cm., i.e. smaller than the nebula if $N = 10^2$. Therefore, they will be re-oriented and share their energies in gas collisions, and not be able to enter the magnetic acceleration mechanism. However, electrons of a few MeV. (e.g. those from nuclear reactions) have free paths $10^{23}/N$ cm., large compared to the nebula. They will be acted on purely by the magnetic field, and need not lose energy to other electrons. While it must be admitted that the collision cross-sections are rough, thermal electrons should not move in straight trajectories, even in an expanding gas mass, and there is no doubt that the range of the MeV. electrons is large, except for the effects of the magnetic fields.

Others have discussed the Fermi acceleration mechanism for cosmic rays. In the case of the electrons in the Crab nebula, starting at MeV. energies, they will be raised to 10^6 MeV. after 14 c/v collisions with magnetized clouds of velocity v/c. For 1000 km./sec. as the velocities of clouds within the Crab nebula, 4×10^3 collisions are required, i.e. four per year. The present size of the nebula is of the order of a light year, the particles traverse it in a straight line in a year, so that the number of separate clouds within the nebula need only be four. Thus the structure required of the magnetic field need not be very fine, or small-scaled. The kinetic energy lost by the clouds is measured by the rate of decay of the relativistic electrons. If we take Oort's estimate of 10^{36} electrons, each of 10^{12} eV., we lose 10^{36} ergs/sec., or 10^{47} ergs in the life of the nebula. With one solar mass and 10^3 km./sec., the kinetic energy of expansion is 10^{49} ergs; it is apparent that a rate of loss much greater than 10^{36} ergs/sec. cannot be maintained. If Oort's mass estimate is used, much too little energy is available, and the relativistic electrons cannot be produced in the nebula, but must come from the star.

REFERENCES

[1] Minkowski, R, and Greenstein, J. L. *Ap. J.* **119**, 238, 1954.
[2] Reber, G. and Greenstein, J. L. *Observatory*, **68**, 15, 1947.
[3] Greenstein, J. L. *A.J.* **54**, 121, 1949; *Sky and Telescope*, **8**, 149, 1949.
[4] Haddock, F. T., Mayer, C. H., Sloanaker, R. M. *Ap. J.* **119**, 456, 1954.
[5] Whitford, A. E. *Ap. J.* **107**, 102, 1948.

[6] Morgan, W. W., Harris, D. L. and Johnson, H. L. *Ap. J.* **118**, 92, 1953.

[7] Sharpless, S. *Ap. J.* **118**, 362, 1953.

[8] Haddock, F. T., Mayer, C. H. and Sloanaker, R. M. *Nature*, **174**, 176, 1954.

[9] Sharpless, S., Osterbrock, D. E. *Ap. J.* **115**, 89, 1952.

[10] Strömgren, B. *Problems of Cosmical Aerodynamics* (Dayton, Ohio, C.A.D.O. 1951), ch. II.

[11] Greenstein, J. L. *Ap. J.* **104**, 414, 1946.

[12] Johnson, H. M. *Ap. J.* **118**, 370, 1953.

[13] Osterbrock, D. E. *Ap. J.* **122**, 235, 1955.

Discussion

Gold: The generation of fast particles seems to be related to fast gas motions in the Sun, the Crab nebula, and probably also in other radio sources. One wonders which phenomenon in rapidly moving gases can be held responsible. Shock waves in magnetized tenuous gases are inadequately understood. They are places where energy would be available in a well-organized form for some subsidiary process, and where indeed most of the kinetic energy of the gas is finally dissipated.

Biermann: It seems that magnetic fields have only been mentioned in connexion with the Schwinger mechanism. I should like to draw attention, however, to the importance of magnetic fields for the emission by plasma oscillations: the ideal plasma oscillation does not radiate, but a superposed magnetic field gives coupling leading to emission. This is perhaps the most probable mechanism under astrophysical conditions.

HYDROGEN EMISSION NEBULAE
AS RADIO SOURCES

F. T. HADDOCK

Naval Research Laboratory, Washington, D.C., U.S.A.*

This talk is based largely upon the centimetre-wave observations of hydrogen emission nebulae made with the Naval Research Laboratory 50-ft. paraboloidal reflector at 3·15 cm. [1], at 9·4 cm. [2], and at 21 cm. [3]. After the detection of the first individual bright galactic nebulae at 9·4 cm. a systematic search was made to detect other hydrogen emission nebulae, principally with the aid of the catalogue of emission nebulae obtained by Sharpless [4] from 48-inch Schmidt plates at Mount Wilson and Palomar Observatories. An interesting correlation between radio detectability and nebular classification by optical size and brightness was found, in spite of the fact that optical extinction was not taken into account. The Sharpless catalogue lists 140 classified emission nebulae, of which sixty-five of the brightest were scanned with the radio antenna beam. Of these, twelve nebulae were detected and measured. They can be grouped as follows.

The catalogue contains seventeen emission nebulae in the optically brightest class, of five classes, of which the five largest nebulae were detected as radio sources. Their average optical angular size was 50' and their average centimetre-wave emission measure was 48,300 parsecs.cm.$^{-6}$ averaged over the 26' circular antenna beam. It is assumed that 72 % of the true brightness temperature is measured when a uniformly bright source covers the main lobe and the neighbouring side lobes because of the spill-over effect, and that $A_1(2) = 19\cdot2$ and $T_e = 10^4$ degrees in the free-free emission formula.

The catalogue contains nine nebulae in the second brightest class and the five largest of these were also detected. Their average size was 110' and their average emission measure was 27,000 parsecs.cm.$^{-6}$.

The two remaining nebulae that were detected are in the third brightness class and have an average size of 150' and average emission measure of 10,600 parsecs.cm.$^{-6}$.

* Now at the University of Michigan Observatory, Ann Arbor, Michigan, U.S.A.

The maximum size of nebulae in the 4th and 5th brightness classification is only 60' and none were detected.

We see here systematic relationships between optical brightness class, angular size, and radio flux density (which is proportional to the beam emission measure). The brighter the nebula the smaller is the critical size required for radio detectability; and the brighter the nebula the greater is the radio flux density in spite of smaller average size.

These relations indicate that a common radio emission process is responsible for the observed radio flux; the proportionality between the radio and optical brightnesses suggests a thermal emission process.

Additional support for the radio-optical proportionality for emission nebulae was found in the correlation between the logarithm of the radio flux density and the magnitude in hydrogen-alpha light for seven nebulae measured by Shajn, Hase and Pikelner [5] and corrected by us for optical extinction.

Boggess [6] has measured calibrated intensity isophotes of four galactic emission nebulae (NGC 6523, 6514, 6611 and 6618) in several colours including H_α light from which he predicted the centimetre-wave thermal flux densities at the earth after making corrections for optical extinction and assuming an electron temperature in the nebulae of 10,000° K. When these were compared with the measured 9·4 cm. flux densities the latter were too large by factors of 1·5, 1·1, 1·9 and 2 for the four nebulae, respectively. He attributed these discrepancies to inadequate extinction data.

A revised evaluation, following the method of Boggess, has been made of the comparison between the H_α measurements by Boggess and the measured 9·4 cm. wave-length flux densities under the assumption of purely thermal radio emission. The chief features of this new evaluation are that the H_α brightness and emitting area was averaged over the antenna beam rather than using the entire nebula or the bright inner 'nucleus', and that Dr Nancy Roman has kindly derived for us absorption corrections for these nebulae based on extinction data recently published [7].

Furthermore when a 65% antenna gain efficiency (which includes the 72% spill-over effect mentioned above) is used and an allowance is made for a variation in electron temperature between the limits of 10,000 and 15,000° K., and for the 2° uncertainty in the radio antenna temperature measurements (which introduces uncertainties of 30, 50, 40, and 8 % in radio flux density for the above four nebulae, respectively) we find the range of values in the predicted and observed radio flux values in M.K.S. units multiplied by 10^{25} given in Table 1.

The upper and lower predicted flux densities correspond to electron temperatures of $1·5 \times 10^4$ and $1·0 \times 10^{4°}$, respectively (see Boggess, thesis, for the effect of temperature). The agreement is good except for NGC 6523, which fails to accord by a factor of $1·2$. This is the nebulae for which Boggess obtained good agreement but he considered only the bright inner 'nucleus' which is appreciably smaller than the antenna beam and the entire nebula, the largest of the four, is larger than the beam but radiates 75% of its H_α flux from a region within 12' of its central brightness peak. Nevertheless this 20% discord is too small to be considered as significant evidence against purely thermal radio emission, and for the other three nebulae the accord is excellent.

Table 1

NGC	Predicted	Observed
6514	6– 7	5–14
6523	26–31	12–22
6611	9–10·5	8–19
6618	52–61	53–62

Boggess did not measure the Orion Nebula, NGC 1976, because of the difficulty of eliminating the effect of the strong continuous radiation due to the presence of interstellar dust within the nebula. It is possible to estimate the expected upper limit of its radio emission from its H_α magnitude of $5·8$ compared to $8·2$ for NGC 6618 as measured by Shajn, Hase and Pikelner. However, it is estimated that the absorption correction for NGC 6618 is $2·5$ magnitudes and is small for NGC 1976. Therefore, it is not unexpected that both at $9·4$ cm. and $3·15$ cm. wave-length the radio flux density is about 40% greater for NGC 6618.

The evidence presented so far for a purely thermal radio emission from galactic emission nebulae has been based upon the theoretically expected proportionality between H_α and centimetre-wave intensities and, as has been discussed above, this evidence is favourable. However, it is possible to present additional evidence, independent of optical measurements, in the form of radio spectra. Fig. 1 shows the measured radio flux density in M.K.S. units for NGC 1976 at five different wave-lengths embracing a 100 to 1 range. The three short wave-length points were measured at the Naval Research Laboratory, the point at $1·2$ metres was obtained by Kraus and Ko (private communication) and at $3·5$ metres by Mills (reported by Pawsey [8]). The symbol at $1·4$ metres indicates the upper limit set by Baldwin from the failure to detect this nebula [9]. The double set of marks at $3·15$ and $9·4$ cm. wavelength indicate the observed and corrected flux densities. The correction was made for the fact that the antenna beam is smaller than or comparable to the radio source size. The amount of the correction is determined directly from the percentage broadening of the drift curve (obtained as the 13′ diameter source drifts through the 9′5 or 26′ antenna beam), and is not too sensitive to the assumed shape of the radio brightness distribution.

The solid curve is the theoretical thermal spectrum based entirely upon the apparent source size estimated from the broadening of the drift curves

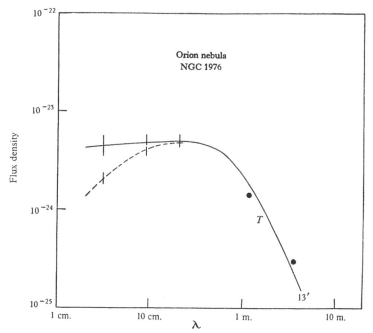

Fig. 1. Observed and computed spectrum of the Orion nebula.

and the measured flux densities at 3·15 or 9·4 cm. wave-length. The good agreement at the longer wave-lengths is strong evidence favouring a thermal model of the Orion nebula. The emission measure at 3·14 cm. averages 465,000 parsecs.cm.$^{-6}$ over its 9′5 beam using the same antenna parameters as used with the 9·4 cm. beam mentioned above and with $A_1(2) = 17$. This is of the same order as the optically determined values [10].

Theoretical thermal spectra have also been obtained for NGC 6618 and 6357, based upon a source diameter of 8′ and 17′, respectively, and are in fair agreement with observed flux densities at four wave-lengths embracing a 50 to 1 range.

The radio source Cygnus X has been mapped at 9·4 cm. with a 26′

beam. It has been found to consist of two principal maxima, one located near the star γ Gygni which is surrounded by hydrogen emission nebulosity and the other maximum at about 2°5 farther east and 0°5 farther north located at a faint un-catalogued emission nebulosity in the heavily reddened Cygnus rift region. The flux densities from the γ Cygni nebulosity at 9·4 and 21 cm. give similar emission measures of 18,600 and 13,600 parsecs.cm.$^{-6}$ respectively, when averaged over their respective beams, thereby indicating thermal emission (the 55′ beam at 21 cm. measures a somewhat lower value as would be expected from the isophotes found with a 26′ beam). Piddington and Minnett [11] obtained a flat radio-spectrum for Cygnus X at longer wave-lengths and suggested it was due to the thermal emission from the γ Cygni nebulosities or from the whole spiral arm in the direction of Cygnus. Their flux density is too high for either the γ Cygni region or the spiral arm alone because their broad antenna beam included both maxima in Cygnus X as well as part of the underlying background from the spiral arm.

REFERENCES

[1] Haddock, F. T. and McCullough, T. P. *A.J.* **60**, 161, 1955.
[2] Haddock, F. T., Mayer, C. H. and Sloanaker, R. M. *Ap. J.* **119**, 456, 1954. *Nature,* **174**, 176, 1954.
[3] Hagen, J. P., McClain, E. F. and Hepburn, N. *Proc. I.R.E.* **42**, 1811, 1954.
[4] Sharpless, S. *Ap. J.* **118**, 362, 1953.
[5] Shajn, G. A., Hase, W. F., and Pikelner, S. B. *Astr. Zhur.* **31**, 105, 1954.
[6] Boggess, A., Thesis, University of Michigan, 1954.
[7] Morgan, W. W., Harris, D. L. and Johnson, H. L. *Ap. J.* **118**, 92, 1953. Morgan, W. W., Code, A. D. and Whitford, A. E. *Ap. J.* Supp. Ser. II, no. 14, 1955.
[8] Pawsey, J. L. This publication, paper 18, p. 130.
[9] Baldwin, J. E. *Observatory,* **73**, 155, 1953.
[10] Greenstein, J. L. This publication, paper 32, p. 179.
[11] Piddington, J. H. and Minnett, H. C. *Aust. J. Sci. Res.* A, **5**, 17, 1952.

POLARIZATION AND THE RADIATING
MECHANISM OF THE CRAB NEBULA

J. H. OORT AND T. WALRAVEN [1]

University Observatory, Leiden, Netherlands

Photo-electric measurements made in Leiden in the first part of 1955 and extended by one of us at the Observatoire de Haute Provence in October 1955, show that the light of the Crab nebula is strongly polarized. The presence of polarization was first suggested by Vashakidze [2] and was firmly established by photo-electric measures by Dombrovsky [3] in 1954. The present observations, which are much more detailed, show that 17 % of the light coming from the central part, with a mean radius of 0ʹ8, is linearly polarized. It may be seen that over the entire bright part of the nebula the polarizations are sensibly parallel, while in the outer parts they tend to be oriented at random.

As is well known from the observations by Baade and Minkowski, the Crab nebula consists of two parts of entirely different nature, namely, an 'amorphous' mass which emits continuous radiation and is rather strongly concentrated towards the centre, and a great number of emission filaments which form a thick, irregular shell surrounding the bright amorphous part. These filaments radiate like ordinary emission nebulae. It is the amorphous mass which emits the polarized light. A closer analysis indicates that wherever a small volume of this mass can be isolated from its surroundings, the light coming from such a volume is totally polarized.

These facts give strong support to the theory advanced by Shklovsky [4] that the continuous light comes from electrons of extremely high energy spiralling in a magnetic field. Plates recently obtained by Baade with the 200-inch Hale telescope through a polaroid screen and through filters to eliminate the emission shell, give a still stronger confirmation of this theory. They show a great amount of detailed structure in the 'amorphous mass'. Whenever the structure runs parallel to the direction of the analyzer, it disappears; this indicates that the light is 100 % polarized and that the structural features represent concentrations of magnetic lines of force.

Radiation by electrons moving in large orbits under the influence of a magnetic field has been first observed in synchrotrons. We propose, therefore, to call it synchrotron radiation. It is strongly concentrated in the direction in which the electron moves, and is totally polarized with the electric vector parallel to the radius of curvature of its orbit.

Consideration of the energy density of the electrons required to explain the observed radiation sets a lower limit to the magnetic field. An upper limit is set by the condition that the electrons must not loose a large fraction of their energy in one circuit through the nebula. From these considerations we find that the average field strength should be near 10^{-3} gauss. The median energy of the electrons emitting the optical radiation is estimated to be 2×10^{11} eV. The orbital radii are of the order of 10^{12} cm., which is very much smaller than the radius of the nebula (3×10^{18} cm.).

The electrons emit radiation over a very wide range of frequencies, extending far beyond the frequencies used in radio-astronomical observations. The relative strength of the radiation emitted at various wavelengths is determined by the field strength and by the numbers of electrons with different energies. As Shklovsky has pointed out, an energy spectrum like that of cosmic rays reproduces approximately the observed ratio of radio-frequency to optical radiation. It is tempting to suppose that the radio-frequency radiation of the Crab nebula is due to the synchrotron mechanism. It may be noted that the idea that the radiation of radio stars might come from particles of cosmic-ray energies caught in a stellar magnetic field, is due to Alfvén and Herlofson [5], who had suggested this as early as 1950.

If we want to compare the observed ratio of radio-frequency to optical radiation with the ratio computed from some assumed energy spectrum the comparison should be based on the radiation emitted from the same volume, rather than from the entire nebula. Measurements by Baldwin have indicated a large size for the radio source. A similar though somewhat smaller difference in scale was found from observations of an occultation of the Crab nebula by the Moon, recently made by Seeger at a wavelength of 75 cm. These accurate measurements show that the scale of the radio source is 1·7 times that of the optical source. In the following comparison this factor has been taken into account.

If the average field strength is 10^{-3}, the electrons responsible for the optical radiation will lose by this radiation half of their energy in 180 years. There are indications that new high-energy electrons are being injected into the nebula by the central star in sufficient amount to make up for the radiation losses. If we assume that the energy spectrum of the injected electrons is like that of the primary cosmic rays we find that this gives a ratio of radio frequency to optical radiation that is about thirty times higher than the observed ratio. In order to obtain the observed ratio we have to take a somewhat less steep energy spectrum. The following distribution will reproduce the observed ratio:

$$E > 15 \cdot 5 \times 10^9 \text{ eV.} \qquad n_0(E) \sim E^{-2 \cdot 3}$$
$$1 \cdot 55 \times 10^9 < E < 15 \cdot 5 \times 10^9 \qquad n_0(E) \sim E^{-1 \cdot 3}$$
$$0 \cdot 155 \times 10^9 < E < 1 \cdot 55 \times 10^9 \qquad n_0(E) \sim E^{-0 \cdot 3}$$

$n_0(E)$ is the number of electrons of energy E injected into the nebula per unit of time. The above exponents are algebraically about 0·5 higher than the exponents that have been assumed for cosmic rays.

Using the spectrum just given we obtain the following radiation per unit of frequency:

ν	$J(\nu)$
10^8	$1 \cdot 19\alpha$
10^9	$1 \cdot 88\alpha$
10^{10}	$2 \cdot 16\alpha$

α being a constant. The emission varies little over this range, in agreement with available observations. The small increase between $\nu = 10^8$ and 10^9 can easily be made to disappear by a slight increase in the slope of the energy spectrum at low energies.

The average strength of the magnetic field probably decreases in the outer parts of the nebula. If we have an energy spectrum like that indicated above, the optical radiation will decline more rapidly with decreasing magnetic field than the radio emission. We have computed that with constant electron density the following relation will hold between light-emission, u_l, and radio emission, u_r:

$$u_l = a u_r^{3 \cdot 2},$$

where a is a constant. If the electron density should decrease proportionally with the magnetic field H, the exponent of u_r would be 1·9.

We can thus understand why the optical radiation decreases more rapidly with increasing distance from the centre. There are indications that in the outer regions the optical radiation diminishes still more rapidly relative to the radio emission than indicated by the above expressions. This may indicate that the highest-energy particles are escaping from these parts.

The theory predicts that the radiation emitted at radio frequencies should also be polarized. The observation of such polarization is hampered by two circumstances. In general we can only observe the radiation of the radio source as a whole. Because of the wider distribution of the radio emission the contribution of the outer parts, where the resultant optical polarization is practically zero, outweighs that of the polarized central part. We have estimated that the resulting polarization of the entire radio source will not exceed 1 %. Except at centimetre waves this polarization will, moreover, be largely effaced by Faraday effects in the filamentary shell. We estimate that at 20 cm. the electric vector of polarized light passing through this shell will be turned over an average angle of the order of 10 radians.

An attempt to measure polarization at 22 cm. wave-length was made by Mr Westerhout at Kootwijk[6]. He found that the polarization must be smaller than 1 %.

It is not unlikely that many other radio sources radiate by the same process. The conditions required are a mechanism for producing high-energy particles and a magnetic field. In most sources the magnetic fields or the particle energies are apparently insufficient to give observable optical radiation of the type discussed. In Virgo A (M 87), however, the bright wisp near the centre of the nebula may radiate by the same process as the Crab nebula. The fact that the radio source is very much larger than the luminous wisp would be due to the fact that weak magnetic fields extend through the entire volume of this galaxy. Measurements of optical polarization of the wisp are still inconclusive.

If the radio emission is due to the synchrotron mechanism the distribution of radio emission through a radio source may show little resemblance to that of optical radiation. Striking differences of this kind have indeed been observed in several sources.

REFERENCES

[1] Oort, J. H. and Walraven, T. *B.A.N.* **12**, 285, no. 462, 1956.
[2] Vashakidze, M. H. *Astr. Circ. U.S.S.R.* no. 147, 1954.
[3] Dombrovsky, V. A. *Proc. Acad. Sci. U.S.S.R.* **94**, 1021, 1954.
[4] Shklovsky, I. S. *Proc. Acad. Sci. U.S.S.R.* **90**, 983, 1953.
[5] Alfvén, H. and Herlofson, N. *Phys. Rev.* **78**, 616, 1950.
[6] Westerhout, G. *B.A.N.* **12**, 309, no. 462, 1956.

Discussion

Baade: I am quite relieved to hear that the Crab nebula does not have to have a large mass. Its mass has to be fairly small as it is a supernova of type I, i.e. a star belonging to population II. There are other objects in which polarization might be expected, as the mechanism of radiation may be the same, e.g. in M 87.

Oort: We have attempted to measure polarization in the jet of M 87 in Leiden, but it is very difficult to measure with a small telescope.

Hanbury Brown: Should we expect any component of circular polarization?

Oort: No, the theory predicts strictly linear polarization.

OPTICAL EMISSION FROM THE CRAB NEBULA IN THE CONTINUOUS SPECTRUM

I. S. SHKLOVSKY

Sternberg Astronomical Institute, Moscow, U.S.S.R.

All conclusions concerning the nature of the Crab nebula up to the present have been based upon the interpretation of its continuous spectrum given by Baade and Minkowski in their well-known studies ([1,2]; see also [3,4]), It was suggested that its continuous emission is a thermal radiation caused by the extremely hot amorphous mass of this nebula. A consequence of this interpretation is that the mass of the envelope is of the order of 10–20 solar masses and the kinetic temperature is extremely high. The filamentary part of the nebula possesses more or less ordinary characteristics.

This interpretation of the continuous optical emission meets with insurmountable difficulties. The discovery of radio emission from the Crab nebula has brought new facts, which lead to altogether new views concerning the nature of the optical emission of the Crab nebula. Several years ago a number of authors have independently proved that the radio emission from the Crab nebula cannot be considered as an extension of its continuous optical emission [3,5]. This is true for any supposition about the kinetic temperature and density of the amorphous mass of the Crab nebula. Thus the character of the radio emission from the Crab nebula must be of a non-thermal nature.

If, however, the radio emission of the Crab nebula cannot be considered as a continuation of its thermal emission, the optical emission from this nebula may be a continuation of its radio emission. Thus, both the optical and the radio emission of the Crab nebula are caused by the same non-thermal mechanism [7].

In the interval of more than six octaves (from $\lambda = 750$ cm. to $\lambda = 9\cdot4$ cm.) the spectral density of the flux of radio emission from the Crab nebula decreases only by a factor of $2\cdot5$. Radio emission from the Crab nebula at a wave-length of $3\cdot2$ cm. was recently discovered in the U.S.S.R. [6]. The value of the flux constitutes 70 % of the total flux at $\lambda = 9\cdot4$ cm. Thus in the interval of 8 octaves the spectral density of the flux decreases by a factor 3·5 only. It would be rather naïve to believe that the radio emission from the Crab nebula comes abruptly to an end at $\lambda = 3·2$ cm. Beyond doubt the flux of radio emission continues towards the region of shorter wave-lengths, where it may reach optical frequencies. According to the optical observations the spectral density for $\lambda = 5·10^{-5}$ cm. is about 1000 times less than in the metre range. If in the interval of 8 octaves the spectral density has decreased 3·5 times, in the interval of the 15 octaves that remain up to the optical range, it may decrease about 300 times. In

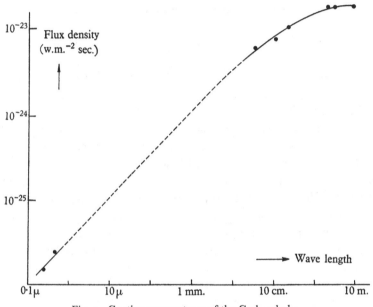

Fig. 1. Continuous spectrum of the Crab nebula.

the optical region the spectral density of the flux is approximately pro-portional to the frequency, the spectrum falls much steeper than in the radio region (Fig. 1).

The non-thermal mechanism of emission in radio and optical frequencies can only be emission by relativistic electrons in magnetic fields. This mechanism was discussed by the author [5, 7] some time ago in connexion with the problem of cosmic-ray generation during outbursts of supernovae and novae.

It may be shown that there must exist in the Crab nebula magnetic fields with intensities of $H \sim 10^{-4}$ gauss, the orientation of which is more or less random [8]. According to the theory of emission by relativistic electrons in a magnetic field the energy of the electrons responsible for radio emission from the Crab nebula ranges from $3 \cdot 10^7$ to $3 \cdot 10^9$ eV. But the energies of electrons emitting optical frequencies range from $3 \cdot 10^{11}$ to $3 \cdot 10^{12}$ eV.

From the observed flux of emission, the composition of its spectrum and the distance of the Crab nebula the concentration of relativistic electrons and their energy spectrum may be calculated. The concentration of the comparatively soft relativistic electrons causing radio emission is found to be of the order of 10^{-5} cm.$^{-3}$ and of the harder 'luminescent' relativistic electrons approximately 10^{-9} cm.$^{-3}$. The differential energy spectrum has to be

$$dN(E) \sim E^{-1\cdot5}dE(3\cdot10^7 < E < 3\cdot10^9 \text{ eV.}) \text{ and } E^{-3}(3\cdot10^{11} < E < 3\cdot10^{12} \text{ eV.}).$$

The total energy of all relativistic electrons of the Crab nebula is about 10^{48} ergs.

If the emission is caused by relativistic electrons in magnetic fields, the theory predicts that it is fully polarized. Dombrovsky [9] was the first to discover this polarization after it had been predicted theoretically. The polarization is of a fairly regular character, which is somewhat unexpected. A faint regular field is obviously superposed upon the magnetic fields of random orientation.

Thus the optical emission of the amorphous mass of the Crab nebula is caused by a comparatively small amount of relativistic particles, the total mass of which is negligibly small. This nebula may therefore be pictured as a 'soap bubble'. Only the expanding network of filaments must be considered as the 'real' nebula, the gases that were formed as a result of an outburst of the supernova of the year 1054. All the difficulties that were met when the old interpretation was accepted arise no longer in this case.

In so far as only the filamentary system constitutes the 'real' nebula the mass ejected as a result of the outburst of the supernova hardly exceeds $0\cdot1$ $M_{\odot}$. Quite a number of difficulties, connected with the supernova problem can thus be avoided.

Most important is the question concerning the cause of acceleration of a comparatively small number of particles up to extremely high energies. This question was investigated in detail by some authors [10]. Obviously a considerable part of the kinetic energy of the gases, ejected during the outburst, is transferred into the energy of a comparatively small number of relativistic particles, the magnetic field playing the role of a 'driving belt' in this process.

REFERENCES

[1] Baade, W. *Ap. J.* **96**, 188, 1942.
[2] Minkowski, R. *Ap. J.* **96**, 199, 1942.
[3] Greenstein, J. L. and Minkowski, R. *Ap. J.* **118**, 1, 1953.
[4] Barbier, D. *Ann. Astr.* **8**, nos. 1–2, 35, 1945.
[5] Shklovsky, I. S. *A.J. U.S.S.R.* **30**, 15, 1953.
[6] Kajdanovsky, N. L., Kardashev, N. S. and Shklovsky, I. S. *Dokl. Akad. Nauk. U.S.S.R.* **104**, 517, 1955.
[7] Shklovsky, I. S. *Proc. Acad. Sci. U.S.S.R.* **90**, 983, 1953.
[8] Shklovsky, I. S. *Proc. Acad. Sci. U.S.S.R.* **91**, 475, 1953.
[9] Dombrovsky, V. A. *Proc. Acad. Sci. U.S.S.R.* **94**, 21, 1954.
[10] Ginzburg, V. L., Pikelner, S. B. and Shklovsky, I. S. *A.J. U.S.S.R.* **32**, no. 6, 1955.

ON THE NATURE OF THE EMISSION FROM THE GALAXY NGC4486

I. S. SHKLOVSKY

Sternberg Astronomical Institute, Moscow, U.S.S.R.

It is well known that the radio galaxy NGC4486 has as a striking peculiarity a small and very bright 'jet' in its central part. It seems to us that the key to the understanding of the nature of this radio galaxy is the purely continuous spectrum of the jet, where not even a slight trace of emission or absorption lines is present.

Also in the case of the Crab nebula we meet intense continuous optical emission accompanied by powerful radio emission. Some time ago we gave a new interpretation of the continuous optical emission from the Crab nebula. This emission is caused by the same mechanism of a non-thermal character that causes radio emission, namely by relativistic electrons in magnetic fields [1].

There are reasons to suggest that the nature of the continuous optical emission of the 'jet' of NGC4486 is similar. From our calculations (Shklovsky, 1955) [2] it follows that the intensity of the magnetic field in the region of the 'jet' is of the order of 10^{-4} gauss, i.e. approximately the same as in the 'amorphous' part of the Crab nebula. As a consequence the energy of the relativistic electrons responsible for the optical emission of the 'jet' must be of the order of 10^{11}–10^{12} eV., and their concentration about $5 \cdot 10^{-9}$ cm.$^{-3}$. The electrons with energies of the order of 10^9–10^{10} eV. should be much more numerous in this case. The region of the 'jet' thus has to be a powerful generator of relativistic particles. The unusual conditions prevailing there are apparently favourable for the acceleration of particles. An obvious assumption is that the acceleration is caused by a Fermi statistical mechanism.

Relativistic electrons formed in the central part of NGC4486 will diffuse into the surrounding space and in the course of 1 to 2 million years (the time during which the 'jet' is formed) they will fill a considerable part of the volume of NGC4486. Wandering through the weak interstellar magnetic fields of that galaxy, the relativistic electrons will radiate in the range of radio waves, which is the cause of the anomalously high radio emission of NGC 4486

The quantitative theory developed by us shows that the mean concentration of relativistic electrons with energies E greater than 5.10^8 eV. in NGC 4486 is of the order of 5.10^{-8} cm.$^{-3}$, and their differential energy spectrum $dN(E) = KE^{-3}dE$. The total energy of the relativistic electrons is about 5.10^{56} ergs over the whole radio galaxy NGC 4486. The number of particularly energetic 'luminous' relativistic electrons with E about 10^{11}–10^{12} eV. located in the region of the 'jet' is millions of times less than the number of less energetic relativistic electrons, that fill a considerable part of the volume of NGC 4486 and cause its radio emission.

Supposing that the optical emission of the 'jet' continues to exist for 10^6 years with the observed intensity, we find that during this time-interval the relativistic electrons lose about 2.10^{55} ergs of their energy by radiation. This gives a new estimate of the energy needed for the process that is the initial cause of the anomalous phenomena going on in NGC 4486. This energy is one-and-a-half order lower than the total energy of relativistic electrons in NGC 4486, estimated above. It may be considered as the lower limit of the energy needed for the formation of the 'jet'.

We may point out that the radio emission of NGC 4486 in 10^6 years, integrated over the spectrum, will be about 10^{54} ergs, i.e. considerably less than the amount of energy lost by the electrons of the 'jet' in the course of the same interval of time. This may be explained by the fact that the relativistic electrons in the interstellar medium of NGC 4486 are losing their energy extremely slowly. They will cause radio emission for at least 5.10^8 years. The 'jet' will become dispersed in the course of that time and its visible traces will disappear.

An extremely important conclusion follows from this suggestion: there may be observed radio galaxies with radio emission entirely similar to NGC 4486, but without any peculiarities in their optical radiation.

It may be considered that all relativistic electrons that fill the volume of NGC 4486 were formed in result of a *single* outburst. It is more natural, however, to suggest that the 'jets' in NGC 4486 are a recurrent phenomenon, and that the relativistic electrons that have filled NGC 4486 were formed as a result of 10–20 outbursts.

What is the nature of a 'jet' (or of 'jets')? Two hypotheses may be suggested.

(1) The 'jet' has originated as the result of a certain enormous explosion in the central part of NGC 4486. In this case it should be assumed that the energy emitted during such an explosion is enormous: it is hundreds of

millions times greater than during an outburst of a supernova. Similar phenomena are unknown in modern physics or astrophysics. The development of such a hypothesis raises numerous difficulties.

(2) The anomalous conditions in the central part of NGC 4486 are caused by collisions of massive aggregates (of the type of large globular clusters) containing interstellar gas. These aggregates, extremely numerous in the spheroidal galaxy NGC 4486, must have velocities of about 500–800 km./sec. and 'frontal' collisions between them may, in a similar fashion as in the case of the colliding galaxy Cygnus A, be the cause of a generation of considerable numbers of relativistic particles.

In [2] it was predicted that a considerable polarization of the optical emisssion of the 'jet' is expected. Special observations of this phenomenon are very desirable.*

REFERENCES

[1] Shklovsky, I. S. *Proc. Acad. Sci. U.S.S.R.* **90**, 983, 1953.
[2] Shklovsky, I. S. *Astr. Zj.* **33**, N 3, 1955.
[3] Baade, W. *Ap. J.* **123**, 550, 1956.

* Editor's note (added in proof): This polarization has since been observed by W. Baade[3].

PART III

GALACTIC STRUCTURE AND STATISTICAL STUDIES OF POINT SOURCES

GALACTIC RADIO EMISSION AND THE DISTRIBUTION OF DISCRETE SOURCES

INTRODUCTORY LECTURE BY

R. HANBURY BROWN

Jodrell Bank Experimental Station, University of Manchester, England

At wave-lengths greater than about 1 metre the majority of the radio emission which is observed from the Galaxy cannot be explained in terms of thermal emission from ionized interstellar gas. This conclusion is widely accepted and is based on observations of the equivalent temperature of the sky and the spectrum of the radiation. The spectrum at metre wave-lengths is of the general form:

$$T_A \propto \lambda^n$$

where T_A is the equivalent black-body temperature of a region of sky and λ is the wave-length. The exponent n varies with direction but lies between about 2·5 and 2·8, and is thus significantly greater than the value of 2·0 which is the maximum to be expected for thermal emission from an ionized gas. Furthermore, the value of T_A is about $10^{5°}$ K. at 15 metres and thus greatly exceeds the electron temperature expected in H II regions.

At centimetre wave-lengths it is likely that the majority of the radiation observed originates in thermal emission from ionized gas; however, the present discussion is limited to a range of wave-lengths from about 1 to 10 metres where the ionized gas in the Galaxy is believed to be substantially transparent and where the origin of most of the radiation is believed to be non-thermal.

I. SOME FEATURES OF THE GALAXY AT METRE WAVE-LENGTHS

(a) The general background radiation

Early surveys [1, 2, 3, 4, 5] of the sky showed that the general radiation is, broadly speaking, concentrated in latitude about the galactic plane and in longitude about the galactic centre. On the basis of these surveys, it was concluded [6] that the sources of emission in the Galaxy, whatever they may be, have a space distribution like that of the common stars. Irregularities in the distribution, which are increasingly pronounced as the wavelength is decreased, were attributed to the effects of spiral structure in the Galaxy.

The early surveys were made with rather wide beams and it is now known that much important detail was lost. Thus a recent high resolution survey by Scheuer and Ryle [7] and also some unpublished work by Mills suggest that the true distribution of intensity normal to the galactic plane is made up of perhaps three component distributions. Two of these components are narrow and have widths of about 2 and 10°. It is difficult, without more extensive surveys, to be sure of their independent existence and this is a problem which requires a considerable amount of further study. The third component appears to be much broader and to have a width of the order of 120°. The variation of these components with galactic longitude is not yet known satisfactorily and is clearly complicated by irregularities in galactic structure; nevertheless it is known that the narrow distributions show a marked concentration towards the galactic centre, while the broader component appears to be concentrated to a lesser extent in that direction.

The space-distribution of the sources of the broad component presents a particularly interesting question, and it is difficult to escape the conclusion that the Galaxy has a radio corona which extends to great distances. In the earlier interpretations of the isophotes it was found that agreement with the general distribution of mass in the Galaxy could be obtained by assuming that a large fraction, about two-thirds, of the total radiation was isotropic and probably of extra-galactic origin. It seems likely, as has been suggested by Shklovsky [8], that a substantial fraction of this isotropic component must now be attributed to the Galaxy and associated with the broad distribution.

(b) *The discrete sources at metre wave-lengths*

The first surveys of the discrete sources, or radio stars, were made with interferometers and its was concluded from these results that the distribution of sources with direction from the sun is isotropic [9]. While this is still believed to be true of the majority of sources, it is now recognized that the resolving power of the interferometers was so high that important sources of large angular diameter were missed. It appears that there are at least two classes of source.

Class I sources, which form a minority of the total, are of relatively high intensity and show a pronounced concentration into the galactic plane [10, 11].

A surprising feature of these sources is that many of them are known to have apparent angular diameters greater than one degree [12].

Class II sources, which form the majority, appear to be uniformly distributed over the sky. The angular diameters of many of these sources are not yet known and it is important that they should be measured; however, the few results which are available suggest that, for the most part, their diameters are of the order of a few minutes of arc or less.

2. THE ORIGIN OF THE RADIO EMISSION FROM THE GALAXY AT METRE WAVE-LENGTHS

Any discussion of the origin of the radiation from the Galaxy must be highly speculative, since recent work has shown that our knowledge of the distribution of the background radiation is seriously incomplete. Furthermore, data on the spectra and angular diameters of the sources are confined to a few of the most intense.

The present evidence suggests that any theory may have to account for both the broad and narrow distributions although it is by no means clear whether these distributions can be regarded as independent.

Two components of the narrow distributions are the thermal radiation from ionized gas in H II regions and Class I sources. At metre wave-lengths the thermal radiation cannot account for the total intensity observed and it is tempting to ascribe the remainder to a population of sources which lie close to the galactic plane. However, this cannot be done, since so little is known about the Class I sources; for example it is not known whether they are a homogeneous population, nor how they are distributed in the Galaxy. Until more data are available we must be prepared to find that some other mechanism, as yet unknown, is responsible for the majority of the radiation. For example, it has been suggested [13, 14, 15, 16, 17] that the non-thermal radiation might be due to cosmic-ray electrons in interstellar magnetic fields.

The nature of the known Class I sources is a fascinating problem. They appear to be rare bodies with a space-density in the neighbourhood of the sun, which we may compare, solely for the purpose of illustration, with that of planetary nebulae. The spectra and apparent surface temperatures of a few of these sources are known and it is clear that some of them are radiating by a non-thermal mechanism. The large angular diameters of several of these sources [12], coupled with the few photographic identifications which have been made [12, 18, 19, 20], suggest that they are associated with extended nebulosities. These nebulosities are of low photographic brightness and some of them have been found to contain filaments which are apparently moving at very high speeds. The nature of these nebulosities is controversial. At least two, if not three, of the sources have been identified with the remnants of supernovae [21, 22, 23], and it has been suggested [24, 25] that many of the other sources arise in the same way.

The physical mechanism by which these nebulosities radiate is also unknown. It seems likely that plasma oscillations cannot be invoked since the plasma frequency in the medium is too low, and the current idea is that the radiation arises from the deflexion of relativistic electrons in magnetic fields [26]. The magnetic fields are presumed to be generated by turbulence in an ionized medium, and the fast electrons to be accelerated by the Fermi mechanism, by shock waves, or by some other process.

The origin of the broad distribution is also a challenging problem and it is an urgent task of observation to establish beyond doubt the shape and spectrum of this distribution. It has been suggested [8, 27] that, whereas the narrow distributions are apparently associated with populations concentrated into the galactic plane, the broad distribution arises in an extended halo which is roughly spherical and extends to radial distances of the order of 10,000 parsecs. It has also been proposed that the generation of the energy in this halo occurs in a very rarefied medium and is due to the deflexion of fast electrons in magnetic fields.

The origin of the majority of the discrete sources, the Class II sources, may be extragalactic. Recent work [23] has shown that their distribution is remarkably isotropic and it is difficult to associate them with any of those components of the background radiation which are clearly of galactic origin.

3. EXTRAGALACTIC SOURCES

A small number of radio sources have been identified with external galaxies, and on the basis of these results it appears that, as far as radio emission is concerned, we must recognize at least two major classes of galaxy, *normal* galaxies and *peculiar* galaxies [28].

(a) Normal galaxies

A study of six type-Sb galaxies shows that they radiate roughly the same ratio of radio to light flux [29, 30]. If the intensity of radio emission at the earth is I w.m.$^{-2}$ (c./s.)$^{-1}$ then, at a wave-length of about 1·9 metres, it has been found that the photographic and radio magnitudes may be taken as roughly equal if we define the radio magnitude (m_R) by the equation:

$$m_R = -53\cdot4 - 2\cdot5 \log I.$$

It is clearly of great interest to know how the value of $m_R - m_{pg}$ varies with the type of galaxy and this is one of the major observational problems of radio astronomy. At the present moment the available evidence is insufficient to draw any definite conclusions about this question.

(b) Peculiar galaxies

A few radio sources have been identified with peculiar objects of which the best known is the pair of colliding galaxies in Cygnus [18]. These are to be described later in greater detail by Dr Minkowski and so will not be discussed here.

(c) Our own Galaxy

A comparison of the radio emission and the light from M31 suggests that, as far as radio emission is concerned, it behaves as a normal Sb galaxy [29]. Furthermore a comparison of our own Galaxy with M31 indicates that the two systems are similar in respect to their total radiation at metre wave-lengths. Thus the total radiation from our Galaxy [31] at 1·9 metres is of the order of $5·0 \times 10^{20}$ watts $(c./s.)^{-1}$ steradian^{-1}, which is in close agreement with that found for M31. In conclusion the extended distribution of radiation around M31, which has been reported by Baldwin, suggests that both our own Galaxy and M31 may possess large radio coronas.

REFERENCES

[1] Reber, G. *Ap. J.* **100**, 279, 1944.
[2] Reber, G. *Proc. I.R.E.* **36**, 1215, 1948.
[3] Hey, J. S., Parsons, S. J. and Phillips, J. W. *Proc. Roy. Soc.* A, **192**, 425, 1948.
[4] Bolton, J. G. and Westfold, K. C. *Aust. J. Sci. Res.* A, **3**, 19, 1950.
[5] Allen, C. W. and Gum, C. S. *Aust. J. Sci. Res.* A, **3**, 224, 1950.
[6] Westerhout, G. and Oort, J. H. *B.A.N.* **11**, 323, no. 426, 1951.
[7] Scheuer, P. A. G. and Ryle, M. *M.N.R.A.S.* **113**, 3, 1953.
[8] Shklovsky, I. S. *Astr. Zh.* **29**, 418, 1952.
[9] Ryle, M., Smith, F. G. and Elsemore, B. *M.N.R.A.S.* **110**, 508, 1950.
[10] Mills, B. Y. *Aust. J. Sci. Res.* A, **5**, 266, 1952.
[11] Hanbury Brown, R. and Hazard, C. *M.N.R.A.S.* **113**, 123, 1953.
[12] Hanbury Brown, R., Palmer, H. P. and Thompson, A. R. *Nature*, **173**, 945, 1954.
[13] Alfvén, H. and Herlofson, N. *Phys. Rev.* **78**, 616, 1950.
[14] Kiepenheuer, K. O. *Phys. Rev.* **79**, 738, 1950.
[15] Ginsburg, V. L. *Dok. Akad. Nauk. U.S.S.R.* **76**, 377, 1951.
[16] Hutchinson, G. W. *Phil. Mag.* **43**, 847, 1952.
[17] Hoyle, F. *Nature*, **173**, 483, 1954.
[18] Baade, W. and Minkowski, R. *Ap. J.* **119**, 206, 1954.
[19] Baldwin, J. E. and Dewhirst, D. W. *Nature*, **173**, 164, 1954.
[20] Hanbury Brown, R. and Walsh, D. *Nature*, **175**, 808, 1955.

[21] Bolton, J. G. and Stanley, G. J. *Aust. J. Sci. Res.* A, **2**, 139, 1949.
[22] Hanbury Brown, R. and Hazard, C. *Nature*, **170**, 364, 1952.
[23] Ryle, M. *Observatory*, **75**, 137, 1955.
[24] Shklovsky, I. S. *Dok. Acad. Nauk. U.S.S.R.* **94**, 417, 1954.
[25] Hanbury Brown, R. *Observatory*, **74**, 185, 1954.
[26] Twiss, R. Q. *Phil. Mag.* **45**, 249, 1954.
[27] Baldwin, J. E. *Nature*, **174**, 320, 1954.
[28] Baade, W. and Minkowski, R. *Ap. J.* **119**, 215, 1954.
[29] Hazard, C. *Occasional Notes, Roy. Ast. Soc.* **3**, 74, 1954.
[30] Mills, B. Y. *Aust. J. Phys.* **8**, 368, 1955.
[31] Hanbury Brown, R. and Hazard, C. *Phil. Mag.* **44**, 939, 1953.

Discussion

Ryle: The interferometric measurements by Scheuer [1] of the distribution of intensity at low galactic latitudes had sufficient resolving power to give the true latitude distribution without effects associated with 'aerial smoothing'. The curves obtained show with absolute certainty the presence of the two components; a narrow belt having a width to half-intensity of the order of $\pm 1°\!.2$, and the 'second component' which for longitudes between 345° and 30° falls to half-intensity at about $\pm 10°$.

Steinberg: Our observations at 33 cm. also clearly show the presence of a narrow component and a much broader one.

Westerhout: Another proof of the existence of a corona around the galactic system may possibly be obtained from 21-cm. line observations. Recent observations at Kootwijk show the presence of very long wings to line profiles at high galactic latitudes. Neutral hydrogen with a very small density and high velocities thus seems to be present at large distances from the galactic plane.

Burbidge: A possible explanation of the origin of the halo, corona, or aura, of radio emission observed in M 31 and in our Galaxy by Mr Baldwin may be obtained by extending the ideas of Pikelner, Shklovsky and Ginzburg who first suggested that there is an extended distribution of gas and magnetic field extending normal to the spiral planes. By using the conditions for the emission of synchrotron radiation by relativistic electrons, for the stability of the spheres of diffuse gas (with radii of the order of 10–15 kiloparsecs), and for the energy density of the electrons to be much less than the cosmic-ray energy density in the planes, I have concluded [2] that the most plausible values of the parameters are

$\bar{H} \approx 10^{-6}$ gauss,
ρ (mean density in the diffuse spheres) $\approx 10^{-26} - 10^{-27}$ g./cm.^{3},
v (turbulent gas velocities) ≈ 200 km./sec.,
E (electron energies) $\approx 2 \times 10^9$ eV.

A fairly uniform distribution of radio brightness away from the planes is obtained since the electrons lose energy by radiation but continuously gain it by collisions with turbulent magnetic cloud elements (Fermi mechanism) whose 'cores' must be about 100 parsecs apart.

The origin of the high-energy electrons is rather uncertain. I have considered that pair production in collisions between high-energy protons of cosmic radiation and thermal photons may be important. Another possibility is that electrons will be the end products following meson decay after the mesons have been produced in high-energy nucleon-nucleon collisions.

REFERENCES

[1] Scheuer, P. A. G. and Ryle, M. *M.N.R.A.S.* **113**, 3, 1953.
[2] Burbidge, G. *Ap. J.* **123**, 178, 1956.

THE CAMBRIDGE SURVEY OF
RADIO SOURCES

J. R. SHAKESHAFT

Cavendish Laboratory, Cambridge, England

A survey of radio sources at a wave-length of 3·7 metres has been carried out with a large interferometric radio telescope (Ryle and Hewish, 1955) [1] which has a receiving area of about 5000 square metres. Four parabolic troughs are arranged at the corners of a rectangle 600 metres east–west by 50 metres north–south. The reception polar diagram of each, $\pm 1°$ by $\pm 7°$ to half-power points, is thus filled with interference fringes in the north–south plane as well as the east–west plane. Sources are observed at transit, the time of which gives the right ascension, while the declination is obtained by comparing the observed intensity on successive days as the phase of the north–south pattern is altered.

The four aerials have been used together in two different ways. In the first the whole system is equivalent to a phase-switching interferometer of aperture 157 wave-lengths and so it can detect only sources with angular diameter less than about 20'. The positions of 1906 such sources have been found (Shakeshaft *et al.* 1955) [2] and they are plotted on the accompanying map (Fig. 1) in equal-area galactic co-ordinates. In the second case, the two north aerials are connected and used as a phase-switch interferometer against the two south aerials. This gives an effective aperture of 14 wave-lengths, and sources of larger angular diameter may be detected. In this way 30 extended sources, marked as open circles, have been found with angular diameters between 20' and 120'. They are concentrated towards the galactic plane, though the less intense also occur near the poles. The latter do not appear to coincide with clusters of extra-galactic nebulae and are therefore probably galactic. The small-diameter sources are isotropically distributed except for a concentration of a few bright sources near the galactic anti-centre, and areas where confusion with the most intense sources is important. There is thus little evidence to relate the majority of the sources to the general galactic structure.

About 500 of the positions are believed to be known, each within an area of 0·05 square degrees, and it was hoped that with the new list many identifications would be made. An intense source coincides with the remnants of Kepler's supernova of 1604, so it is rather surprising that another, already tentatively identified as Tycho Brahe's supernova of 1572 by Hanbury Brown and Hazard (1952) [3], in fact seems to be 6' of arc away in right ascension, which is considerably greater than the estimated error. IC 443 has been identified as a source and the nebulosity in Auriga (Hanbury Brown, Palmer and Thompson, 1954) [4] confirmed as one. No significant coincidences occur with novae, globular clusters, magnetic stars or flare stars.

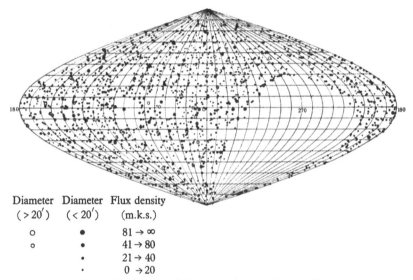

Diameter (> 20′)	Diameter (< 20′)	Flux density (m.k.s.)
○	●	81 → ∞
○	●	41 → 80
·	●	21 → 40
	·	0 → 20

Fig. 1. Distribution of the sources in galactic co-ordinates.

Of normal extra-galactic nebulae M 31 and M 51 are observed and it is not to be expected that many more would be seen owing to the upper limit of angular size set by the interferometer. The positions of two sources previously identified with M 33 and M 101 by Ryle, Smith and Elsmore (1950) [5] were found not to agree with those of the optical objects and in each case the angular diameter is less than 7', thus not supporting the original conclusion.

Seventy-five of the best positions were first examined by Dewhirst, with the 17-inch Cambridge Schmidt, and then by Minkowski, using the 48-inch Sky Survey plates, with little success except for a coincidence with NGC 2623 which shows distorted spiral arms and may be another case of colliding galaxies like NGC 1275.

The most startling result of the survey is the number-magnitude distribution. If a curve is plotted of log N against log I, where N is the number of sources per steradian having a flux density greater than I, the slope is steeper than -1.5, and this indicates that there are more faint sources than would be expected for a uniform distribution. Mr Ryle discusses this point in the following paper.

REFERENCES

[1] Ryle, M. and Hewish, A. *Mem. R.A.S.* **67**, 97, 1955.
[2] Shakeshaft, J. R., Ryle, M., Baldwin, J. E., Elsmore, B. and Thomson, J. H. *Mem. R.A.S.* **67**, 106, 1955.
[3] Hanbury Brown, R. and Hazard, C. *Nature*, **170**, 364, 1952.
[4] Hanbury Brown, R., Palmer, H. P. and Thompson, A. R. *Nature*, **173**, 945, 1954.
[5] Ryle, M., Smith, F. G. and Elsmore, B. *M.N.R.A.S.* **110**, 508, 1950.

THE SPATIAL DISTRIBUTION OF RADIO STARS

M. RYLE

Cavendish Laboratory, Cambridge, England

I. THE OBSERVATIONS

The Cambridge survey of radio stars [1], which has been described by Shakeshaft, revealed some 1900 sources of small angular diameter, which appeared to be distributed nearly isotropically. An examination of the number–magnitude distribution shows, however, that they cannot be accounted for in terms of a uniform spatial distribution of sources [2].

The results are summarized in Fig. 1 which shows for seven areas of sky a plot of log N against log I, where N represents the number of sources per unit solid angle having an intensity greater than I. A uniform distribution of point sources would lead to a slope of $-1 \cdot 5$; it can be seen that for all the areas the slope is substantially greater than this, and that for the five areas away from the plane where the observations are unaffected by the presence of rare galactic sources, the slope is approximately $-3 \cdot 0$.

Before considering the significance of this result it is necessary to examine all possible instrumental defects or errors in the interpretation of the records which might lead to an apparent increase in the number of faint sources.

(i) *Subsidiary maxima in the primary reception pattern*

The reception pattern has been measured by a technique which allows the determination of subsidiary maxima with great precision. It was found that except near the two principal planes the sensitivity was less than 10^{-5} of that in the forward direction. It is therefore very simple to eliminate errors due to this cause.

(ii) *Accuracy of reading the records*

The signal–noise ratio is more than 5 : 1 even on the weakest sources, and resultant errors in determining the relative intensities of the sources are unimportant.

(iii) *Difficulty of interpreting the records in presence of confusion*

It is clear that the confusing effects of adjacent sources will limit the detection of faint sources; the flattening of the log N–log I curve for small I

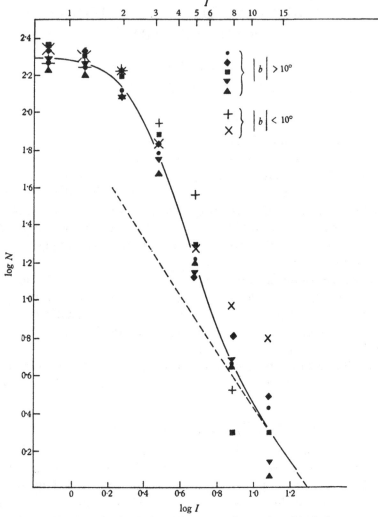

Fig. 1. Curve of log N against log I where N represents the number of 'point' sources per unit solid angle having an intensity greater than I.

is predominantly due to this cause. By making an analysis of the effect it is possible to show that when there are less than 0·25 sources per beam-width, the errors in the deduced number of sources is very small; this result

222

indicates that in the curve of Fig. 1 there should be negligible error for $I > 3.5 . 10^{-25}$ M.K.S. units.

Fortunately, however, the possibility of the large apparent slope being due to the effects of confusion can be eliminated entirely by using an independent, statistical method of analysing the records. In this method no account is taken of individual sources, but the probability distribution of the interference pattern amplitude, D (see Fig. 2), is determined.

In the absence of intense sources which could be resolved individually, the record would be composed of the random addition of the components

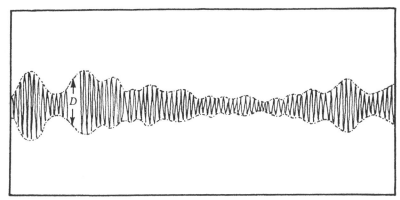

Fig. 2. Section of record showing how the probability distribution $P(D)$ is derived from measurements of the amplitude (D) of the envelope of the interference pattern.

due to a large number of weak sources; under these circumstances the probability distribution $P(D)$ would have a Rayleigh distribution. The presence of intense sources would lead (for the case of a uniform spatial distribution) to a curve which tended to $D^{-\frac{5}{2}}$ for large D. The complete probability curve which would be produced by any particular assumed spatial distribution of the sources can be computed, and that for a uniform distribution is shown in Fig. 3. The observed probability curve obtained for areas away from the galactic plane is also given, and this shows important differences from the computed curve; the observations again indicate an excess of faint sources or a lack of intense ones.

(iv) *The presence of a large number of extended sources*

The main survey of radio stars was designed to ignore sources having angular diameters greater than 20 minutes of arc. If an appreciable fraction of the radio sources had angular dimensions of this order, then the nearest, most intense ones would be missed, or recorded as of smaller intensity, whilst similar sources at greater distances would not be so affected; in this way, an apparent lack of intense sources might arise. Such an interpretation is, however, incompatible with the result of the second survey described by Shakeshaft in which a different arrangement of the aerials was employed, to allow the detection of sources having angular diameters of up to 3°. The sensitivity of this survey was not as good as in the main survey, but over most of the sky it is believed to include the majority of sources having a flux greater than $4 . 10^{-25}$ M.K.S. units.

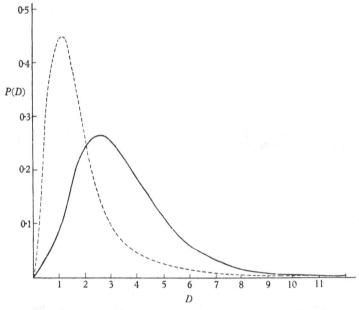

Fig. 3. The probability distribution $P(D)$ of the amplitude of the record D. The full curve represents the observations whilst the theoretical curve for a uniform spatial distribution of sources is shown dotted.

If the apparent lack of intense sources is to be explained in terms of their partial resolution by the main survey it is possible to predict the number of extended sources which should exist in any range of intensities. By comparing this number with that observed it is easy to show that the number of extended sources is quite inadequate for this explanation; for example, for sources having a flux density greater than $12 \cdot 5 \times 10^{-25}$ M.K.S. units, where the second survey must be regarded as effectively complete, a total of six extended sources were found in the area investigated. The number expected on this interpretation would be forty-six.

Similar but more complicated arguments may be used to show that the effect cannot be due to clustering of the sources.

(v) *Dispersion in the absolute luminosity*

It can also be shown that any reasonable dispersion in the absolute luminosity of the sources cannot explain the steep slope of the log N–log I curve; at least one class of object must have a spatial density which increases with distance.

From the above considerations it is concluded that the effect is real, and that either the spatial density or the luminosity of the sources shows a progressive increase with distance. Such an increase cannot extend indefinitely because both the amplitude of the recorded interference pattern and the

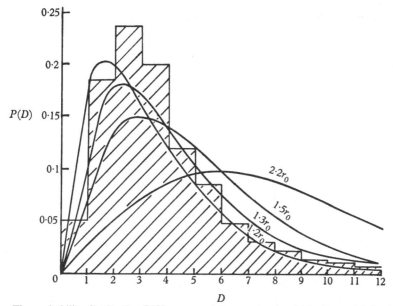

Fig. 4. The probability distribution $P(D)$ computed for a region in which the spatial density of sources increases as r^3 to a distance r_2 (expressed in terms of r_0, the distance of the farthest resolvable source). The observations are shown by the shaded area.

integrated radiation would diverge. A study of the former by the statistical method of analysis already discussed leads to the most sensitive method of deriving the extent of the divergent region in terms of the distance r_0 of the weakest sources which can be resolved individually.

Fig. 4 shows a number of theoretical curves of $P(D)$ derived for a region in which the spatial density of sources is assumed to increase as r^3 out to a distance r_2; four values of r_2 are shown, each expressed in terms of r_0. The shaded area represents the observations. From these results it appears that the region of increasing spatial density or luminosity does not extend much beyond the limit of the present survey.

2. THE INTERPRETATION OF THE RESULTS

(i) *In terms of galactic sources*

If it were supposed that the majority of the radio sources were within the Galaxy, it would be necessary to conclude that the solar system were situated in a region of abnormally low density, in which the density increased with distance equally in all directions. By considering the integrated radiation which would be produced by such a distribution of sources it can also be shown that the region would have to show effectively spherical symmetry out to radial distances from the sun of at least 2 kiloparsecs.

If on the other hand the radio sources were supposed to occur mainly in a spherical shell surrounding the Galaxy, the radius would have to be of the order of 80–100 kiloparsecs or departures from symmetry in directions towards the centre and anti-centre would be detectable.

Both of these models present considerable difficulties.

(ii) *Extragalactic sources*

If it is supposed that the majority of radio stars are extragalactic, it is not possible to account for the isotropic increase of apparent density in terms of random clustering. By applying similar arguments relating to the integrated extragalactic radiation, it can be shown that the radius of the region of increasing density must be at least 10^8 parsecs.

The very large scale which must be adopted in this case suggests that the explanation may be found in effects associated with large red-shifts. Owing to the relatively flat spectrum of the radio stars the spectral term is relatively unimportant even for red-shifts considerably beyond the limit of optical observation, where other effects might become appreciable. The observation that the intense source in Cygnus lies at a distance of 66.10^6 parsecs indicates that other sources of similar luminosity would indeed produce a detectable flux at distances considerably beyond the optical limit.

If this interpretation is correct, the observations give clear evidence that the distant regions differ from those in the neighbourhood of the Galaxy; such a result is incompatible with the predictions of the steady-state theories of cosmology proposed by Bondi and Gold [3] and Hoyle [4] but might be interpreted in terms of evolutionary theories. If indeed the majority of radio stars are of the Cygnus type, resulting from collisions between spiral galaxies, then a progressive decrease in the spatial density of radio sources with time might be expected, especially if collisions between such galaxies remove the interstellar matter, as has been suggested by Spitzer and Baade [5].

Some independent support for the suggestion that the majority of radio stars are similar to that in Cygnus has been provided by recent observations. Minkowski has reported that the intense source in the constellation of Hercules has been identified as another faint extra-galactic object showing a recession velocity of 26,000 km./sec. This observation indicates that its radio emission must be about 20 % of that of the Cygnus source.

The recent observations made by Palmer and Thompson [6] have shown that three of the four unidentified intense sources they have studied have angular diameters less than 50 sec. of arc; unless these sources are some invisible stellar object within the Galaxy, it seems probable that they are of galactic dimensions. The observed intensity and angular diameter then allow a lower limit to be set to their radio luminosity; their luminosity must be at least 10 % of that of the Cygnus source.

These results are therefore compatible with the suggestion that the majority of the radio stars are of comparable luminosity to that of Cygnus, and that their local spatial density at the present time is of the order of 2.10^{-26} parsecs^{-3}. In this way it seems possible to explain the observed log N–log I curve in terms of evolutionary theories of cosmology, whilst the difficulty of observing related optical objects as discussed by Mr Shakeshaft has a natural explanation; only some tens of the main class of radio star would be within reach of the 200-inch telescope.

REFERENCES

[1] Shakeshaft, J. R., Ryle, M., Baldwin, J. E., Elsmore, B. and Thomson, J. H. *Mem. R.A.S.* **67**, 106, 1955.
[2] Ryle, M. and Scheuer, P. A. G. *Proc. Roy. Soc.* A, **230**, 448, 1955.
[3] Bondi, H. and Gold, T. *M.N.R.A.S.* **108**, 252, 1948.
[4] Hoyle, F. *M.N.R.A.S.* **108**, 372, 1948.
[5] Spitzer, L. and Baade, W. *Ap. J.* **113**, 413, 1951.
[6] Palmer, H. P. and Thompson, A. R. This publication, paper 28, p. 162.

PRELIMINARY STATISTICS OF DISCRETE SOURCES OBTAINED WITH THE 'MILLS CROSS'

J. L. PAWSEY

*Division of Radiophysics, Commonwealth Scientific and Industrial Research Organization,
Sydney, Australia*

The statistics of the discrete sources observed in Cambridge and the interpretation given by Ryle and his colleagues constitute one of the most interesting items of recent astronomy. It is therefore of great importance to check the observational data and this can be done from the independent results being obtained in Sydney by Mills and his colleagues with the 85 Mc./s. Mills Cross. With this in mind Ryle sent me some two months ago a pre-publication account of the Cambridge work. (Now published, Ryle and Scheuer, 1955[1].) The currently available observations with the Mills Cross are not yet sufficient to give a decisive answer, but those available disagree with the Cambridge ones. Because of the importance of the subject it seems desirable to give here an interim account of these observations. The general position of the observations is discussed in a separate paper (paper 18, Pawsey). As stated there observations to date have been aimed at the study of known objects. The beam in each case was adjusted to the appropriate declination and an extended record, including the selected object in a small section, was taken. Most of these records have been examined for discrete sources and such sources listed with their intensities when sufficient records at adjacent declinations were available to delineate them. The list was restricted to sources which, within the 50' limits of resolution of the equipment, appeared to be discrete point sources. Extended sources were neglected. This method gives an irregular coverage of the sky so that the sampling must be watched.

At the time of Ryle's letter some 550 sources had been listed over a solid angle in the sky of roughly one steradian. The region concerned included an unduly large proportion of sky adjacent to the Milky Way. For these sources the statistical distribution of flux densities is shown in the form used by Ryle by the black dots in Fig. 1. Here ρ_S is the number of sources per steradian with flux density greater than S. The vertical dotted line through each point shows the limits $\pm\sqrt{N}$, where N is the number of sources in the actual sample, and indicates the probable statistical error. As discussed by Ryle a uniform space density of sources should give a line of slope] $-3/2$ and the straight line has this slope. It is clear that the original sample

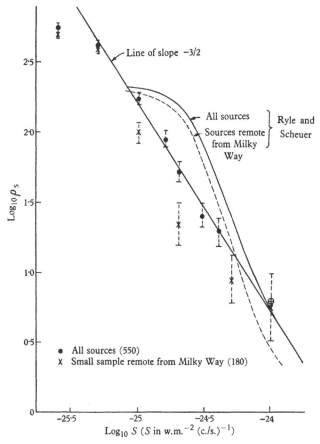

Fig. 1. Intensity distribution of first 550 discrete sources listed from 85 Mc./s. Mills Cross records. ρ_s is the number of sources per steradian with flux density greater than S. The highest intensity group, small circle, includes sources from other parts of the sky previously located by interferometers. Ryle and Scheuer's 81·5 Mc./s. distributions are shown for comparison.

showed no significant departure from the $-3/2$ distribution except for intensities less than about 5×10^{-26} w.m.$^{-2}$ (c./s.)$^{-1}$ where instrumental limitations might be expected.

In a first attempt to exclude the influence of galactic sources those sources remote from the Milky Way were selected and the resulting distribution is shown by the crosses. Unfortunately, the sample, 180 sources, is unduly small and the statistical errors great. This sample does show a suggestion of an excess of faint sources over the number expected on the uniform space density hypothesis but when the distribution is compared with those obtained by Ryle and Scheuer, which are also shown, it is seen that the excess, corresponding to the steeper slope, occurs at substantially

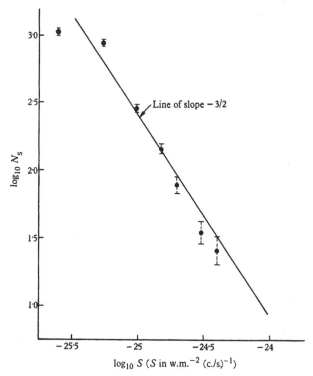

Fig. 2. Intensity distribution of first 1030 Sydney sources. N_s is the number of sources with flux density greater than S.

greater intensity in Ryle and Scheuer's case. This difference is not accounted for by the very slight difference in frequency of observations (85 and 81 Mc./s.).

In the short interval available before this symposium a considerable number of further sources were listed in regions remote from the Milky Way. This gave a total of 1030 sources in areas much less biased towards the Milky Way and the intensity distribution of these sources is shown in Fig. 2.

On considering the three Sydney curves it appears that none show a deviation from the $-3/2$ law which we can be sure is significant. There is a tendency for an increase in steepness of the curves at intensities just short of the survey limit, but we should not like at this stage to exclude possible instrumental or other extraneous effects at such intensities. The important point is that in the intensity range in which the Cambridge workers found excess steepness the Sydney results do not show such an effect. The essential difference in the results is that in the range about

$$S = 2 \times 10^{-25} \text{ w.m.}^{-2} \text{ (c./s.)}^{-1}$$

Ryle and Scheuer report two or three times as many sources despite the fact that the Sydney sensitivity limit is several times lower than theirs.

It is clear that the details of individual sources, positions, intensities and sizes, in areas common to the Cambridge and Sydney surveys, should be compared in order to elucidate the nature of the discrepancy. But this has not yet been possible because none of the detailed Cambridge information has been available to us.

There is thus a substantial disagreement between the Cambridge and the preliminary Sydney results and it seems best to withhold judgment on the most interesting interpretation put forward by Ryle and Scheuer until the Sydney observations are complete. At that stage quite definite conclusions should be reached because the pencil-beam technique used is substantially free from confusion. In the intensity range of interest for the comparison, sources stand out unambiguously as illustrated by the record of the source NGC 253 shown in Fig. 2 of paper 18 (p. 125). The flux density of this source, $S = 1 \cdot 1 \times 10^{-25}$ w.m.$^{-2}$ (c./s.)$^{-1}$ is at the lower end of the intensity range where the Cambridge and Sydney results disagree.

REFERENCE

[1] Ryle, M. and Scheuer, P. A. G. *Proc. Roy. Soc.* A, **230**, 448, 1955.

Discussion

Gold: The fact that the steepening of the slope occurs in both Ryle's and the Australian survey, in each case near the limit of the instrument but at a different level for the two methods, suggests that this is an experimental effect. In the case of Ryle's survey it is clear that weak sources with angular diameters $> 20'$ are missed near the galactic plane. A similar cut-off might have been operative at high latitudes for still weaker extended sources which are therefore perhaps missing from the lower end of Ryle's curve.

Another way in which a steepened curve could be brought about is by the erroneous judgment of intensity of some of the faint sources. When there are several sources in the beam, it might frequently occur that one is recorded of greater than the correct intensity. This would produce an increase in the number in one range of the curve at the expense of a proportionally much smaller decrease in a higher section of the curve. An interpretation of that sort would imply that the Cambridge survey is much more liable to such an error, and already at a higher intensity than the Australian one.

Ryle: The steepening of the slope in the Cambridge results does *not* occur near the confusion limit; the slope is significantly greater than $1 \cdot 5$ for an intensity of $5 \cdot 10^{-25}$ M.K.S. units where the number of sources per beam width is about $0 \cdot 06$. The errors caused by confusion for such a small value of sources per beam-width is readily computed and is quite unimportant in the present case.

It is also worth mentioning that the limit of detection in the Australian survey is determined by sensitivity and not confusion; it would be a remarkable coincidence if two such entirely different factors should produce a steepening at about the same part of the curve.

The possibility of the high slope being due to extended sources has already been discussed at some length and shown to be incompatible with the Cambridge survey of low resolving power. It would also be remarkable if such an explanation could account for the similar increased slope of the Australian survey, where a much lower resolving power was used.

Bondi: I wish to make three points:

(1) The arguments given by Ryle to show that finite size and dispersion of luminosity are *separately* unable to affect the results are invalid when the two effects are considered together.

(2) Has any allowance been made for the influence of clustering?

(3) In a confusion-limited instrument like the Cambridge interferometer it is very hard to tell what the quantity designated as intensity actually measures in the case of faint sources. If this quantity contains any admixture of differentiation with respect to angular position, as is only too likely, a substantial steeping of the log N–log I curve would follow. The inverse cube law arises if half an order of angular differentiation is introduced in both directions.

Ryle: We have already discussed the effects produced by each of these possibilities independently; it is not clear why a combination of them should be any more effective in producing an increased slope of the log N–log I curve without becoming apparent on the survey of low resolving power.

Further, in connexion with Pawsey's communication, I would like to point out that a number of extended sources were found which do not appear on the main survey. Their number is too small to allow of an explanation of the increased slope in terms of partial resolution of the intense sources, but they would be sufficient to modify the slope found in a survey made with a lower resolving power such as that of Mills' aerial (50'). The discrepancy in the slopes of the two log N–log I curves ($-3 \cdot 0$ for Cambridge, and $-2 \cdot 2$ for the area containing 180 sources away from the plane in Mills' survey), may be due to such a cause.

THE SPHERICAL COMPONENT OF THE GALACTIC RADIO EMISSION

J. E. BALDWIN

Cavendish Laboratory, Cambridge, England

As part of the programme of observations with the large Cambridge radio telescope, a survey of the integrated radio emission has been made using one of the four elements of the interferometer. At a wave-length of 3·7 metres this aerial has beam-widths to half-power points of 2° in right ascension and 15° in declination. The use of a long wave-length makes it possible to obtain accurate measurements of the brightness temperature of the sky in regions away from the galactic plane. It is with the radiation from these regions that this paper is primarily concerned.

Westerhout and Oort (1951) [1] showed, from a study of previous surveys, that the background radio emission could be conveniently divided into two main components: (*a*) a distribution similar to that of the distribution of mass in the Galaxy, and (*b*) an almost isotropic distribution which might arise from the integrated radiation from extra-galactic sources. Distribution (*a*) contributes the greater part of the radiation at latitudes less than 30°, whilst at higher latitudes (*b*) is the principal source of radiation. On the basis of this classification, the areas of minimum brightness of the sky should lie within a few degrees of the galactic poles. An examination of the present survey shows that these areas lie at latitudes of ± 45° between longitudes 100° and 215°, i.e. in directions away from the galactic centre. This suggests immediately that much of the radiation at high galactic latitudes may be galactic in origin. Shklovsky (1952) [2] has already proposed that the source of most of the radiation from the Galaxy is distributed throughout a large spherical region and this suggestion has received support from observations of the Andromeda nebula made at Cambridge (Baldwin, 1954) [3] which showed the existence of such a distribution in M 31.

In order to investigate in detail this distribution in the Galaxy, a comparison has been made between the observed distribution of radiation at latitudes greater than 30° and that expected on the basis of a simple model. We shall first take one in which

(1) there is no contribution from extra-galactic sources,

(2) the distribution is spherical in form, of radius R, centred on the Galaxy, and having

(3) uniform emission per unit volume, σ, throughout the sphere.

The observed distribution of brightness at latitude $+50°$ as a function of galactic longitude is shown in the form of a polar plot in Fig. 1, together

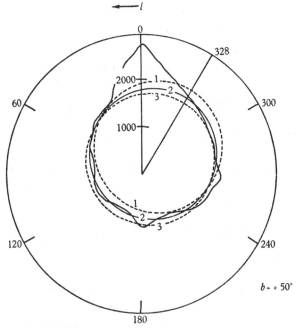

Fig. 1. The distribution of radio emission at $b = +50°$. The observations are shown in comparison with the computed curves for three spherical models having different values of R: (1) 12 kiloparsecs, (2) 16 kiloparsecs, (3) 20 kiloparsecs.

with the calculated curves for three different values of the radius R of the spherical model—12, 16 and 20 kiloparsecs. In each case the value of σ has been adjusted to give the best fit to the experimental curve. It may be seen that a value of R of 16 kiloparsecs gives the best agreement with the observations. The corresponding value of σ is $1\cdot8 . 10^8$ watts ster.$^{-1}$ (c./s.)$^{-1}$ parsecs^{-3}. Using these figures, the distributions of brightness expected at other galactic latitudes have been calculated and are shown in comparison with the observed distributions in Fig. 2. The fit is good except in the neighbourhood of $l = 0°$ where a bright arm of radiation

234

exists extending from the equator towards the pole, a feature which has been observed in several previous surveys.

Although the assumption of a model with uniform emission per unit volume leads to good agreement between theory and observation, there

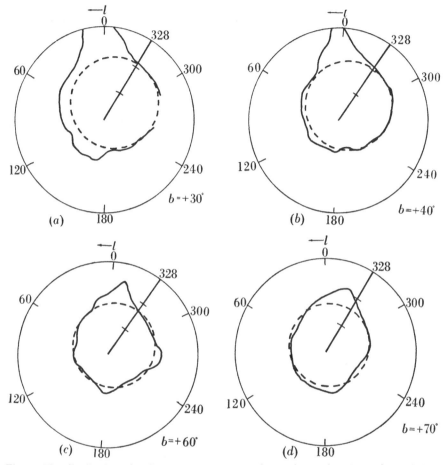

Fig. 2. The distribution of radio emission at $b = +30°$, $+40°$, $+60°$ and $+70°$. Full line: observed curve. Dotted line: computed curve for a spherical model having a radius of 16 kiloparsecs.

are two regions in which appreciable variations in the value of σ would be undetectable.

(1) The region within 4 kiloparsecs of the galactic centre. This region lies at galactic latitudes less than 30° where the radiation associated with the disk of the Galaxy is superimposed on that from the spherical distribution.

235

(2) Beyond 10 kiloparsecs from the centre of the Galaxy. The difficulty of interpretation in this region is that a spherical shell of emission whose radius is appreciably greater than the distance of the sun from the centre gives a brightness distribution over the sky which is almost isotropic as seen from the neighbourhood of the sun. It is thus not possible to determine the variation of the spatial density of the source of emission with radius in the outer parts of the sphere, and in the limit it is impossible to distinguish between emission from a very large galactic shell and that from extra-galactic sources.

A range of models may therefore be derived containing different contributions from extragalactic sources, all of which fit the observations equally well. The essential constants of these models are given in the first three columns of Table 1.

Table 1

Assumed extragalactic radiation (° K.)	R (kps.)	σ watts ster.$^{-1}$ (c./s.)$^{-1}$ ps.$^{-3}$	Axial ratio
0	16		> 0·75
150	14·5	$1\cdot8 \times 10^8$	> 0·7
300	13		> 0·6
500	11		> 0·5

In the previous discussion only spherical models having a uniform emission per unit volume have been considered. An investigation has been made of distributions in the form of flattened ellipsoids of revolution, their minor axes lying perpendicular to the galactic plane, and of the effect of assuming the emission per unit volume to be a function of the radial distance from the centre. None of these models fit the observations as closely as do the simple spherical models. From these calculations it has been possible to place a lower limit on the axial ratio of the ellipsoid appropriate to any assumed value of the extragalactic emission and these limits are given in the last column of Table 1. It has also been shown that the value of σ cannot vary by more than a factor of 2 between 4 and 10 kiloparsecs from the centre of the Galaxy.

The range of models may be narrowed still further by using estimates of the extra-galactic emission derived in other ways. The recent survey of radio stars has provided evidence that the extragalactic component must be at least 150° K. at this frequency. The radius of the sphere must there-fore lie in the range 11–14·5 kiloparsecs. This radius is comparable with the radius of 18 kiloparsecs found for the similar distribution in M31. In this latter case too, the value of σ is sensibly constant throughout the spherical region, although its actual value is some six times smaller in M31 than in the Galaxy. The respective values are

Galaxy $\quad$ $1\cdot8 \times 10^8$ watts ster.$^{-1}$ (c./s.)$^{-1}$ ps.$^{-3}$
M31 $\quad$ $0\cdot24$–$0\cdot35 \times 10^8$ watts ster.$^{-1}$ (c./s.)$^{-1}$ ps.$^{-3}$

For M31 it has been shown that at least two-thirds of the total radiation originates in the spherical distribution and this fraction could be much greater. The range of values of σ quoted for M31 correspond to this uncertainty.

REFERENCES

[1] Westerhout, G. and Oort, J. H. *B.A.N.* **11**, 323, no. 426, 1951.
[2] Shklovsky, I. S. *Ast. Zh.* **29**, 418, 1952.
[3] Baldwin, J. E. *Nature*, **174**, 320, 1954.

RADIO ASTRONOMY AND THE ORIGIN OF COSMIC RAYS

A. UNSÖLD

Director of the Institute for Theoretical Physics and of the Observatory, Kiel, Germany

At present we can *observe* the origin of only the solar component of cosmic rays. The sun emits cosmic rays in connexion with flares and probably also continuously on a smaller scale. Thus the emission of cosmic rays appears to be connected much more closely with non-thermal radio emission than with thermal radiation of light and heat.

The large cosmic ray bursts are often not exactly coincident with the H_α flare itself, but are retarded by about half an hour and more. The coincidence is much better with the 'second part' of the radio outburst, e.g. at 200 Mc./s. Both are evidently connected with streams of high-speed particles penetrating the outer parts of the solar atmosphere. Expressing the flux of the radio-frequency radiation for 100 Mc./s. in w.m.$^{-2}$ (c./s.)$^{-1}$ and of cosmic rays (energy range $\sim$ 0·5 to 5 BeV) in units of the total cosmic radiation we obtain the observed ratio as given in the Table.

Ratio of radio frequency radiation to cosmic radiation

Flare of 25 July 1946: Major burst	$3\cdot0 \times 10^{-18}$	
Second part	$0\cdot1 \times 10^{-18}$	
Moderate flares (Firor, neutron pile)	$1\cdot7 \times 10^{-18}$	2×10^{-18}
Solar component (Firor, Simpson, Treiman)	$1\cdot8 \times 10^{-18}$	
Galactic radio-frequency radiation/total cosmic radiation		$3\cdot6 \times 10^{-21}$

In the last line the 'isotropic component' or 'halo radiation' of the galactic radio-frequency radiation has not been included. Assuming it to be galactic too would increase the ratio by about a factor two, which is still within the range of uncertainty.

We should be aware that in our Galaxy the (charged) cosmic-ray particles are stored in an interstellar magnetic field of $\sim 5 \times 10^{-6}$ gauss. Their mean free path Λ (measured along the coiled orbits) must be somewhat shorter than that for the destruction of heavy nuclei by collisions with interstellar protons. In this way Morrison, Olbert and Rossi obtain

$\Lambda \approx 1 \cdot 2 \times 10^6$ parsecs. The average distance from which light or radio waves reach us is only $l \approx 500$ to 1000 parsecs. This means, the cosmic-ray particles are 'stored' in the Galaxy by a factor $\Lambda / l = (1 \text{ to } 2) 10^3$ and the ratio for the *production* of the two non-thermal radiations in the Galaxy,

$$\text{radio-frequency radiation/cosmic rays}$$
$$= 1 \cdot 5 \times 10^3 . 3 \cdot 6 \times 10^{-21} = 5 \times 10^{-18},$$

is practically *equal* to that found for the sun. Furthermore, the radio-frequency spectra of the non-thermal galactic radiation and of the average non-thermal solar radiation are very similar, the radiation temperature T_ν varying as $\nu^{-(2 \cdot 7 \text{ to } 2 \cdot 5)}$ and $\nu^{-2 \cdot 3}$ in the two cases.

On the other hand we know from the radio-spectrum observations by Wild *et al.* that the non-thermal radio-frequency radiation on the sun is produced by plasma oscillations which are excited by fast-moving matter, or—one might just as well say—corpuscular radiation. As we have seen already, cosmic-ray particles are accelerated in the same medium, probably by induction effects of fast-moving magnetized matter. Although neither the physical processes producing the radio-emission nor those producing the high-energy particles are known in detail, the observations presented above are best summarized by the hypothesis that:

On the sun as well as in galactic sources non-thermal radio-frequency radiation and cosmic rays (at least up to 5 BeV) are produced together in the same ratio in violently moving plasma, probably having also magnetic fields.

The question, whether particles of energies up to 10^{18}–10^{19} eV. in the general cosmic radiation are produced from the BeV particles by interstellar acceleration or by step-wise acceleration in the powerful sources themselves, might be left open for the present. Since already the 1 to 5 BeV-particles on the sun are probably produced by acceleration of less energetic particles which in turn have been accelerated magnetically, the latter possibility appears not unreasonable.

Looking for galactic sources which might account *simultaneously* for the major part of the non-thermal radio-frequency radiation and of the cosmic rays we realize that the types of radio sources which have so far been identified with optical objects do not come into account.

As to more or less *interstellar sources* one might first think of supernova shells, but with present data this leads into difficulties concerning the energy of cosmic rays. About one supernova with $\sim 10^{49}$ ergs every 300 years would produce all over the Galaxy $\sim 10^{39}$ ergs/sec. On the other hand a volume of $\sim 2 \times 10^{11}$ parsecs^3 (a cylinder of 10 kiloparsecs radius and 600 parsecs thickness) ought to be refilled with cosmic radiation of

239

$\sim 10^{-12}$ ergs/cm. every $\sim 4 \times 10^6$ years, thus requiring $\sim 4 \times 10^{40}$ ergs/sec. With ordinary novae one finds a similar discrepancy.

In spite of this result, the following considerations show that the hypothesis of an interstellar origin still merits further investigation. (a) Present estimates of the frequency of supernovae of type I and especially those of type II (resembling violent ordinary novae) might be too low. (b) In the central parts of the Galaxy the interstellar hydrogen—according to recent work of the Dutch group—has high velocities whose mechanism is not yet understood. (c) The 'galactic halo', supposedly producing the nearly isotropic radio-frequency radiation, is still quite enigmatic.

Looking for *stellar sources* we have called attention to the extremely cool dwarf stars which are very numerous and show many signs of solar-type 'activity' on a very enhanced scale (bright $H + K$ lines as in the plages faculaires; huge spots producing light-variation; violent flares). Also theoretically we should expect that the hydrogen convection zone, which ultimately drives all these mechanisms, is more strongly developed in these stars than in the sun. The requirements that such sources produce the observed radio emission per parsec^3 *and* that individual sources of this type have not been identified with the largest existing radio telescopes can be accounted for by postulating about 7×10^{-2} radio stars per parsec^3 with $M_{\text{phot.}} \approx 18$, $M_{\text{vls.}} \approx 14$ and $M_{\text{bol.}} \approx 9$.

The volume density fits very well with the luminosity function for the nearest stars. The cosmic-ray output would have to be about 4 % of the total light plus heat radiation. Since all our assumptions are astrophysically very reasonable, the hypothesis that the galactic radio-frequency radiation *and* cosmic rays originate in enormous numbers of extremely faint, cool stars has much to recommend itself. It is obviously not essential to postulate only *one* type of galactic objects involving highly turbulent and magnetic plasmas.

The alternative hypothesis, that the galactic radio-frequency radiation be produced by high energy electrons spiralling in magnetic fields, cannot be discussed here in detail. But it should be clear that on this hypothesis the agreement of the radio spectra and of the ratio of radio frequency to cosmic ray emission between the active sun and the Galaxy would be only a matter of chance.

BIBLIOGRAPHY

Kosmische Strahlung (ed. W. Heisenberg) (Berlin-Göttingen-Heidelberg, 1953).
Pawsey, J. L. and Bracewell, R. N. *Radio Astronomy* (Oxford, 1955).
Proc. Internat. Cosmic ray conference (Guanajuato, 1955), to be published.
Unsöld, A. *Physik der Sternatmosphären* (2nd ed.) (Berlin-Göttingen-Heidelberg, 1955).

SOME PROBLEMS OF META-GALACTIC RADIO-EMISSION

I. S. SHKLOVSKY

Sternberg Astronomical Institute, Moscow, U.S.S.R.

The observed cosmical radio-emission has to be divided into the galactic and metagalactic components. The separation of these components is a problem of first importance. It was shown by us (Shklovsky, 1952 [1]) that at least 75% of the observed intensity near the galactic poles is caused by sources located in our stellar system. These sources form an almost spherical sub-system. Another model of the distribution of the cosmical radio-emission sources (Westerhout and Oort, 1951 [2]) is popular in the western countries. The so-called 'isotropic component' of metagalactic origin used in that model was considered responsible for the greatest part of the intensity near the galactic poles.

Recent observations of the distribution of the intensity of radio-emission in M 31 (Baldwin, 1954 [3]) give a new and most important argument in favour of our model of the distribution of the sources of cosmical radio-emission. The temperature of the metagalactic emission at $\lambda = 3$ metres thus lies between 100° and 200°.

Optical observations do not permit us to estimate the integral brightness of the Meta-Galaxy, because the faint light from the latter is lost in the much brighter glow of atmospheric, interplanetary and galactic origin. The fundamental cosmological characteristics entering into the theory of the integral photometric effect of the Meta-Galaxy are the average density and the red-shift constant. This demonstrates the decisive importance of radioastronomical methods in cosmology, which was underlined by us in 1953[4]. It was proved (Shklovsky, 1953 [4]) that the phenomenon of the red-shift exists also for the meta-galactic radio-emission. The so-called 'photometric radius of the Meta-Galaxy', i.e. the radius of an imaginary sphere of constant density, in which the radio sources (similar to those in the surroundings of our Galaxy), when uniformly distributed, give a radio intensity equal to the observed, was estimated. This radius was found about 500 megaparsecs (on the new scale), which confirms the theoretical calculations, based upon relativistic cosmology.

Radio-emission from several clusters of galaxies and the so-called 'Super-Galaxy' was discovered recently. The question arises whether this radio-emission is caused by all 'radio galaxies' contained in this cluster (the 'discrete' hypothesis) or localized in the intergalactic space within the cluster ('continuous' hypothesis). The 'discrete' hypothesis was until now accepted by the majority of investigators. However, we have shown that this hypothesis meets considerable difficulties (Shklovsky, 1954[5]) and that the 'continuous' hypothesis is more acceptable. If radio-emission is caused by relativistic electrons in magnetic fields, then, according to our estimates, the concentration of the relativistic electrons in metagalactic space should be of the order of 10^{-13} cm.$^{-3}$ and the strength of the magnetic field H $\sim 10^{-7}$ gauss. This also gives us a new aspect for the problem of the origin of cosmic rays.

The inter-galactic medium in clusters of galaxies may also cause the radio line of 21 cm. The frequency of this emission from the given cluster will be strongly shifted and will be concentrated in a wide band of the order of 1 to 2 Mc./s. The brightness temperature may reach several tens of degrees. Special instruments are being developed in the Moscow university for the solution of this most important problem.

Up to 2000 discrete sources of $F_\nu > 7.5 \cdot 10^{-26}$ watt per m.^{2} per herz ($\lambda = 3.7$ metres) are known at present. The majority of them are of a meta-galactic origin. The question arises to what type of radio galaxies the majority of these sources must be identified. What causes the observed radio intensity from the Meta-Galaxy: clusters of galaxies or discrete sources? The total flux emitted by all the known discrete sources equals about $5 \cdot 10^{-22}$ watt per m.^{2} per herz, which corresponds to an average brightness temperature of about $15°$ ($\lambda = 3.7$ metres), or about 5 % of the total brightness of the Meta-Galaxy.

If these faintest sources should be similar to Cygnus A then it may be found by means of methods used in stellar astronomy that their spatial density should be 10^8 times smaller than the density of late-type spirals. The mean distance between them should be of the order of 300 megaparsecs and the faintest of them should be located at an enormously large distance of $4 \cdot 10^9$ parsecs. However, we did not take into account the red shift, which will cut off the emission of the objects, located at distances over 800 megaparsecs, rather strongly. Consequently, unless recourse is taken to very special cosmological models, the faintest of the known sources can hardly be identified with objects like Cygnus A. If the sources should be identified with objects like NGC 4486 then their spatial density should constitute about 1/100 of the spatial density of late spirals. The sources with $F_\nu = 7 \cdot 5 \times 10^{-26}$ watt per m.² per herz should then be located at distances of about 80 megaparsecs. But while the 'photometric radius' of the Meta-Galaxy equals 500 megaparsecs, these sources, located in different parts of the Meta-Galaxy, will produce only about 35% of the observed surface brightness of the Meta-Galaxy. This may signify that the brightness of the Meta-Galaxy is determined to a considerable extent by the integral effect of clusters of galaxies.

Finally, if the faintest known sources are identified with radio galaxies of comparatively low luminosity of the type of NGC 5128 or NGC 1316, the partial density in respect to the late spirals must be about 1/10. The main part of the observed brightness of the Meta-Galaxy can be explained by the integral radiation from these sources.

In this connexion attention may be paid to the fact that from seventeen galaxies brighter than 10ᵐ, three (NGC 5128, 4486 and 1316) are radio galaxies. This may signify that these three belong to a class of objects met comparatively often in the Meta-Galaxy. If the majority of faint discrete sources are considered similar to NGC 5128 and NGC 1316, then these sources should be identified with galaxies of about 14 to 15ᵐ. If the faint discrete sources are caused by objects of the type of NGC 4486 they must be identified with galaxies of about 16ᵐ. The total number of such galaxies in the sky is about 10⁵. Thus, the problem of identification of metagalactic discrete sources with galaxies becomes extremely important.

REFERENCES

[1] Shklovsky, I. S. *A.J. U.S.S.R.* **29**, 418, 1952.
[2] Westerhout, G. and Oort, J. H. *B.A.N.* **11**, 323, no. 426, 1951.
[3] Baldwin, J. *Nature*, **74**, 320, 1954.
[4] Shklovsky, I. S. *A.J. U.S.S.R.* **30**, 495, 1953.
[5] Shklovsky, I. S. *A.J. U.S.S.R.* **31**, 533, 1954.

COMPARISON BETWEEN RADIO AND OPTICAL SURFACE BRIGHTNESS DISTRIBUTIONS IN THE MAGELLANIC CLOUDS

G. DE VAUCOULEURS

Yale-Columbia Southern Station, Mount Stromlo, Canberra, Australia

I

The distribution of radio surface brightness in the Large Cloud has been measured at 3·5 metres by Mills [1], and in the 21-cm. line by Kerr, Hindman and Robinson [2], in Sydney. A comparison with the distribution of the optical surface brightness in red light ($\lambda\lambda$ 5600–6600) [7] and with the surface density of stars brighter than $m_{pg} = 14\cdot0$ ($M_{pg} = -4\cdot7$) [3], determined by the author at Mount Stromlo, is made in this paper.

Fig. 1 shows the observed distributions of red light (i.e. faint stars), bright stars, neutral hydrogen and 3·5 metres radiation.* At 21 cm. the velocity separation makes it easy, at least in principle, to separate-out genuine Cloud features from the foreground galactic radiation. The distinction is still fairly definite in the photometry in red light where the unresolved background radiation, measured on small-scale photographs between the brighter stars ($m < 11$) and corrected for a smooth galactic gradient across the field, must refer mostly to the Cloud. For the bright-star counts confusion by the galactic field stars becomes serious and it is often difficult to decide whether an excess of star density in the outer parts of the field indicates a significant Cloud structure or merely a fluctuation in the foreground. At 3·5 metres the large irregularities of the background and calibration difficulties make most uncertain the separation of the radiation associated with the Cloud. In all cases a rather conservative attitude has therefore been adopted in determining the extent of the excess brightness (or density) associated with the Cloud. The accuracy and reliability of the data is probably best for light and poorest for 3·5 metres, with hydrogen and bright stars intermediate; but all the present data should be considered as preliminary and subject to revision.

* The 21-cm. map differs slightly from that originally published as it includes some subsequent minor revisions of the data.

For comparison with the radio contours the optical distributions were in each case smoothed with the appropriate reception diagram of the radio-telescopes whose beam-widths between half-power points are $0°8 \times 1°0$ (at 3·5 metres) and $1°5 \times 1°5$ (at 21 cm.). Fig. 2 shows by comparison with Fig. 1 the effect of smoothing with the 21-cm. beam. A key map [3] of the major structural features of the Cloud as shown by direct photography is added for reference in Fig. 4. The complexity or amount of structure shown by the various elements, after smoothing, is greatest for the bright stars, least for 3·5 metres and light, with 21 cm. intermediate. The different methods of reduction used tend, however, to exaggerate the details of the 21-cm. picture and to smooth out any 3·5 metres fine structure.

Bearing this in mind, Figs. 1 and 2 indicate the main characters of the *absolute* distributions as follows:

(1) The faint stars are very strongly concentrated in the axial 'bar' and only moderately so in the spiral arms, especially in the inner, asymmetrical arm near $5^h 30^m$, $-67°$. There is also some concentration in the bright region near 30 Doradus ($5^h 40^m$, $-69°$).

(2) The bright stars are very strongly concentrated in the spiral arms, especially the inner, asymmetrical arm, and in the bright region around and preceding 30 Doradus. Their near absence in the axial bar is a conspicuous and probably very significant phenomenon [4].

(3) Neutral hydrogen shows a pronounced concentration in the bar and in the 30 Doradus area; there are also indications of relatively high density in the inner, asymmetrical arm.

(4) The 3·5 metres radiation has a sharp maximum around 30 Doradus, superimposed on a smooth distribution which bears little or no resemblance to other structural details of the Cloud. There is definitely no concentration in the bar.

3

In order to study more closely the *relative* distributions, contour maps showing the variations of the ratio of each element (with the appropriate smoothing) to each of the three others were established and compared with the main types of structure found in the various parts of the Cloud. By analogy with the usual colour index this ratio was expressed in magnitudes with an arbitrary zero point. Only the variations of the ratio are of interest here. Contours of constant ratio index at half-magnitude intervals have been drawn for a number of combinations. The accuracy of the data does not warrant closer intervals. The most interesting example is reproduced in Fig. 3.

The main points which emerge from the comparisons are as follows:

(1) *Ratio of* 21 *cm. to light* (Fig. 3). There is a large excess of red light compared to neutral hydrogen in the core. As reference to Figs. 1 and 2 indicates, this is due to the fact that although both elements are markedly

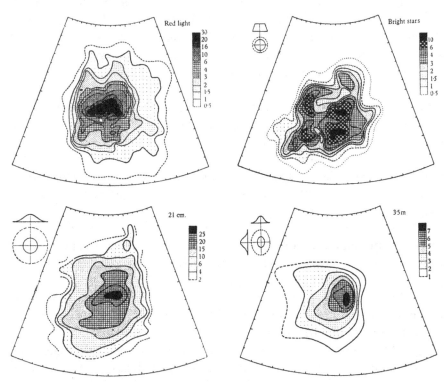

Fig. 1. Observed distributions of red light, bright stars ($m \leqslant 14$), neutral hydrogen (21 cm.) and 3·5 metres radiation in the Large Cloud. Intensities are on arbitrary scales. The beam profiles and diffraction disks are shown on the same scale.

concentrated in the axial bar, the degree of concentration is much higher for red light, i.e. faint stars, than for neutral hydrogen. Another area of excess light relative to hydrogen is in the following section of the Cloud, along R.A. 6^h 20^m, where it is due mostly to the sharp decrease of hydrogen density in this direction—a check on the original data seems to confirm the reality of this hydrogen deficiency in the weak outer spiral arm or loop.

On the contrary there is an excess of hydrogen compared to light in the preceding section of the Cloud, especially near 4^h 40^m, $-67°$, but this

246

may be apparent only as there are signs of local (foreground) obscuration in this area. Hydrogen is also in excess relative to light in the regions occupied by the inner spiral arms, in particular near $5^h 30^m$, $-67°$; that is, in these spiral arms there is a greater degree of concentration for the gas than for the dwarf stars. Further, hydrogen seems to be also in relative

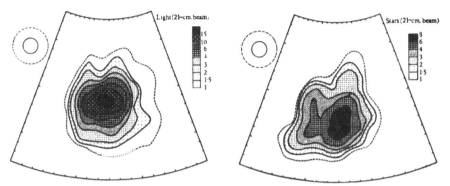

Fig. 2. Contours of the distribution of red light and of bright stars smoothed with an aerial beam of $1°5$ diameter in order to make them comparable to the observed radio isophotes shown in Fig. 1.

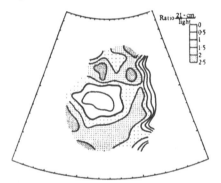 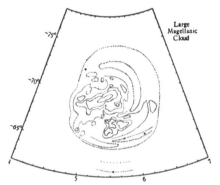

Fig. 3. Sample contour map of ratio index. This map shows the ratio of neutral hydrogen (21 cm.) to faint stars (red light).

Fig. 4. Key map showing the main structure of the Large Magellanic Cloud appearing on photographs.

excess in the dark inter-arm spaces, for example near $5^h 15^m$, $-67°$ and $5^h 30^m$, $-73°$, where there is some evidence for the presence of dark matter in the Cloud [5].

(2) *Ratio of 21 cm. to bright stars.* There is a pronounced excess of hydrogen relative to bright stars in the axial bar and to the south of it which, as shown by Figs. 1 and 2, is due to both a real concentration of hydrogen and a marked deficiency of bright stars in this section. Conversely the bright stars are in excess in the inner spiral arms which they

delineate; that is, the concentration of hydrogen in these spiral arms is much less marked than that of supergiant stars. In the outer spiral arm both hydrogen and bright stars seem to be rare or absent.

(3) *Ratio of bright stars to light.* There is a very large excess of light, that is faint stars, relative to bright stars in the axial bar; this results from the combined effect of the concentration of faint stars and the deficiency of bright stars in this section. Conversely there is a marked excess of bright stars relative to faint stars in the inner, asymmetrical spiral arm and to a lesser extent in the other sections of the inner spiral structure.

(4) *Ratio of 3·5 metres to light.* There is a large excess of light in the core where the 3·5 metres radiation is not concentrated, and in the outer loop where reference to the original records seems to confirm the near absence of radio emission. There is a slight relative excess of 3·5 metres radiation north and south of the core and a greater excess in the preceding section, along R.A. 4^h 40^m, but the radio isophotes are rather uncertain in this section and optical obscuration may also be present. On the whole the relative distribution is to a large extent a reflexion of the optical distribution with which the radio distribution shows little correlation.

(5) *Ratio of 3·5 metres to bright stars.* The relative distribution picture is essentially a reflexion of the bright-star distribution. Hence it is probably of little significance.

(6) *Ratio of 3·5 metres to 21 cm.* There is a slight excess of the 21-cm. radiation in the bar and in the sections occupied by the inner spiral structure; this is mainly due to the real concentration of neutral hydrogen in these regions, but in the bar a slight deficiency of 3·5 metres radiation may also be a factor. There is a greater relative excess of 21-cm. radiation in the following section, beyond R.A. 6^h 00^m, that is, in the outer loop, where both 21 cm. and 3·5 metres are weak compared with light, but apparently 3·5 metres drops off more sharply than 21 cm. in this direction. The 3·5 metres radiation is in excess in the 30 Doradus area as could be expected from the higher degree of ionization in the giant H II region.

4

The main results of the comparisons in sections (2) and (3) may be summarized by the following table.

Scale: ** very strong concentration; * strong concentration; o some concentration; † element present, but not conspicuously concentrated; †† element very weak or absent.

The degree of relative variability, as measured by the range of the ratio index, is lowest between 21 cm. and 3·5 metres (range 1 mag.) and highest between bright stars and light (range 3 mag.) with bright stars/3·5 metres, bright stars/21 cm., light/21 cm. and light/3·5 metres intermediate.

Region in L.M.C.	Light	Bright stars	21 cm.	3·5 metres
Bright core (axial bar)	**	††	*	†
Bright region around 30 Dor.	*	**	*	**
Bright inner spiral arms	*	**	*	o
Dark inter-arm spaces	†	††	*	o
Faint outer arms	*	†	†	††

The effective radii defined by $A = \pi r_e^2$, where A is the area of the (smoothed) isophote enclosing half the total energy (or numbers of stars), are listed below for both the 3·5 metres and 21-cm. beams:

	Beam	Light	Bright stars	21 cm.	3·5 metres
Effective	(3·5 metres)	$2°6$	$2°7$	—	$3°0$
Radii r_e	(21 cm.)	$2°8$	$2°9$	$3°1$	—

Hence the gas and the source of the 3·5 metres radiation are somewhat less concentrated than either light (faint stars) or bright stars.

5

These results taken as a whole tend to indicate that the space distribution of the source of the 3·5 metres radiation has, except in the H II regions, little in common with the distribution of stellar or interstellar matter. Hence the hypothesis of its originating in a tenuous electronic corona surrounding the main body of the stellar system [2, 6] is not inconsistent with the present evidence.

The existence of a definite concentration of neutral hydrogen in the bar where bright supergiant stars are rare or absent should be regarded as significant. It may be noted that, while ordinary giants (shown by star counts beyond $m = 14$) and dwarfs (as indicated by the red light distribution) are highly concentrated in the bar, dust clouds are rare and well localized in it.

In the inner spiral arms where supergiants are most abundant, there is a definite concentration of both faint stars and neutral hydrogen, but it is more pronounced for the latter which also seems concentrated in some dark (obscured) inter-arm sections.

In the faint outer arms both the brighter supergiants and the gas seem to be rare or absent. The same appears to be true for the outermost extensions, in particular for the 'anti-galactic' arm [3]. This may have evolutionary significance.

In the Small Cloud the data do not warrant as yet detailed comparisons, but if one allows for the smaller size, lower intensity, distortion and tilt of the system, it appears that the same elements bear the same general relations to each other as in the Large Cloud [1].

The generous co-operation and assistance given by Messrs F. J. Kerr and B. Y. Mills in supplying the radio data and in their discussion is gratefully acknowledged.

REFERENCES

[1] Mills, B. Y. *Aust. J. Phys.* **8**, 368, 1955.
[2] Kerr, F. J., Hindman, J. V. and Robinson, B. J. *Aust. J. Phys.* **7**, 297, 1954.
[3] Vaucouleurs, G. de. *A.J.* **60**, 126, 1955.
[4] Vaucouleurs, G. de. *Irish Astr. J.* **4**, 13, 1956.
[5] Shapley, H. and Nail, V. McK. *Proc. Nat. Ac. Sc., Washington,* **37**, 133, 1951.
[6] Shklovsky, I. S. *A.J. U.S.S.R.* **30**, 15, 1953.
[7] Vaucouleurs, G. de. *A.J.* **62**, 1957 (in the Press); see also Eggen, O. J. and Vaucouleurs, G. de, *P.A.S.P.* **68**, 421, 1956.

PART IV

THE QUIET SUN

THE QUIET AND ACTIVE SUN

INTRODUCTORY LECTURE BY

C. W. ALLEN

University of London Observatory, England

It is both convenient and realistic to regard the solar radio-emission as being composed of a quiet permanent part coming from the whole of the sun's surface, together with a superimposed variable part emitted from centres of activity. The variable part can be further subdivided into various components to which separate reference will be made. The components are almost independent and add to one another to give the solar emission recorded by radio telescopes. For our understanding of the phenomena it would be a great advantage if the components could be completely and reliably segregated but this is not always possible. The results from a wide network of observational stations are recorded in the *Quarterly Bulletin of Solar Activity*, from which the intensities and characteristics of most of the components may be extracted.

I. QUIET SUN

Radiation from the quiet sun can be detected, and is to be expected, at all usable radio wave-lengths. Observations plotted in Fig. 1 show that the probable errors of smoothed intensities are less than 10 % at most wave-lengths. Fig. 1 also shows that the intensity measurements for frequencies below 100 Mc./s. are not yet satisfactory and it is uncertain whether there is a maximum of the apparent temperature T_a near this frequency.

The measurement of the change in quiet sun radiation throughout the sunspot cycle is complicated by the superimposed active components, particularly the slowly varying component (Piddington and Davies, 1953 [1]; Christiansen and Hindman, 1951 [2]; Waldmeier, 1955 [3]). The corrected data for the curves T_a against frequency f are shown in Fig. 1 for both sunspot maximum and sunspot minimum. The sunspot maximum data are from Allen (1951) [4] and sunspot minimum data are extracted from the *Quarterly Bulletin of Solar Activity* for 1953 and 1954. We find that the curves are nearly parallel with a shift of 0·17 in log f. This is the type of change to be expected if the coronal distribution and temperature remained constant during the cycle but the electron density varied by a factor of 1·5. This is smaller than the factor 1·8 derived by van de Hulst (1950) [5] from brightness variation. At frequencies near 100 Mc./s. the values of T_a should be closely related to the coronal temperature, which we would judge from Fig. 1 to drop 10 or 20 % from sunspot maximum to sunspot minimum. This is in quite good agreement with Waldmeier's (1952) [6] estimation of temperature change. It is interesting to see, however, that the minimum intensity on metre waves comes well before sunspot minimum. In Fig. 2 we plot the third lowest recorded intensity (avoiding the

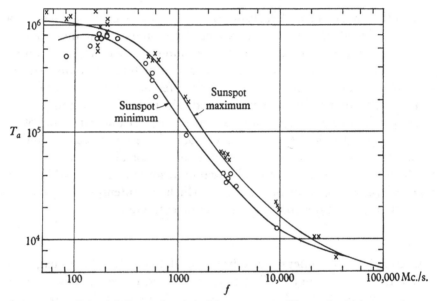

Fig. 1. Values of T_a as a function of frequency. ×, Observations at sunspot maximum. ○, Observations at sunspot minimum.

very lowest for reasons of statistical stability) in each month from 1951 to 1954. Although the sunspot minimum is about May 1954 there is a distinct increase of intensity during 1954 on nearly all metre-wave recordings. Fig. 2 would suggest that the minimum of coronal temperature was about a year before sunspot minimum. Line intensity measurements, on the other hand (Waldmeier, 1952 [6]), indicate a polar temperature minimum one year before sunspot minimum and an equatorial temperature one year after sunspot minimum.

In recent years considerable efforts have been made to determine the distribution of intensity across and beyond the sun's disk. The conclusion appears to be that eclipses alone do not usually give definitive results although they provide useful information. On the other hand in favourable cases interferometric measurements can be analyzed to give a complete curve of T_e, the effective temperature, against r, the radial distance from the sun's centre in terms of the sun's radius. From the curves published (Christiansen and Warburton, 1955 [7]; O'Brien, 1953 [8], 1955 [9]; Covington and Broten, 1954 [10]; Aoki, 1953 [11]; Alan, Arsac and

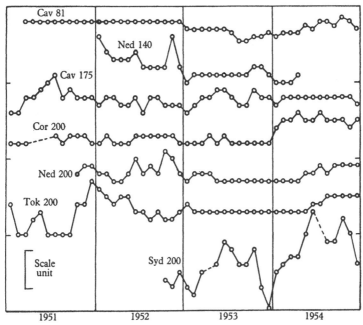

Fig. 2. Variation of metre-wave flux near sunspot minimum. Symbols representing the various observing stations are from the *Quarterly Bulletin of Solar Activity*. The scale unit is 1×10^{-22} w.m.$^{-2}$ (c./s.)$^{-1}$ for Cav 81 and 5×10^{-22} w.m.$^{-2}$ (c./s.)$^{-1}$ for the other stations.

Steinberg, 1953 [12]; Priester and Dröge, 1955 [13]) I have extracted, when possible, two factors which seem best to express the nature of the curves without giving too much detail.

The first factor is T_c/T_a, where T_c is the value of T_e at the centre of the sun's disk. The distribution measurements have been converted to T_c/T_a and are shown in Fig. 3 b. When measurements in the equatorial direction differ from the polar direction a mean has been taken. However, there is a general tendency to use an equatorial run for the observations and then to assume circular symmetry for interpreting the run of T_e. This tends to give too small a value of T_c/T_a. Even if allowance is made for the observa-

255

tional errors there appear to be some significant differences between the observations and the theoretical calculations of Smerd (1950) [14] (using a 10⁶ ° K. model), Reule (1952) [15] (isothermal corona and $Q\Theta = 3$), and Hagen (1951) [16] (special model). The reason why Reule's results are in

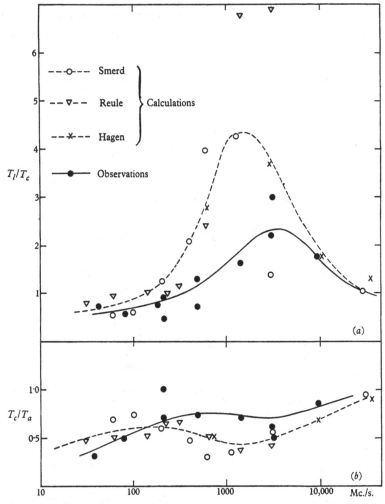

Fig. 3. Comparison of calculated and observed parameters characterizing brightness distribution across the sun. Upper graph: ratio of limb temperature to centre temperature. Lower graph ratio of centre temperature to apparent temperature, defined as average over one solar disk.

better agreement with observations than Smerd's is that we chose $Q\Theta = 3$ for Reule and effectively 1 for Smerd. $Q = (\overline{N_e^2})^{\frac{1}{2}}/\overline{N_e}$ (with N_e the electron density and the bar signifying the mean over a large volume) is a measure of the inhomogeneity of the corona, and $\Theta = (10^6/T)^{\frac{3}{2}}$. $T = 700,000°$ K. and $Q = 3$ give about the best fit. The fact that Hagen's calculations fit into those of Reule can only show how insensitive the calculations are to the model used. The Reule and Hagen models differ considerably.

Another factor which gives a rather better indication of the limb brightening is T_l/T_c, where T_l is the peak value of T_e *near* the limb—it is taken precisely *at* the limb if there is no evidence of limb brightening. Observations of T_l/T_c are shown in Fig. 3 a and compared again with the calculations by Smerd, Reule and Hagen. The smallness of the observed ratio as compared with theory can easily be ascribed to instrumental resolution. The Reule calculations with $Q\Theta = 3$ are again better than Smerd (with $Q\Theta = 1$) up to 1000 Mc./s. The Hagen results give reasonable agreement at high frequencies.

Table 1. *Observed T_a and T_c*

		T_a			T_c	
λ (cm.)	f (Mc./s.)	Sunspot maximum (° K.)	Sunspot minimum (° K.)	T_c/T_a	Sunspot maximum (° K.)	Sunspot minimum (° K.)
600	50	1,100,000	600,000	0·42	460,000	250,000
300	100	1,000,000	800,000	0·55	550,000	440,000
150	200	900,000	770,000	0·67	600,000	520,000
60	500	580,000	350,000	0·75	440,000	260,000
30	1,000	230,000	130,000	0·71	163,000	92,000
15	2,000	90,000	59,000	0·69	62,000	41,000
6	5,000	31,000	22,000	0·75	23,000	16,500
3	10,000	17,000	12,000	0·83	14,100	10,000
1·5	20,000	10,000	9,000	0·92	9,200	8,300
0·6	50,000	6,400	6,400	0·99	6,300	6,300

Many attempts have been made to deduce the structure of the sun's chromosphere and corona with the help of radio observations. The most useful data for such purposes are the curves for T_c as a function of radio frequency. The data compiled in this article are given in Table 1. Unfortunately there are no data from which we could find the change of T_c/T_a with the solar cycle and it has been assumed not to vary. In Fig. 4 the observed T_c is plotted against f at sunspot maximum and sunspot minimum and compared with some calculations from coronal models (Woolley and Allen, 1950 [17]; Hagen, 1951 [16]; Piddington, 1954 [18]; and Woltjer, 1954 [19]). This diagram illustrates the difficulty of deriving a model atmosphere directly from radio observations. The Piddington and Hagen models have a smooth temperature change from chromosphere to corona, the Woolley and Allen model has a sudden change, while

Woltjer's model has two sets of conditions at each level. All are in fair agreement with the observations.

It should be possible to estimate the value of QN_e for the outer corona from metre-wave flux measurements at distances 2 or 3 radii from the sun. The measurements available at present (O'Brien, 1953, 1955) [8,9] show rather large variations which are probably experimental. However, the tendency is to show greater intensity than calculations based on the van de Hulst (1950) [5] model, and suggest either that outer electron densities are higher than the model or that the effective value of Q is high.

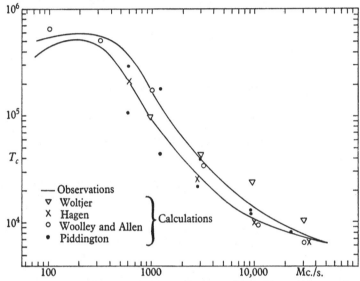

Fig. 4. Comparison of observed and calculated values of T_c. Observations are from Table 1. Calculations are from Hagen (1951), Woolley and Allen (1950) with coronal temperature 700,000° K., Piddington (1954) for sunspot maximum and sunspot minimum, and Woltjer (1954).

2. SLOWLY VARYING COMPONENT

The monthly and daily variations of this component and eclipse evidence all show clearly that it is emitted from the neighbourhood of sunspots. The steadiness of the radiation indicates that it is probably thermal, and the size of the emission areas as determined at eclipses leads to local temperatures of about 5×10^6 ° K. in agreement with coronal active area temperatures from Doppler broadening.

In order to segregate the component one may use the method of plotting daily radio flux against sunspot area or number and determining the mean linear relation (Christiansen and Hindman, 1951 [2]; Covington and Medd, 1954 [20]). Since emission from various centres of activity must be additive the relation must be linear unless it is dominated by an unusually active sunspot. The difficulty is to allow for the delayed action of the radio emission as compared with the visible sunspots (Piddington and Davies, 1953 [1]; Waldmeier, 1955 [3]). It is important to discriminate between radiation from centres of activity which are no longer attended by sunspots and quiet radiation from the whole sun that is in excess of the zero-sunspot quiet sun. Waldmeier's recent analysis comes close to this but would be improved by the inclusion of the sunspot minimum.

The explanation of the slowly varying component as the thermal radiation from a hot coronal condensation gives reasonable agreement with the observations. However, the proposed condensations (e.g. Waldmeier and Müller, 1950) [21] appear rather too condensed to fit observations of the white corona (Lyot, see Kiepenheuer, 1953) [22], and they would not be in pressure equilibrium.

3. ENHANCED RADIATION (NOISE STORMS)

Enhanced radiation is so variable that some of its characteristics are difficult to measure and express. The *Quarterly Bulletin* tabulations of median flux, variability, and polarization are concerned mainly with enhanced radiation and are an attempt to overcome these difficulties. At metre wave-lengths the median flux is practically a measure of noise-storm emission; the variability can express either the general activity or (when the flux is high) the ratio of emission of individual bursts to the steadier component; and the polarization expresses the sense of the magnetic field in the neighbourhood of the source.

Noise-storm emission can be located on the sun by the 'swept lobe' method (Payne-Scott and Little, 1951) [23] and is always found to come from the neighbourhood of large sunspot groups. There is, however, conflicting evidence on the question: which groups produce noise? Thus Payne-Scott and Little (1951) [23] find all groups with a single spot greater than 400 millionths of the sun's disk (i.e. surface magnetic field > 2000 gauss) produce noise, while Hatanaka and Moriyama (1952) [24] and Owren (1954) [25] find all storms are preceded by an outburst. These two findings are not entirely compatible since flares are not regularly associated with large sunspots. Becker and Denisse (1954) [26] find strong evidence that magnetic storm particles are emitted from sunspots that emit storm noise and not others. This seems at first sight to agree well with the prevalent belief that particle emission is associated with flares. However, Simon

(1955) [27] has shown fairly clearly that flares have very little to do with particle emission. Apparently there are three types of particle emissions: (*a*) those associated with great flares and causing great storms; (*b*) those associated with enhanced radiation and presumably far more frequent than (*a*); and (*c*) the recurrent streams from M regions which avoid sunspots.

4. OUTBURSTS

The *Quarterly Bulletin* tabulations of Outstanding Occurrences are designed mainly for outbursts. These should usually carry the type CD. It would be most helpful if the type designation could make it clear whether the observer regards the occurrence as an outburst or not. A fairly elaborate classification of outburst type has been made by Dodson, Hedeman and Owren (1953) [28] on the basis of 200 Mc./s. radiation but, as emphasized by Smerd (1954) [29], single-frequency recordings do not give a unique discrimination between types. The realization that outbursts are sometimes (if not always) the cause of noise storm commencement makes the segregation between these two components more difficult.

5. NON-POLARIZED BURSTS

It is difficult to recognize non-polarized bursts from single-frequency recordings. Their occurrence is often sporadic but also they occur in clusters, sometimes introducing an outburst. These points make it difficult to systematize non-polarized burst information which is therefore not available in the *Quarterly Bulletin*.

Two questions are prompted by these phenomena. Firstly, is all non-thermal solar radio-emission in the form of bursts and if so what are the physical differences between the polarized and non-polarized types? Secondly, are the double-hump bursts caused by a reflexion or are they harmonics?

6. ABSORPTION PHENOMENA

Since the absorption of radio waves by cosmic material is well understood, any clear indications of absorption may lead to useful conclusions. Three examples of solar radio absorption phenomena have been reported up to the present. (*a*) Observations of the excess outburst-flare association on the east side of the sun have been interpreted by Hey and Hughes (1955) [30] to indicate absorbing material ejected from flares. (*b*) Observations of the disappearance of the Taurus source as it approaches the sun (Hewish, 1955) [31] give an indication of the electron density in the outer corona. However, the effect is one of refraction rather than absorption and indicates only the maximum electron density along the line of sight. (c) Covington and Dodson (1953) [32] have found that decimetre-wave radiation can be decreased by absorption (probably of a dark floculus) after certain outbursts.

REFERENCES

[1] Piddington, J. H. and Davies, R. D. *M.N.R.A.S.* **113**, 582, 1953.
[2] Christiansen, W. N. and Hindman, J. V. *Nature*, **167**, 635, 1951.
[3] Waldmeier, M. *Z. Ap.* **36**, 181, 1955.
[4] Allen, C. W. *7th report of Commission on Sol. Terr. Relations*, p. 63, 1951.
[5] van de Hulst, H. C. *B.A.N.* **11**, 135, no. 410, 1950.
[6] Waldmeier, M. *Z. Ap.* **30**, 137, 1952.
[7] Christiansen, W. N. and Warburton, J. A. *Observatory*, **75**, 9, 1955.
[8] O'Brien, P. A. *M.N.R.A.S.* **113**, 597, 1953.
[9] O'Brien, P. A. *Observatory*, **75**, 11, 1955.
[10] Covington, A. E. and Broten, N. W. *Ap. J.* **119**, 569, 1954.
[11] Aoki, K. *Tokyo Astr. Obs. Reprints*, no. 106, 1953.
[12] Alan, I., Arsac, J. and Steinberg, J. L. *C.R.* **237**, 300, 1953.
[13] Priester, W. and Dröge, F. *Z. Ap.* **37**, 132, 1955.
[14] Smerd, S. F. *Aust. J. Sci. Res.* **3**, 34, 1950.
[15] Reule, A. *Z. Naturf.* 7a, 234, 1952; Mitt. Tübingen, no. 3.
[16] Hagen, J. P. *Ap. J.* **113**, 547, 1951.
[17] Woolley, R. v. d. R. and Allen, C. W. *M.N.R.A.S.* **110**, 358, 1950.
[18] Piddington, J. H. *Ap. J.* **119**, 531, 1954.
[19] Woltjer, L. *B.A.N.* **12**, 165, no. 454, 1954.
[20] Covington, A. E. and Medd, W. J. *J. R.A.S. Canada*, **48**, 136, 1954.
[21] Waldmeier, M. and Müller, H. *Z. Ap.* **27**, 58, 1950.
[22] Kiepenheuer, K. O. *The Sun*, ed. Kuiper, see p. 421, 1953.
[23] Payne-Scott, R. and Little, A. G. *Aust. J. Sci. Res.* A, **4**, 508, 1951.
[24] Hatanaka, T. and Moriyama, F. *Tokyo Astr. Ob. Repr.* no. 99, 1952.
[25] Owren, L. *Radio-Astron.* Rep. no. 15, Cornell, 1954.
[26] Becker, U. and Denisse, J. F. *J. At. Terr. Phys.* **5**, 70, 1954.
[27] Simon, P. *C.R.* **240**, 940, 1056, 1192, 1955.
[28] Dodson, H. W., Hedeman, E. R. and Owren, S. *Ap. J.* **118**, 169, 1953.
[29] Smerd, S. F. 'On the distribution of world-wide data on solar radio emission', Mimeographed Notes, July, 1954.
[30] Hey, J. S. and Hughes, V. A. *Nature*, **173**, 771, 1954; *M.N.R.A.S.* **115**, 605, 1955.
[31] Hewish, A. *Proc. Roy. Soc.* A, **228**, 238, 1955.
[32] Covington, A. E. and Dodson, H. W. *J. R.A.S. Canada*, **47**, 207, 1953.

Discussion

Pawsey: In connexion with the resolution of the individual bright-regions which are the source of the 'slowly varying component' Christiansen and Warburton have taken an extended series of observations on 21 cm. with their multiple-element interferometer which has a beam-width of 3 minutes of arc. They find that the majority of these bright regions are resolved. The regions have, typically, an extent of from 3 to 5 minutes of arc, and agree closely in size and position with the associated optical features: Ca or H plages, regions of intense coronal emission, and Babcock's regions of magnetic field. It is expected that this work will be published in the *Australian Journal of Physics*.

Miss Dodson: The individual areas of enhanced radio emission have also been resolved by Covington at 10·3 cm. The agreement with the calcium plages on the sun has been pointed out in the literature (Dodson, H. W. *Ap. J.* **119**, 564, 1954).

THE STRUCTURE OF THE SOLAR CHROMOSPHERE FROM CENTIMETRE-WAVE RADIO OBSERVATIONS

J. P. HAGEN

Naval Research Laboratory, Washington, D.C., U.S.A.

The atmosphere of the sun is transparent to visible radiation, is nearly transparent to millimetre and centimetre radio radiation, and becomes opaque to the metre and longer wave radiation. Information about the chromosphere can then be given by observing the radiation from the sun at short radio wave-lengths. In its outer part, the atmosphere of the sun is highly ionized. Absorption in any region is directly proportional to the square of the density and the wave-length squared and inversely to the temperature to the three-halves power

$$\kappa \propto \frac{n^2 \lambda^2}{T^{\frac{3}{2}}}.$$

This is the familiar equation for the absorption of radio waves in an ionized medium. By consequence of this, the longer wave radiation is absorbed in the outer layers of the sun's atmosphere and can escape only from these outer regions. The shorter wave-length radiation is absorbed very little in the outer part of the solar atmosphere where the density is quite low, and hence radiation from the chromosphere escapes as centimetre and millimetre radio waves. In fact, the principal radiation from the sun in the centimetre and millimetre region comes from the chromosphere.

Since the sun subtends an angle small compared with most radio antenna beam-widths, it is necessary to resort to unusual methods to resolve the sun if the variation in brightness across its surface is to be obtained. At 8 mm. wave-length, this can be done with the N.R.L. 50-ft. antenna which has a beam-width of about 3 minutes of arc at this wave-length. A better way, however, is to observe the sun at the time of a total solar eclipse when the advancing edge of the moon provides the necessary resolution at any wave-length. Eclipses are particularly useful in the centimetre region where the radio diameter of the sun is not too much larger than the visual and where the moon will still effectively eclipse the sun. At metre wave-lengths the radio size of the sun is so large that at totality more than half of the sun is seen around the moon, especially at the longer wave-lengths.

A consideration of early measurements of the radio flux from the sun at many wave-lengths revealed that the radio measurements were an effective means of defining the temperature gradient in the sun's atmo-sphere. The curve marked (1) in Fig. 1 shows the results of such an analysis.

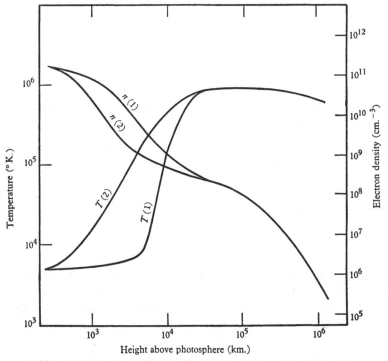

Fig. 1. Temperature and density distributions in the spicules (1) and in the inter-spicule gas (2).

Note that there is a steep temperature rise in the chromosphere. It is one of the consequences of the eclipse measurements that the location of the steep rise in the chromosphere may be better defined. This model of the sun requires limb brightening at all short radio wave-lengths and recent measurements here and abroad have now amply confirmed that limb brightening does exist.

The most effective means in the centimetre region for measuring the radio brightness distribution has been to measure the changing flux through the course of the total eclipse. Our most recent measurement was at

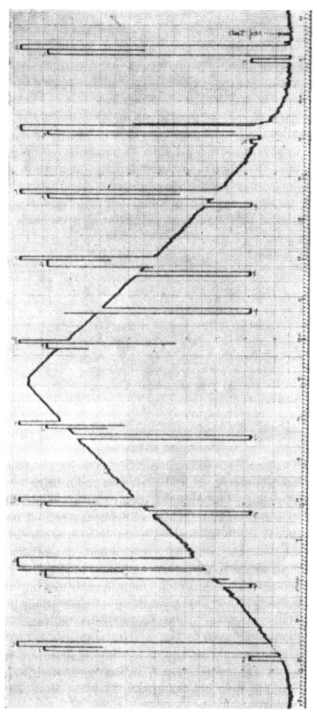

Fig. 2. Record of the solar eclipse obtained at Oskarshamn at $\lambda = 8 \cdot 65$ mm.

Oskarshamn, Sweden, in 1954 where measurements at 8 mm. and 10 cm. wave-length were made. Both measurements improved and confirmed measurements made in Khartoum in 1952. The 10 cm. results show a clear case of limb brightening and reduce to a model atmosphere similar to those of Fig. 1 with a temperature rise occurring near that of Model (1) of that figure.

I will describe the 8 mm. results, obtained with Gibson, Coates and McEwan, in somewhat more detail since they pertain to the most important aspect of my talk. The antenna used at 8 mm. had an aperture of 16 ft. × 2 ft. producing a fan beam 1° wide in one dimension and 8 minutes wide in the other. The antenna beam was so oriented that the elliptically shaped cross-section was centred on the sun and aligned so that the path of the moon

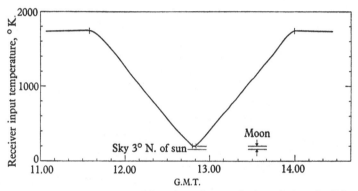

Fig. 3. Eclipse curve corrected for gain and atmospheric variations, λ=8·65 mm., Oskarshamn, 30 June 1954.

was along the major axis of the pattern. In this way we watched a central slice of the sun being eclipsed and derived a radio brightness curve for the equatorial region. The antenna consists of a cylindrical parabola, 16 ft. × 2 ft., which illuminates along a line focus a second smaller cylindrical parabola 2 ft. × ¼ in. in cross-section; this in turn brings the radiation to a focus at a point. This antenna was mounted equatorially and driven by a synchronous motor to follow the motion of the sun.

With this equipment, the flux of the sun was measured during the entire course of the eclipse. The record was obtained on a recording milliammeter and the raw data are shown in Fig. 2. In this curve, you will see many breaks for calibrations of sky temperature, the temperature of a calibrating hot load, and ambient temperature. While the weather was quite cloudy at Oskarshamn, it is apparent even from this curve that there was little effect on the radio data by the clouds.

This curve has been corrected for variations in receiver gain, for deviations from linearity, for effects due to antenna beam shape and for atmospheric absorption. In its corrected form it appears as Fig. 3. It is to be noted that there is a very small residual radiation at totality, less than half of 1 % of the total flux of the sun, and in addition, that at totality the principal radiation entering the receiver appears to be thermal radiation from the moon. Earlier measurements on the moon have allowed this radiation to be evaluated.

The radio brightness distribution derived from this 8 mm. curve differs markedly from that predicted on the basis of the simpler model. The model shows a relatively flat distribution with a narrow bright ring at the edge of the disk. The derived curve, Fig. 4, shows limb brightening but in

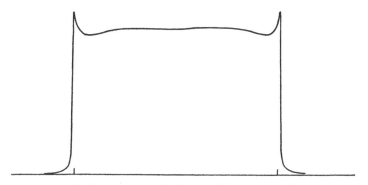

Fig. 4. Radial brightness distribution of the sun at $\lambda = 8.5$ mm.

addition shows a pronounced amount of brightness at the centre. A model of the chromosphere that could produce this effect has been constructed and this is what I would like to report today.

It was earlier suggested by Giovanelli that the chromosphere may not be uniform but may have a grass-like structure. Roberts and others have observed spicules extending through the chromosphere. I have assumed that the chromosphere is made up of spicules blending into a gas which smoothly joins the photosphere with the corona. The spicules would have a random or gaussian distribution about a given size and would also be random in duration and occurrence. For simplicity of analysis, it is assumed that all spicules are pyramidal in shape, have the same base width, 3 seconds of arc, and have the same height, 10,000 km., which is assumed to be the average height of all spicules. These dimensions were arrived at by trial and error to obtain a fit with the radio radial brightness distribution curves. The derived temperatures and pressure gradients in the spicules (1) and in the inter-spicule material (2) are shown in Fig. 1. It is seen that the spicules are cooler, and hence more dense than the ambient gas. Equality of pressure across the face of the spicules was assumed. Schematically the chromosphere will appear as shown in Fig. 5.

Qualitatively, one can see that such a model will fit the radio results. At millimetre waves where the penetration is deep, the central ray will emerge from the region high in the spicule and low in the inter-spicule material. The equivalent temperature will be the average of the two and will be high. For the ray emerging nearer the limb, the spicules will be

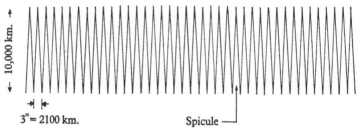

Fig. 5. Schematic shape of the spicules.

seen more nearly face-on and since their absorption coefficient is higher than that for the inter-spicule material they will contribute more to the escaping radiation than will the inter-spicule material. Hence, the equivalent temperature of this ray will be lower. In this way centre-brightening is seen. At the extreme limb the escaping ray must pass tangentially through the atmosphere and absorption will occur at greater heights where temperatures are higher and thus limb brightening will be observed. In this way the radial brightness distribution of Fig. 4 can be reconstructed from the spicule model. At the longer centimetre wavelengths, where absorption and radiation by the corona plays a larger part, the penetration is such that the spicules will play little part and one would expect strong limb brightening with only a mild centre-brightening at 10 cm. wave-length.

It is thus seen that radio measurements predict a two-fluid chromosphere with cooler spicules surrounded by a more tenuous, hotter gas.

OBSERVATION OF THE SOLAR ECLIPSE OF 30 JUNE 1954 AT 9·4 CM. WAVE-LENGTH

C. H. MAYER, R. M. SLOANAKER AND J. P. HAGEN

Naval Research Laboratory, Washington, D.C., U.S.A.

The observation of the total solar eclipse of 30 June 1954 at 9·4 cm. wave-length was one of the experiments carried out by the Naval Research Laboratory eclipse expedition to Sweden. The apparatus was set up in the city of Oskarshamn at a point about 5 km. north of the centre-line of totality. At this location the eclipse was total, with magnitude 1·035, at $12^h 48^m$ U.T.

A Dicke-type radiometer using a paraboloidal reflector 1·83 metres in diameter was mounted on a polar axis and motor driven to track the sun during the eclipse. The angular width of the antenna pencil-beam was about 3°5. The superheterodyne receiver had an average noise factor of 7 and the band-width of the intermediate-frequency amplifier was 5·5 Mc./s. The power from the antenna was interrupted thirty times a second with an absorber and the resulting modulation was detected in a coherent detector circuit. The output time constant was about one second. For the eclipse measurement, the modulation-frequency output was amplified and detected in two separate channels. The output of each channel was displayed by two recording meters. Four simultaneous, semi-independent records of the receiver output were obtained.

During three weeks preceding the eclipse the characteristics of the apparatus and the operating conditions at the site were evaluated. On the day of the eclipse, a set measurement procedure was followed from one hour before first optical contact until one hour past fourth optical contact. The radiation from the sun was monitored continuously except for intervals of about two minutes duration when reference points were inserted into the output records. The reference points were of four types: the output when a thermal noise source at about 400° K. was substituted for the antenna, the output when a thermal noise source at about 295° K. was substituted for the antenna, and the outputs when the antenna was pointed to declinations 8° above and below that of the sun. The four reference-point measurements were repeated at 15 or 20 minute intervals and were spaced from each other by 2-minute measurements of the radiation from the sun.

The information obtained during the eclipse is illustrated in Fig. 1, which shows the data from one of the recorders. The variation in the gain of the receiver was calculated from the measurements of the two thermal noise sources and is plotted in the upper box. The uncertainty in the gain measurement is of this same order. The receiver output information is plotted in the lower box. The heavy line represents the uncorrected output meter readings when the antenna was pointed at the sun. The measure-

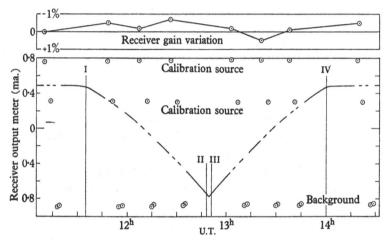

Fig. 1. Total eclipse of 30 June 1954 at 9·4 cm. wave-length (uncorrected data—recorder no. 1).

ments of the thermal noise sources and of the background level above and below the sun are enclosed in circles. The data through which the eclipse curve was plotted are smooth and symmetrical to within 1 % of the uneclipsed solar flux level.

The data from the four simultaneous records of the eclipse agreed to less than 1 %. The radiation from the sun as measured by all four recorders is plotted in Fig. 2. The two halves of the eclipse curve have been normalized separately and superimposed. The dots represent points from the first half of the eclipse curve and the crosses the second half. It was not possible to represent all the points with strict accuracy, but a good indication is given of the agreement between the four records and the smoothness of the change in solar flux during the eclipse. The solar radio radiation decreased smoothly during the eclipse to a minimum value of about 7 % of the flux measured before and after the eclipse.

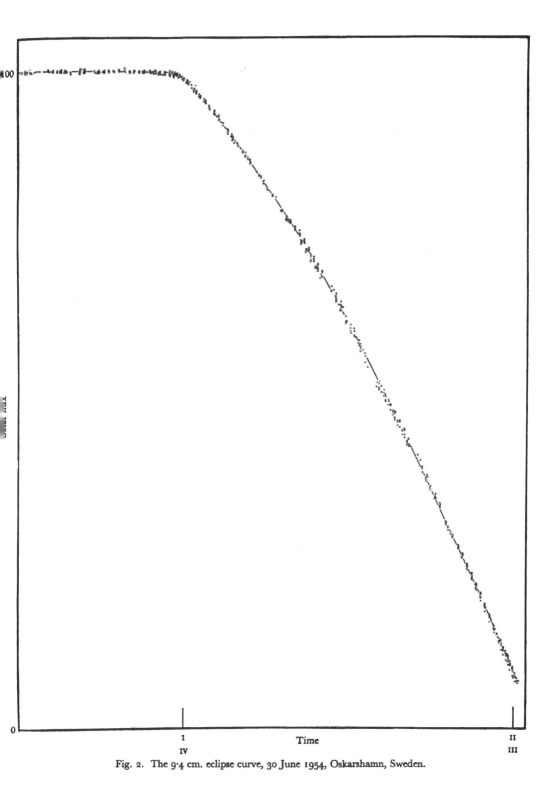

Fig. 2. The 9·4 cm. eclipse curve, 30 June 1954, Oskarshamn, Sweden.

The smoothness and symmetry of the eclipse curve indicate that no appreciable part of the radiation from the sun originated in a localized region emitting intense radio radiation. This indication of a very low level of solar activity is supported by the absence of active regions on the visible hemisphere of the sun, as reported by the solar observatories, and by the measurements of solar radio radiation. The low level of solar activity increases the value of a measurement of this eclipse as a control over models of the undisturbed solar chromosphere.

A radial brightness distribution for the sun at 9·4 cm. wave-length has been calculated from the eclipse measurement under the assumption of spherical symmetry. The curve showing this distribution is reproduced in Fig. 2 of the following paper (p. 276). It has the form with limb-brightening predicted by quiet-sun theory under the same assumption of symmetry. The eclipse curve corresponding to the brightness distribution, shown on p. 276, agrees with the measured eclipse curve to within one percent. A redistribution of a small percentage of the total radiation to the equatorial limb regions gave a slightly improved comparison to the measured eclipse curve. A definite brightening near the limb of the sun is shown in both cases.

THE RADIAL BRIGHTNESS DISTRIBUTION
OF THE SUN AT 9·4 CM.

F. T. HADDOCK

*Naval Research Laboratory, Washington, D.C., U.S.A.**

The Naval Research Laboratory has supported four eclipse expeditions (in the years 1947, 1950, 1952 and 1954) under the direction of Dr J. P. Hagen. The principal purpose was to find the variation of the solar microwave radiation during a total optical eclipse of the sun. During the last two eclipses the sun was sufficiently inactive to enable us to derive the centre-to-limb brightness distribution at a wave-length of 9·4 cm., on the assumption that the distribution was circularly symmetrical. Since these two eclipses were of the same optical magnitude (within 0·12 %) and were measured with the same equipment located at each eclipse near the centre-line of mid-totality, it is of interest to compare their results. The 1954 eclipse observation was described in the last paper by Mayer, Sloanaker and Hagen. The 1952 observation has already been described [1].

The 30 June 1954 eclipse occurred at a time of exceptionally low solar activity as evidenced by both optical and radio observations. The unexpectedly high degree of symmetry and smoothness of the eclipse curve described in the following article indicates a regular, unchanging, quiet solar atmosphere. It is unlikely that within the next several decades the sun will again be as quiet during an eclipse.

The 25 February 1952 eclipse also occurred when there were no visible sunspots. However, a bright calcium plage was seen on each limb near the solar equator. Sudden changes in the observed eclipse curve corresponded in time to the immersion and emersion of these plage regions. No other sudden changes or bright plages were noted. In order to derive an eclipse curve representing the quiet sun it was necessary to subtract the excess radiation from these localized active regions. This was done by attributing differing amounts of excess emission to the two regions until the four sudden changes were smoothed out and the resulting eclipse curve was symmetrical about mid-totality. The amount of excess flux required to do

* Now at the University of Michigan Observatory, Ann Arbor, Michigan, U.S.A.

this was from 7·5 to 10 % of the total uneclipsed flux. The larger value agrees well with the 12 % excess radiation estimated from the low daily flux level of the undisturbed sun a few days after the eclipse to that on the eclipse day.

Fig. 1 shows the corrected eclipse curve for 1952 and the observed eclipse curve for 1954, each with their maximum flux normalized to 100. Actually the uneclipsed flux for 1952, after subtraction of the excess flux, was about 15 % greater than the 1954 value. This point will be mentioned later in connexion with the difference between the minimum flux values.

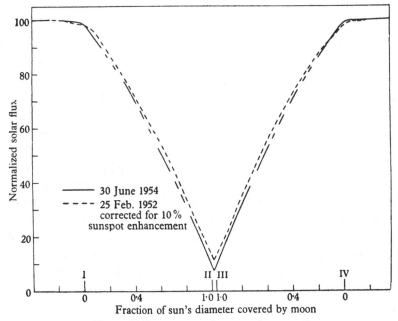

Fig. 1. Eclipse curves at λ 9·4 cm. for two eclipses.

The curves agree well in shape, both having fairly straight sides and a sharp V-shaped minimum. The disagreement in the regions of I and IV contact should be expected because the 1952 curve was corrected for excess emission in those regions and, furthermore, the curves must cross to be consistent with their minimum values. The 1952 V-shaped minimum was felt to be artificial when it was first derived but it was found difficult to deviate far from it and still produce a symmetrical eclipse curve. The fact that the 1954 observed eclipse curve displayed a sharp V-shaped minimum supports the validity of the corrections made on the 1952 curve.

It is impossible to derive a two-dimensional projected brightness distribution of the sun from a single one-dimensional eclipse curve (or a single fanbeam scan) without arbitrarily restricting the form of the distribution. For simplicity it is commonly assumed that the sun's atmosphere is spherically symmetrical, producing circular radio isophotes. However, it has been known for several years that elliptical isophotes are better representative of the sun at metre wave-lengths, and recently Christiansen and Warburton [2] have derived a two-dimensional brightness distribution of the quiet sun at 21-cm. wave-length which has quadrant symmetry with pronounced limb brightening in the equatorial regions changing smoothly to limb darkening at the poles. The 9·4 cm. brightness distribution should be similar to this, although its deviation from circular symmetry may be less since the coronal contribution to the brightness is appreciably less.

Radial brightness distributions with circular symmetry have been found giving eclipse curves which agree with the observed curves to within 1 % of the sun's uneclipsed flux. These distributions are not unique, however, even with the assumption of circular symmetry, because there exists a set of radial brightness functions which are not visible in the eclipse curve. That is, they can be added to a given radial brightness distribution without changing the associated eclipse curve. These 'invisible' radial functions can be represented by a linear combination of the zero-order Bessel functions, centred on the sun, which have a maximum or a minimum value at the radial distance which equals the moon's radius (unpublished work by S. Ament). We do not have a good method for eliminating these fictitious functions hidden in derived distributions, although their effect can be reduced by smoothing the data. Their presence accounts for the variation of shape in the set of distributions which were found to satisfy the 1952 eclipse data. The distribution shown in Fig. 2 was chosen from this set because it resembled in shape the theoretical distributions. The 1954 data are more precise and with simple smoothing produced the distribution shown in Fig. 2. It closely resembles published theoretical distributions [3]. Both distributions in Fig. 2 have a peak brightness of about twice the central brightness, with the peak located just beyond the optical limb of the sun.

An indication of the degree of deviation from circular symmetry in the actual brightness distribution can be obtained in the following way. The upper part of the eclipse curve (lunar positions for which the centre of the solar disk is unclipsed) completely determines the lower part of the eclipse curve *if* the brightness distribution has circular symmetry. On the other hand, if for example the distribution has a brighter limb at the equator than at the poles, and the track of the moon's centre over the disk does not make an angle with the sun's equator close to 45°, then the upper part of the eclipse curve would predict, on the basis of circular symmetry, a lower part in discord with observation. For the two curves in Fig. 1 this discordance was comparable with the observational uncertainties from which it was estimated that only a small percentage of solar flux at 9·4 cm. arises from a non-symmetrical component in the actual brightness distribution. Therefore the distributions shown in Fig. 2 are believed to be reasonable representations of the brightness distribution of the quiet sun on the two eclipse days. They are in general agreement with theoretical models of the quiet sun. The fact that the bright limb occurs outside the

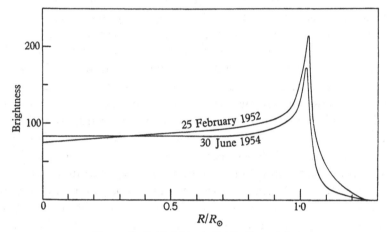

Fig. 2. Radial brightness distributions at λ 9·4 cm.

optical disk is a significant point of agreement with theory since there are notable exceptions to this in published distributions, obtained with fan-beam scans, at wave-lengths of 10·3 cm. [4] and 21 cm. [2] for example.

The principal long-range value of the eclipse observations will undoubtedly lie in the eclipse curves themselves because they can be used as standards against which any future brightness distributions of the quiet sun near sunspot minimum can be tested by generating artificial eclipse curves for a magnitude of 1·036. In fact, any model of the sun's atmosphere from which a 9·4 cm. brightness distribution could be derived could be tested in the same way.

The minimum flux value, measured at mid-eclipse, is of special interest because an eclipse of magnitude 1·036 completely occults the chromosphere and the exposed part of the corona is both optically thin and non-refracting for wave-lengths shorter than about 30 cm. Therefore, if the radiation is thermal, the minimum flux value is the same at these snort wave-lengths.

It is estimated that the mid-eclipse flux density for the quiet corona on 25 February 1952 was 9·4 units (where one unit equals 10^{-22} w.m.$^{-2}$ (c./s.)$^{-1}$) and 5·1 units on 30 June 1954. Using the Allen values for electron densities and an isothermal corona at $10^{6°}$ it can be shown that one-third (or 13·7 units of flux density) of the total coronal radiation (of 53·5 units) is uneclipsed at mid-eclipse (magnitude of 1·036) for all wave-lengths shorter than about 30 cm. On the basis of this model the 4·3 unit decrease of mid-eclipse flux requires a 12·9 unit decrease in total coronal flux. The actual drop in the total solar flux was only 10·6 units. Thus the mid-eclipse flux decrease from 1952 to 1954 was greater in proportion to the total flux decrease than expected from the Allen model. A plausible explanation for this is that the emission parameter ($N_e^2 . T_e^{-1/2}$), decreased a greater percentage in the outer corona than in the inner, thereby decreasing the fraction of uneclipsed flux at mid-eclipse. This greater change at greater heights as sunspot minimum is approached agrees with the changes in electron densities given by van de Hulst [3]. Reasonably expected changes in temperature would have a negligible effect in this situation.

It is of interest to compare the observed mid-eclipse flux densities with theoretical values. Van de Hulst [5] has also pointed out that the radio opacity of the corona should decrease by a factor of 3 from the sunspot maximum to minimum. Since the Allen model applies to the sunspot maximum the calculated mid-eclipse flux density would then decrease from 13·7 to 4·5 units at the minimum. This is in substantial agreement with the observed values of 9·4 in 1952 and 5·1 in 1954.

In 1947, at a sunspot maximum, the N.R.L. eclipse expedition measured at 3·2 cm. a flux density at mid-totality during an eclipse of nearly the same magnitude (1·038) of about 15 units. Since emission from the moon can only account for 3 or 4 units of this value, we again have reasonable agreement with the calculated flux density.

For various reasons, such as the corona being optically thin and non-refracting and the precise geometry at mid-eclipse, it appears that this type of measurement is a good test for a model of the corona; it is also sensitive to the height scale of the model. For example, with the 9·4 cm. equipment it would be possible to detect the change in flux caused by a change in the moon's radius of less than 1′ arc at mid-eclipse.

REFERENCES

[1] Haddock, F. T. *J. Geophys. Res.* **59**, 174, 1954.
[2] Christiansen, W. N. and Warburton, J. A. *Observatory*, **75**, 9, 1955.
[3] Hagen, J. P. *Ap. J.* **113**, 547, 1951.
[4] Covington, A. E. and Broten, N. W. *Ap. J.* **119**, 585, 1954.
[5] van de Hulst, H. C. *Nature*, **163**, 24, 1949.

Discussion

Hey: Observations of partial eclipses can be very useful in indicating deviations from circular asymmetry. We observed the eclipse of 30 June 1954 at 10 cm. wave-length from a site where the eclipse was partial and we found an indication that the intensity distribution across the solar disk must have greater equatorial than polar extent. Is there any such indication in your data?

Haddock: If 5 % of the total flux were distributed in a double sine wave around the limb of the sun a slightly better fit would be obtained.

ON THE UNIFORMITY OF THE LOWER CHROMOSPHERE

RICHARD N. THOMAS

Harvard College Observatory, Cambridge, Mass., U.S.A.

AND

R. G. ATHAY

High Altitude Observatory, Boulder, Colorado, U.S.A.

It is appropriate at a solar session during a radio astronomy conference to report on an analysis of optical observations to infer the properties of a non-spherically-symmetric chromosphere. For the first detailed model of such a non-symmetric chromosphere was that presented by Giovanelli (1949) [1] in an attempt to reconcile apparent contradictions between radio and optical data. Here we summarize some investigations based only on optical data, obtained by the High Altitude Observatory at the 1952 eclipse. Our observations of this eclipse were obtained as part of a joint programme with the Naval Research Laboratory, which conducted radio observations. Dr Hagen reports on the radio material (papers 46 and 47). The optical data in the present paper come from hydrogen and helium alone, the metallic data being still in reduction.

I. POINTS OF CAUTION

(*a*) An analysis for the distribution of n_e and T_e under the assumption of a spherically-symmetric atmosphere is relatively straightforward. Dropping the assumption introduces a wide range of possibilities. If, for example, one adopts the Giovanelli two-component model, it is tempting to identify the two components with spicular and inter-spicular regions, as Hagen (1953) [2] and Woltjer (1954) [3] have done. In the following we adopt this concept of two kinds of regions, homogeneous within themselves at a given chromospheric height, recognizing that it can be only a first approximation.

(*b*) Our use of the data from hydrogen and helium alone reduces the information that could be used, particularly for heights < 2000 km., where the metals would contribute, and for heights > 4000 km., where the radio results contribute. Any final model must include all these data, not available for the present analysis.

(c) Comparison of results from eclipse measurements at different phases of the solar activity cycle demonstrates (Athay and Thomas, 1955) [4] an appreciable variation of chromospheric structure with phase of the cycle. The present results apply only to the 1952 phase, thus near minimum.

2. DATA

There are basically two kinds of data. Type I whose use does not involve the question of whether or not the chromosphere is in local thermodynamic equilibrium, namely, free-bound and scattered light; type II in which the equilibrium question is highly relevant, namely, all line emission. Only the first type gives direct information on T_e and n_e; the second requires an indirect approach through a treatment of the non-equilibrium factors.

Type I data give the average value of two functions of n_e and T_e along the line of sight:

$$\overline{n_e n_p T_e^{-\frac{3}{2}}} = f(h) \quad h < 2500 \text{ km.} \tag{1}$$

$$\overline{n_e + n_e^2 T_e^{-\frac{1}{2}} n_p \alpha_{\mathrm{H}} - e^{\psi/kT_e} c_1} = g(h) \quad h \gtrsim 50,000 \text{ km.} \tag{2}$$

These two relations are sufficient to specify $n_e(h)$ and $T_e(h)$ over a limited height range (Athay, Menzel, Pecker, Thomas, 1955) [5], if spherical symmetry be assumed.

Type II data give certain indirect information, some of which may be used to check the consistency of the above solution, assuming spherical symmetry.

The hydrogen lines provide principally data on self-absorption. Results on self-absorption obtained under the assumption of a spherically-symmetric chromosphere may be applied to a combination of line and continuum data to estimate non-equilibrium factors. These non-equilibrium factors imply T_e values much in excess of those obtained from the continuum model. (Athay and Thomas, 1955 [6].)

Lines of He I appear to be free from self-absorption, and provide for several series the possibility of extrapolating the emission to the series head. From this extrapolation, one obtains the quantity:

$$\overline{n_{\mathrm{HeII}} n_e T_e^{-\frac{3}{2}}} = \phi(h) \quad h \gtrsim 1200 \text{ km.} \tag{3}$$

The total helium abundance obtained from (3) and the spherically-symmetric continuum model exceeds that of hydrogen by a factor 10^6.

Then, the observed helium emission cannot come only from material at greater heights lying along the line of sight because of the absence of the 'shell' effect in a plot of emission vs. height until $h \sim 1100$ km. Thus, the helium data appear to require regions of high T_e at heights above ~ 1100 km. (Athay and Menzel, 1956) [7].

3. THE TWO-COMPONENT MODEL

Tentatively, then, we assume a model of the Giovanelli type; i.e. two types of regions, each homogeneous within itself, and occupying total fractions a_1 and $a_2 = 1 - a_1$ of the distance along the line of sight, specifying region 1 to be the source of He emission.

(a) Equations

We have the three equations already discussed, written for the two regions.

$$a_1 n_{e1}^2 T_{e1}^{-\frac{3}{2}} + a_2 n_{e2} n_{p2} T_{e2}^{-\frac{3}{2}} = f(h), \tag{4}$$

$$a_1 n_{He} {}_n n_{e1} T_{e1}^{-\frac{3}{2}} = \phi(h), \tag{5}$$

$$a_1 n_{e1} + a_2 n_{e2} [1 + n_{e2} n_{p2} T_{e2}^{-\frac{1}{2}} \alpha_H e^{\chi_a/k T_{e2}} c_1] = g(h). \tag{6}$$

There are six unknowns: a_1; n_{e1}, n_{e2}; T_{e1}, T_{e2}; R. The H/He abundance is denoted by R, which we assume satisfies $5 \leqslant R \leqslant 20$. We assume constant pressure across the horizontal boundaries between the two regions to add the equation:

$$2n_{e1} T_{e1} = (2n_{e2} + n_{H2}) T_{e2}. \tag{7}$$

While there are apparently two free parameters in this system of four equations in six parameters, the actual range of solution is relatively small. In the following summary of the solution, we have deliberately picked the widest range of solution that is at all compatible with the data, in order to emphasize the physical implication of the results.

(b) Summary of numerical solution

We find that a_1 increases from 0·05–0·11 at 1500 km. to 0·9–0·99 somewhere between 2500–3500 km.

Further, in the height-range 1500–3000 km., T_{e1} has a value in the range 15,000–30,000° K.; while T_{e2} lies in the range 6000–8000° K. These values represent the range in solution; the change with height is small. Equation (4) rests on direct observations only for $h < 2500$ km.; the equations (4) and (5) become inconsistent for $h \sim 5000$ km. If we are permitted to use the equations (4)–(7) up to 3500–4000–4500 km., the numerical results suggest a considerable rise in T_{e2} at the greater heights.

(c) Comment on solution

We note first that the chromospheric line-emission becomes concentrated into spicule-like structures at ~ 5000 km., and that coronal line-emission begins below 10,000 km. (Athay and Roberts, 1955) [8]. It is then tempting to regard the above numerical results as giving the following properties to the two components.*

Cold component	Hot component
Size: $> 90\%$ at 1500 km., gradually decreasing to $\sim 1\%$ at 5000 km.	$< 10\%$ at 1500 km., gradually increasing to $\sim 99\%$ at 5000 km.
T_e 6000–8000° K. for $1500 < h < 3000$; rises to 15,000–30,000° K. between 3500–5000 km.	15,000–30,000° K. for $1500 < h < 3000$ rises to much higher value between 3500–5000 km.

4. RELATION OF THE RESULTS TO OTHER INVESTIGATIONS

(a) General form of $T_e(h)$ from stability considerations

Any solution for $T_e(h)$ should come from a balance of net radiative emission against mechanical (or other non-radiative) energy input, and should give stability against fluctuations in T_e due to variation in local energy input. The chief sources of emission appear to be H⁻ in the lowest chromosphere, neutral hydrogen in the next higher regions, then ionized helium, and finally either free-free or highly-ionized metals. The computed regions of stability differ for the different sources. Thus, as each source gives rise to the succeeding, there will in general occur a rather abrupt rise in T_e. We find these jumps should occur from somewhere in the interval 7000–10,000° K. to a value near 20,000° K. as neutral hydrogen becomes unstable; and from 20,000–40,000° K. to $\sim 10^5$ as ionized helium becomes unstable (Athay and Thomas, 1956) [9]. The resemblance of these values to the values found in the empirical analysis of the actual chromosphere is striking.

One would assume that the thermodynamic structure of the spicule is somehow fixed by its dynamic state, as in the jet model (Thomas, 1948 and 1950) [10]. The fact that its thermal structure falls within the range suggested by stability considerations emphasizes the need to include a coupling between radiative and kinetic degrees of freedom in treating aerodynamic models in astronomy. The above spicule model (Thomas, *loc. cit.*) is inadequate in this respect. Moreover the interpretation of $a_1(h)$ also rests,

---

* We note that the expressions for emission per cm.³, $(f(h), \phi(h), g(h))$, come from a double differentiation of the observations under the assumption of continuity of variables along, and perpendicular to, the line of sight. Thus, a_1 and a_2 contain implicitly the differential emission gradients of our two regions, and a correction is necessary before a literal interpretation in terms of relative areas is made. In the present rough analysis we simply defer the problems, since the corrections can only be made by successive approximation.

in the actual case, on the statistics of spicule height-distribution, and the associated thermodynamic properties.

(b) *General comment on the 'hot' spicule models*

Woltjer has presented a two-component model, based on the interpretation that spicules are everywhere hotter than the surroundings, and occupying about 2 % of the area along the line of sight. We find (Athay and Thomas, 1956) [4] four objections to his model. First, we are unable to represent our data with such a model and we believe these data to be more extensive than those which were available to Woltjer. Secondly, there are certain difficulties in his photometric data, since he was forced to calibrate observations made with an H_α filter by means of spectroheliographic data, obtained several years apart. It is not clear that the effect of variation of the chromosphere and the effect of scattered light in the spectroheliograph are negligible. Thirdly, we believe the actual choice of the spicule as hot is arbitrary, and cold spicules in a hot medium will satisfy the observations equally well. We believe that spectra, rather than only H_α filter observations, are required to settle the point. Finally, the questions of the stability of a hot spicule against radiative dissipation and of how the spicule can avoid heating the chromosphere must, we believe, be explored before accepting this model.

This work was supported in part by the Air Force Cambridge Research Centre, Geophysics Research Directorate, through contract AF 19(604)–146 with Harvard University, and partially by the Office of Naval Research in co-operation with the Naval Research Laboratory.

REFERENCES

[1] Giovanelli, R. G. *M.N.R.A.S.* **109**, 298, 1949.
[2] Hagen, J. P. Private communication, 1953.
[3] Woltjer, L. *B.A.N.* **12**, 165, no. 454, 1954.
[4] Athay, R. G. and Thomas, R. N. *Ap. J.* **123**, 309, 1956.
[5] Athay, R. G., Menzel, D. H., Pecker, J. C. and Thomas, R. N. *Ap. J. Suppl.* **1**, 505, 1955.
[6] Athay, R. G. and Thomas, R. N. *Ap. J. Suppl.* **1**, 491, 1955.
[7] Athay, R. G. and Menzel, D. H. *Ap. J.* **123**, 285, 1956.
[8] Athay, R. G. and Roberts, W. O. *Ap. J.* **121**, 231, 1955.
[9] Athay, R. G. and Thomas, R. N., *Ap. J.* **123**, 299, 1956.
[10] Thomas, R. N. *Ap. J.* **108**, 130, 1948; *Ap. J.* **112**, 343, 1950.

Discussion

Minnaert: Is there a possibility of distinguishing between hot jets in a cold gas and cold jets in a hot gas?

Thomas: In our model this makes no difference.

OBSERVATIONS OF BRIGHTNESS OVER THE DISK OF THE QUIET SUN AT FREQUENCIES OF 85, 500 AND 1400 MC./S.

J. L. PAWSEY

Division of Radiophysics, Commonwealth Scientific and Industrial Research Organization, Sydney, Australia

Observations of radio emission from the quiet sun, which is believed to be due to thermal emission from the ionized gases of the solar atmosphere, provide information about the electron density and temperature throughout the atmosphere. Comprehensive information requires observations, preferably of the brightness distribution over the solar disk, over a large range of wave-lengths. Because such observations are both difficult and tedious it is desirable that a final model atmosphere should be based on a pool of observations taken by independent methods and observers. The following are recent contributions to this pool from the Radiophysics Laboratory.

I. OBSERVATIONS AT 21 CM.

Christiansen and Warburton (1955) [1] have described observations of the brightness distribution over the solar disk at a wave-length of 21 cm. (the continuum, not the hydrogen line). They used two multiple-element interferometers which scanned the sun strip-wise in various directions. They were able to recognize and reject contributions from the disturbed bright areas by means of the narrow peaks which these areas give on the records. The 'quiet sun' distribution was recognized as the lower envelope of a number of superposed records. The brightness distribution over the disk was deduced from an adaptation of the Fourier method described by O'Brien (1953) [2].

The derived quiet sun distribution is shown in contour form in Fig. 1 and corresponding radial brightness distributions in various directions in Fig. 2. The observations were taken between 1952 and 1954 and, as no substantial change was apparent during this interval, they probably refer to representative sunspot minimum conditions. The angular resolution of the observations varied between 3 and 4 minutes of arc and the results shown in Figs. 1 and 2 have been smoothed where necessary to correspond to a uniform 4 minutes. Limited angular resolution distorts the derived distribution to some extent and this factor is discussed in a paper by Smerd and Wild (paper 51).

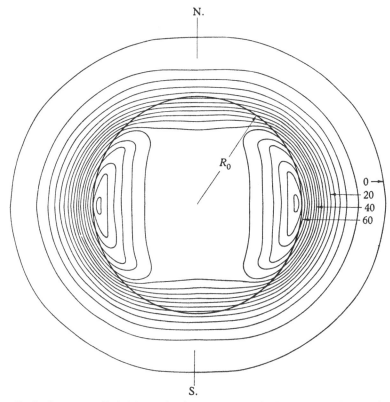

Fig. 1. Derived 21-cm. radio brightness isophotes as observed with 4′ beam. Contour interval $4 \times 10^{3\circ}$ K., central brightness temperature $4 \cdot 7 \times 10^{4\circ}$ K., maximum in peak $6 \cdot 8 \times 10^{4\circ}$ K.

The most interesting feature of the distribution is the crescent-shaped bright areas at the east and west limbs. The northern and southern limits to these areas are also the places where the corona shows the typical changes in structure which differentiate the polar from the equatorial corona at sunspot minimum.

2. OBSERVATIONS AT 60 CM.

Swarup and Parthasarathy (1955) [3] in Sydney adapted the original 32-element east–west Christiansen interferometer for use at a wave-length of 60 cm. so that they could check Stanier's (1950) [4] results for the distribution over the quiet sun using the Christiansen–Warburton method. Stanier, using a multi-spacing two-aerial interferometer, had reached the surprising conclusion that the sun showed no limb brightening at this wave-length. At the same time in Cambridge O'Brien and Tandberg-Hanssen (1955) [5] repeated the observations using Stanier's interfero-

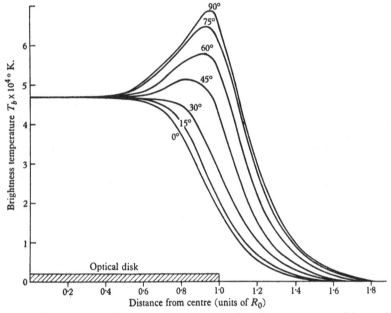

Fig. 2. Radial brightness distributions at 21 cm. in various directions measured from pole of sun. (Note that a corresponding figure published in *Observatory* in 1955 is slightly in error.)

meter method but extending it by using base lines in various directions. The Cambridge observations are the more complete because of the wide range of angles used and those authors have used them to derive a brightness distribution over the disk. However, the Sydney observations, being based on an independent method, give a valuable check.

Figs. 3a and b show a comparison between the Cambridge and Sydney observations for two directions of scan. The agreement is good and confirms the general distribution of brightness derived by the Cambridge workers. Both series conflict with the early Stanier results but it is not yet

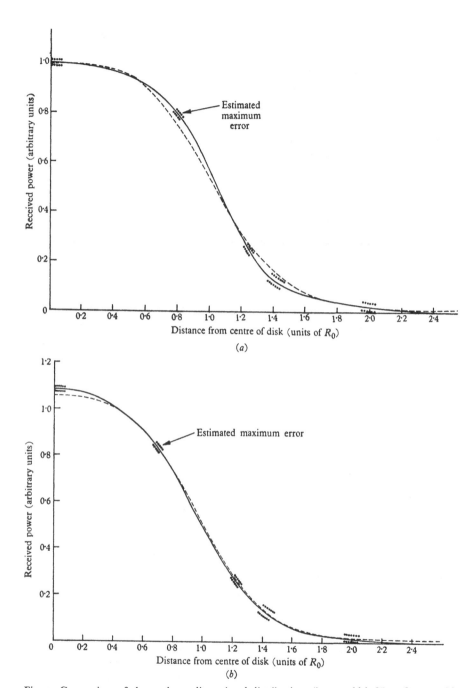

Fig. 3. Comparison of observed one-dimensional distributions (beam-width 8′) at 60 cm. with those derived from O'Brien and Tandberg-Hanssen's two-dimensional distributions by smoothing with the aerial beam. (a) Scanning direction 90° to solar axis; (b) 64°.

certain that the latter were in error since the difference might be due to changes associated with the changing phase of the sunspot cycle.

Turning to details, the Cambridge and Sydney observations show slight divergences which are believed to be outside the experimental errors of the latter. The form of the difference is that which would follow if the Cambridge observations were more 'smoothed' by unsuspected instrumental effects than the Sydney ones. The Sydney equipment had a resolution of 8 minutes of arc.

3. OBSERVATIONS AT 3·5 METRES

A series of observations of the sun were taken by A. G. Little in February and March 1955 using the 85 Mc./s. Mills Cross. This instrument has a pencil beam and in operation five traverses are made effectively simultaneously across the sun at declination intervals of 23′. Unfortunately the beam-width, 50′ in Right Ascension and from 52′ to 58′ in declination (the increase is due to foreshortening effects at low zenith angles) severely limits the derived information. The derived contours at times when the sun was quiet were approximately elliptical and sections through the centre exceeded the beam-width by 16′ in the east–west direction and 7′ in the north–south. It follows that the sun at 3·5 metres is substantially elongated in the east–west direction.

REFERENCES

[1] Christiansen, W. N. and Warburton, J. A. *Aust. J. Phys.* **8**, 474, 1955.
[2] O'Brien, P. A. *M.N. R.A.S.* **113**, 597, 1953.
[3] Swarup, G. and Parthasarathy, R. *Aust. J. Phys.* **8**, 487, 1955.
[4] Stanier, H. M. *Nature*, **165**, 354, 1950.
[5] O'Brien, P. A. and Tandberg-Hanssen, E. *Observatory*, **75**, 11, 1955.

Discussion

Ryle: Our measurements show that at 60 cm. the Fourier components of the distribution are less than $\pm 1 \%$ from a spacing of 200λ to 260λ. An analysis has been made both at this wave-length and at the other wave-lengths of $1\cdot4$, $3\cdot7$ and $7\cdot9$ metres to determine the possible errors caused by the finite resolving power used in the various measurements. The method used was to determine the difference between the adopted distributions, which in Bracewell and Roberts'* nomenclature are the 'Principal Solutions', and the unknown actual distributions. It can be shown that the difference or 'error distribution', if it exists, must have a periodic form having in each of the present measurements

* Bracewell, R. N. and Roberts, J. A. *Aust. J. Phys.* **7**, 615, 1954.

several maxima across the sun's disk. The difficulty of giving physical reality to the distributions found in this way gives considerable confidence in the distributions computed from the observations.

It is worth noting that the possible errors in the derived distribution depend on the amplitude of the highest order Fourier components obtained from the observations. The use of a moving aerial interferometer has in fact considerable advantages over a pencil-beam system of the same total aperture when used for examining a single source such as the sun, because all the Fourier components can be determined with equal weight; in the pencil-beam system the high-order components are necessarily severely attenuated.

Pawsey: The small differences which do exist between the Cambridge and Sydney observations indicate a tendency for the Sydney observations to show finer detail. For example, high-frequency Fourier components exist in the Sydney curves which the Cambridge observers measured but found negligible. Hence one or other of the series of observations must be slightly in error. The only conclusion I should like to draw at this moment is that it would be unwise to base theoretical conclusions on these details until better observations are available.

INTERPRETATION OF SOLAR RADIO-FREQUENCY DISK BRIGHTNESS DISTRIBUTIONS DERIVED FROM OBSERVATIONS WITH AERIALS EXTENDED IN ONE DIMENSION

S. F. SMERD AND J. P. WILD

Radiophysics Laboratory, Sydney, Australia

Several recent papers have dealt with observations of brightness distributions over the solar disk, which were derived either from two-aerial interferometer observations at various spacings and orientations (e.g. O'Brien, 1953) [1], or from multiple-element interferometer fan-beam observations at various orientations (e.g. Christiansen and Warburton, 1954) [2]. In each a two-dimensional distribution is derived from a number of essentially one-dimensional observations by a Fourier synthesis method described by O'Brien. The detail given by these methods must be limited by the finite resolution of the individual observations (limited by the maximum aperture of the aerial system), but the form of the limitation is not obvious, though its knowledge is required when relating the observations to a solar model.

We have found that the 'derived' distribution is identical with that which would be obtained on scanning the true distribution point by point with a hypothetical pencil-beam aerial. This hypothetical aerial has a circular beam such that a strip scan of the power response along any diameter gives the power response of the observing aerial system in its direction of high resolution. This result may be obtained by the following reasoning:

Consider a two-dimensional brightness distribution $f(x, y)$ having small angular extensions, so that a system of *rectangular* co-ordinates (x, y) may be used to specify positions on the celestial sphere. When such a distribution is scanned across the x-direction by a fan-beam parallel to the y-axis and with power response $A(x)$, the instrument registers the profile

$$\int\int A(x - x') f(x, y) \, dx \, dy \qquad (1)$$

(the integrals here and subsequently extending over the entire distribution). When many such profiles, obtained by scanning the distribution in different directions, are combined to derive a two-dimensional distribution (e.g. following the method used by O'Brien), we obtain not the true distribution $f(x, y)$, but a smoothed distribution $g(x, y)$. This distribution must be such that its line integral in any direction corresponds to the observed profile. Thus, considering the x-direction (which is chosen arbitrarily in the first place) we have

$$\int g(x', y') \, dy' = \iint A(x - x') f(x, y) \, dx \, dy. \tag{2}$$

We now consider the distribution g to be formed from the true distribution, f, by smoothing the latter point by point with a pencil beam whose power response is given by $B(x, y)$. By symmetry, the latter must be a circular pattern. So B, f and g are related by

$$g(x', y') = \iint B(x - x', y - y') f(x, y) \, dx \, dy. \tag{3}$$

Substituting (3) in (2), we obtain

$$\iiint B(x - x', y - y') f(x, y) \, dx \, dy \, dy' = \iint A(x - x') f(x, y) \, dx \, dy.$$

This is satisfied if

$$\int B(x - x', y - y') \, dy' = A(x - x')$$

i.e. if

$$\int B(x, y) \, dy = A(x),$$

giving the result stated above.

As $B(x, y)$ depends on x and y only through the function $r = \sqrt{x^2 + y^2}$, the practical interest is centred on finding the function $B(r)$ from the integral equation:

$$2 \int_x^\infty \frac{B(r) \, r \, dr}{\sqrt{r^2 - x^2}} = A(x).$$

The application of this theorem to two cases of particular importance is shown in Figs. 1a and 1b. In either figure the dashed curve represents the power response function of the strip instrument, $A(x)$, and the solid curve represents the radial distribution of the circular equivalent, $B(r)$. Both x and r are measured in units of λ/l radians. Fig. 1a refers to a uniform one-dimensional aerial of length l, so that $A(x)$ is proportional to $\theta^{-2} \sin^2 (l\pi\theta/\lambda)$. Fig. 1b refers to the aerial pattern appropriate to observations made with a two-element interferometer, in which the information obtained using a succession of different aerial spacings between o and l is appropriately combined (Stanier, 1950) [3]. Here $A(x)$ is proportional to $\theta^{-1} \sin (2l\pi\theta/\lambda)$. In both cases it is seen that the side lobes of the equivalent circular pattern are negative as well as positive, and are of somewhat smaller amplitude than those of the observing linear aerial.

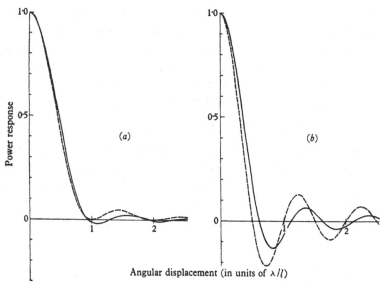

Fig. 1. The power response of aerials having high resolution in only one dimension (dashed curve) and the cross-section of their circular equivalents (full curve).

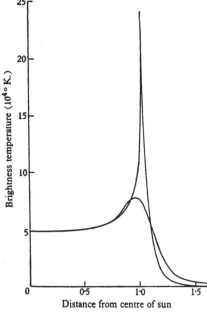

Fig. 2. The thin line shows the radial cross-section of the brightness distribution at $\lambda = 21$ cm. derived from a spherically symmetrical model of the solar atmosphere. The thick line shows the radial cross-section of the distribution which would be derived from strip scans in different directions using a uniform strip-shaped aerial having a beam width of 4 ft. 3 in. between point of half-power.

Using this theorem to compare solar brightness distributions derived from observations with those derived from model solar atmospheres, we find that the finite resolving power of the instrument tends to introduce the following effects:

(1) Broadening, lowering, and shifting towards adjacent bright regions of peaks in the true distribution (e.g. a peak at the limb).

(2) An artificial extension of the outer fringes of the distribution.
Observed values of the central brightness, however, are practically correct.

Numerical examples (e.g. Fig. 2) show effects (1) and (2) to be significant and it seems that in comparing a theoretical with an observed distribution the only reliable procedure is to degrade the theoretical one by smoothing it with the appropriate aerial beam. Current results suggest that it will not be necessary to depart from present ideas on electron densities, etc., in the solar atmosphere in explaining the radio observations which apply near sunspot minimum.

REFERENCES

[1] O'Brien, P. A. *M.N.R.A.S.* **113**, 597, 1953.
[2] Christiansen, W. N. and Warburton, J. A. *Observatory*, **75**, 9, 1954.
[3] Stanier, H. M. *Nature*, **165**, 354, 1950.

BRIGHTNESS DISTRIBUTION OF THE SUN AT 1·45 METRES

J. FIROR

*Department of Terrestrial Magnetism, Carnegie Institution of Washington,
Washington, D.C., U.S.A.*

The brightness distribution of the quiet sun at radio frequencies is of importance in determining the electron density and temperature in the chromosphere and the corona. Calculations made by Smerd, based on parameters derived from optical observations of the sun, indicate that the brightness distributions for wave-lengths near 1·5 metres are sensitive to the assumed coronal conditions and hence are well suited to checking these parameters.

The brightness distribution of the quiet sun at 1·45 metres has been derived from measurements made between December 1953 and July 1955. The measurements were made with a variable-spacing interferometer and several fixed-spacing interferometers. The base-line of the interferometers was either east–west or north–south. With the fixed interferometers the method of measurement consisted of comparing the amplitude and phase of the output record produced by the sun with that produced by a known radio source—either the strong source in Cygnus or the Crab nebula. The measurements with the variable-spacing interferometer gave only the shape of the amplitude-spacing curve.

The measurements were made at times when the sun was quiet, as indicated by lack of burst activity, no large sunspots or new spot groups, and so on. The period during which measurements were made includes the epoch of sunspot minimum. The measurements vary, however, from day to day. In Fig. 1 the solid line gives the average amplitude of the interferometer trace as a function of spacing for the east–west base-line. The plotted points are the individual measurements and are seen to vary from the average curve. On any one day the measured points generally formed a smooth curve. The phase of the solar trace was always near zero for spacings up to about 100 wave-lengths and near 180° for the region 100 to 175 wave-lengths. Only the amplitude was measured in the range 200 to 340 wave-lengths, and in Fig. 1 these points are shown with assumed positive phase.

The results of the measurements made with the north–south base-line are shown in Fig. 2. In this case the maximum antenna spacing was smaller and no phase measurements were made. The overall similarity of the curve to that from the east–west measurements and the manner in which the measurements vary with spacing near 120 wave-lengths both indicate that the proper phase for the second maximum is negative, as plotted in Fig. 2.

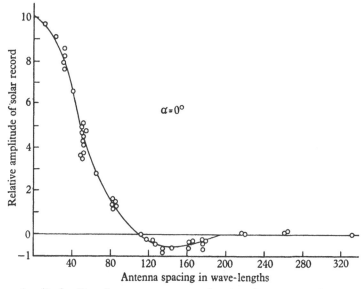

Fig. 1. Amplitude of interferometer output as a function of antenna spacing for the case of the interferometer base-line parallel to the solar equator. The open circles are the measured values and the solid line is the average curve used for computing a brightness distribution. The negative values in the region of 100 to 200 wave-lengths indicate that, for these spacings, the interferometer output was a minimum at the time the centre of the solar disk was on the instrumental meridian.

A single curve, such as the one in Fig. 1, can be used to derive a two-dimensional brightness distribution if brightness contours of some particular form are assumed. With a second curve available, such as Fig. 2, the validity of the assumed contour shape can be partly checked and one parameter adjusted. If, for example, elliptical contours are assumed, the second experimental curve must be the same as the first with the abscissa expanded or compressed, and the parameter, which may be adjusted to fit the measured values, is the eccentricity of the elliptical contours.

For the present measurements it is seen that the north–south curve is not the same as the east–west curve, but has differences in shape rather than in scale. The brightness distribution must therefore have contours more complicated than central ellipses. The measurements are insufficient to determine these contours completely, but they may be reduced to give a distribution consistent with the measurements. This reduction was made by interpolation between the two measured curves to obtain curves for all inclinations of the interferometer fringes to the solar axis. The brightness distribution of the solar disk is then the two-dimensional Fourier transform of the set of amplitudes so obtained.

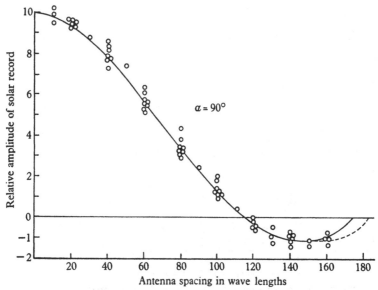

Fig. 2. Amplitude of interferometer output as a function of antenna spacing for the case of the interferometer base-line perpendicular to the solar equator.

The results of this procedure are given in the solid lines in Fig. 3a. The two curves are the brightness distributions along the solar equator and axis. It is seen that both curves exhibit the bright limb and that the equatorial curve is broader than the polar curve. Although both curves are quite similar to the 1·5-metre curve calculated by Smerd and shown in Fig. 3b, it must be remembered that we may add a great variety of distributions to the one obtained and still be entirely consistent with the experimental data. It is only necessary that the distributions to be added contain only higher frequencies, that is, distributions varying more rapidly with angle than are detected by the interferometer at its maximum spacing.

It is also to be noted that the derived brightness along the solar axis becomes negative around 1·6 solar radii. This is a result of the uncertainties in the amplitude of the interferometer trace at large spacings. For example, if the experimental curve in Fig. 2 is extended to zero in a different manner (dotted line) then the dotted curve in Fig. 3a results, which does not become appreciably negative, but also has a more prominent bright limb. As we require a brightness distribution that does not become negative, the dotted line in Fig. 4 is perhaps the more reasonable one to associate with the present measurements.

Because the validity of the derived distribution was seen to depend so strongly on measurements at greater spacings, some points were taken at 216, 260 and 340 wave–lengths with the east–west interferometer. The

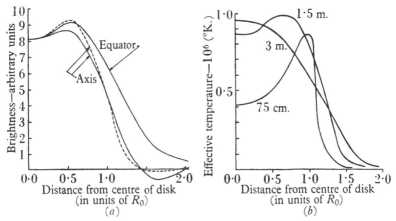

Fig. 3. (a) Brightness distribution of the sun at 1·45 metres wave-length. The solid curves are derived from the solid curves in Figs. 1 and 2. If the dashed curve in Fig. 2 is used, the equatorial curve is not changed appreciably while the brightness along the axis is changed to the dashed curve here. (b) Theoretical brightness distributions calculated by Smerd (1950).

amplitudes only were measured and were sufficiently small so that they do not change any major feature of the distribution derived without them, regardless of which phase is assumed. These points could, if taken into account, vary the details of the distribution, such as shifting the bright limb in or out a small amount, brightening or darkening the centre and so on. But the lack of any major change produced by these points lends confidence to the two main features of the derived distribution: the polar flattening and the bright limb.

Discussion

Unsöld: Why does the maximum brightness occur inside the limb?
Burke: This occurs also in some theoretical computations, e.g. Smerd's.
Pawsey: Moreover, aerial smoothing tends to move it inside.

RADIO OBSERVATIONS OF THE SOLAR CORONA AT SUNSPOT MINIMUM

A. HEWISH

Cavendish Laboratory, Cambridge, England

Previous experiments to determine the distribution of radio brightness across the solar disk at metre wave-lengths using the method of Fourier synthesis have been described by O'Brien (1953) [1]. These results could be explained in terms of a spherically symmetrical corona only if the

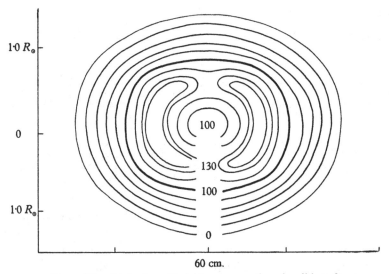

Fig. 1. The distribution of brightness across the solar disk at 60 cm.

electron density and temperature differed appreciably from estimates derived by visual methods (O'Brien and Bell, 1954) [2]. In order to check these measurements a similar series of observations was carried out in 1954 during an extended period of exceptionally low solar activity. Additional experiments, which gave information concerning the ellipticity of the brightness distribution, were made at Cambridge during the partial eclipse of June 1954.

Interferometers of variable spacing in an east-west direction were used on wave-lengths of 1·4, 3·7 and 7·9 metres; on a wave-length of 60 cm. more detailed observations were made in which the direction of the interferometer axis was varied over a wide range of angles. The 60 cm. experiments enabled the complete distribution to be derived assuming symmetry about the rotation axis only, and this assumption was justified by the consistency of the observations over a period of several weeks. The results are shown in Fig. 1 where it is seen that the contours are considerably flattened towards the poles.

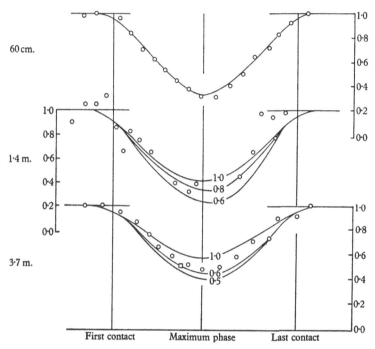

Fig. 2. Observations obtained during the partial eclipse of 1954 at Cambridge. At 1·4 and 3·7 metres the smooth curves denote the results derived theoretically for models of different axial ratio. At 60 cm. the smooth curve is that derived from the distribution shown in Fig. 1.

At the metre wave-lengths it was not possible to derive the complete distribution unambiguously since the direction of the interferometer was fixed. The method adopted was to assume that the contours were elliptical, the degree of ellipticity being derived independently from eclipse measurements.

The eclipse measurements are shown in Fig. 2. At a wave-length of 60 cm. the shape of the eclipse curve was in excellent agreement with that expected from the distribution obtained by the interferometer method. At metre wave-lengths the eclipse results indicated an appreciable degree of ellipticity, and the best fit was obtained for models having an axial ratio of 0·8 at 1·4 metres and 0·6 at 3·7 metres.

Using these results the distributions shown in Fig. 3 were obtained. The curves are largely similar to those reported previously by O'Brien, but the

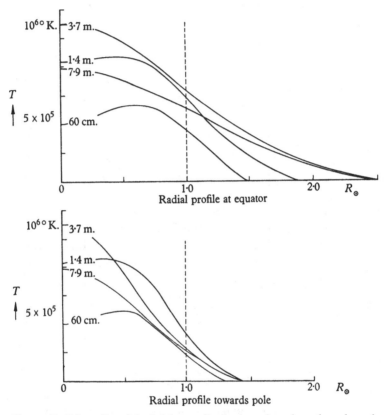

Fig. 3. Radial profiles of the brightness distribution taken along the polar axis and in the equatorial plane.

absolute brightness has been modified on account of the ellipticity of the contours. The radio model of the corona derived by O'Brien and Bell on the basis of spherical symmetry must similarly be modified, and it now seems probable that the radio observations may be explained without the necessity for a rapidly falling temperature in the outer corona. In addition, limb brightening has now been detected at 60 cm. The absence of limb brightening in the earlier measurements of Stanier (1950) [3] could be due to the presence of long-lived regions of enhanced emission which might be a permanent feature on the disk at times other than sunspot minimum.

REFERENCES

[1] O'Brien, P. A. *M.N.R.A.S.* **113**, 597, 1953.
[2] O'Brien, P. A. and Bell, C. J. *Nature,* **173**, 219, 1954.
[3] Stanier, H. M. *Nature,* **165**, 354, 1950.

THE ELLIPTICITY OF THE CORONA AT 80 MC./S. DURING SUNSPOT MINIMUM 1954

J. TUOMINEN

The University, Helsinki, Finland

The solar eclipse of 30 June 1954 was observed in the neighbourhood of Helsinki on the frequency 81·5 Mc./s. [1]. At the maximum of the optical eclipse 90 % of the solar diameter was occulted by the moon. We may accordingly assume that on the wave-length of observation the eclipse was nearly central. Further, the angle between the orbit of the moon and the solar equator was only 12°.

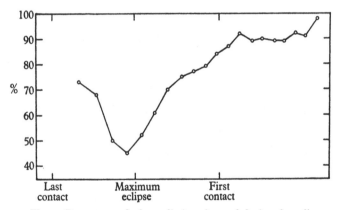

Fig. 1. Percentage of solar radiation observed during the eclipse.

Fig. 1 shows the results obtained. We have used the phase switching system; each circle corresponds to a maximum deflexion of the recorder. At the maximum of the optical eclipse 52 ± 7 % of the radio-frequency radiation is occulted by the moon. At the first contact the corresponding number is 16 ± 4 %. At the last contact the percentage seems to be the same as at the first, although observations could not be continued to the end of the eclipse due to an interfering radio sonde. Assuming symmetry of the curve, the diagram of Fig. 2 represents the observations.

The diagram shows the rate of solar radiation emanating from various areas occulted by the moon. The sum of all these percentage values is about one hundred, the exact figure being 102 %. Accordingly, the observations

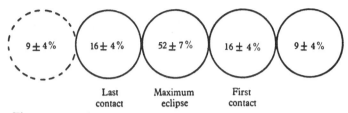

Fig. 2. The percentage in each circle indicates how much of the total solar radiation is taken away by the moon in that position.

indicate that all radiation on 80 Mc./s. emanates from the strip along which the moon passes the sun. It is not possible to deduce from our observations alone the breadth of the active strip in the polar direction or to determine the ellipticity of the corona.

REFERENCE

[1] Tuominen, J., Riihimaa, J. and Tuori, K. *Ann. d'Astrophys.* **18**, 3–6, 1955.

RADIO OBSERVATIONS OF THE ECLIPSE
OF 30 JUNE 1954

M. LAFFINEUR

Institute of Astrophysics, Paris, France

The laboratory of radio astronomy of the Institut d'Astrophysique at Paris has observed the eclipse of 30 June 1954 at Öland (Sweden) and at Meudon (France) [1].

The equipment was analogous to that used at Khartoum for the eclipse of 25 February 1952 [2]. At Högby ($\phi = 57°\ 9'$, $L = -17°\ 2'$) a mirror of

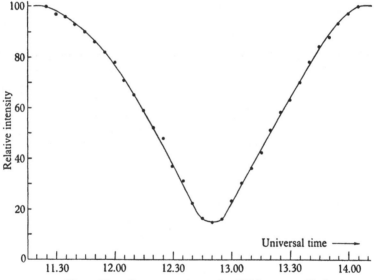

Fig. 1. The eclipse curve observed at Högby at 545 Mc./s.

6 metres diameter and 3·40 metres focal length was used. The beam width was 22° between the first zeros. The receiver at 545 Mc./s. possessed one stage of r.f. amplification and had a band-width of 1·5 Mc./s. The time constant was about 1 sec. A fixed region of the sky served as a standard at 5-minute intervals.

Table 1. *Solar radiation observed during the eclipse*

Högby 545 Mc./s.		Meudon 545 Mc./s.		Meudon 255 Ms./s.	
U.T.	Intensity (%)	U.T.	Intensity (%)	U.T.	Intensity (%)
11h·25	100	11h·17	100	11h·06	100
30	97	27	98	17	100
35	96	37	88	26	99
40	93	43	84	37	96
45	90	48	79	44	91
50	86	56	74	50	86
55	82	12h·02	69	57	83
12h·00	78	09	64	12h·02	75
5	71	16	56	08	71
10	65	21	51	16	65
15	59	28	46	21	58
20	52	36	44	27	54
25	48	44	42	36	49
30	37	56	46	43	43
35	31	13h·06	51	56	42
40	22	15	58	13h·06	54
45	16	24	65	15	63
50	14·7	32	74	24	71
55	16	41	84	32	77
13h·00	23	50	93	42	87
5	30	58	100	51	94
10	36	14h·06	100	59	99
15	42	—	—	14h·17	100
20	51	—	—	—	—
25	58	—	—	—	—
30	63	—	—	—	—
35	70	—	—	—	—
40	78	—	—	—	—
45	84	—	—	—	—
50	88	—	—	—	—
55	94	—	—	—	—
14h·00	97	—	—	—	—
5	100	—	—	—	—

At Meudon the equipment for the daily measurements of solar intensities was used. The antenna is a Wurzburg paraboloid with 7·50 metres diameter having two crossed quarter-wave dipoles at its focus. The beam-width at 545 Mc./s. is 16° in azimuth and 24° in altitude. A pilot computor is used for following the sun.

The records were corrected for the following effects:

(1) The presences of side-lobes; no correction was needed.

(2) The interference by ground reflexion; measured on adjoining days.

(3) The radiation of the comparison field on the sky; measured six months later, leaving an error $< 1\%$ in the eclipse curve.

(4) Sensitivity variations; the residual errors after using the sky comparisons are $< 2\%$.

The corrected results are given in Table 1.

REFERENCES

[1] Laffineur, M., Vauquois, B., Coupiac, P. and Christiansen, W. N. *C.R.* **239**, 1589–90, 1954.
[2] Laffineur, M. *Ann. d'Astrophys.* **17**, 358, 1954.
[3] Laffineur, M. *Bull. Astronomique*, **18**, 1, 1953.

RADIO OBSERVATION OF THE PARTIAL
SOLAR ECLIPSE, 20 JUNE 1955

T. HATANAKA

Tokyo Astronomical Observatory, Mitaka, Tokyo, Japan

The partial solar eclipse of 20 June 1955 was observed at Tokyo and Toyokawa, where regular solar radio observations are conducted. A party was sent to Kagoshima by the Tokyo Astronomical Observatory. The locations, the frequencies observed and the types of aerial are listed in Table 1. The paths of the northern limb of the moon at three stations are shown by dotted lines in Fig. 3.

Table 1

Station (latitude and longitude)	Magnitude of the eclipse	Frequency (Mc./s.)	Type of aerial	Organization
Kagoshima (+31° 37′, E. 130° 32′)	0·38	3000	2 metre dish	Tokyo Astronomical Observatory
Toyokawa (+34° 50′, E. 137° 22′)	0·21	3750	2·5 metre dish	Research Institute of Atmospherics
		4000	Eight 1·5 metre dish interferometer	
Tokyo (+35° 40′, E. 139° 33′)	0·16	3000	10 metre dish	Tokyo Astronomical Observatory

The variations of the total flux at three stations are shown by the curves marked *A* in Fig. 1. The observed variations in flux at the three stations show marked differences. These differences are apparently due to the different magnitudes of obscuration at the three stations of an enhanced region at the position of the sunspot group in the southern hemisphere. At Toyokawa the interferometric observation with eight dishes has been continued during the day. Curves *A* in Fig. 2 give the drift curves before and during the eclipse.

It is possible to determine the location, the size and the brightness distribution of the enhanced region by combining these eclipse curves. The contours for the radio isophotes thus obtained are shown by full lines

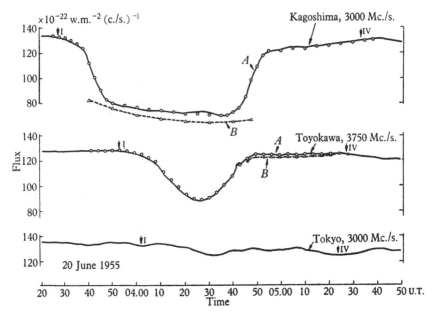

Fig. 1. Variation of flux at three stations. Arrows with I and IV indicate the time of the first and fourth contacts.

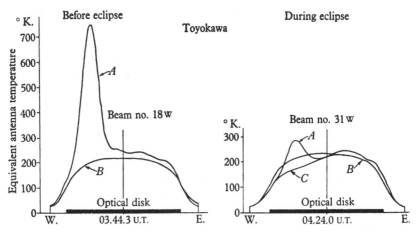

Fig. 2. Drift curves with eight-dish interferometer. A, observed curve. B, lower envelope of the drift curves observed during June 1955. C, calculated drift curve for the basic radiation during the eclipse.

in Fig. 3. The numbers 1, 2 and 3 in the figure give the measurements for the relative brightness and roughly correspond to $10^{6°}$ K., $2 . 10^{6°}$ K. and $3 . 10^{6°}$ K. in brightness temperature. The curve in dotted line gives the calcium plage observed at Mount Wilson kindly made available to us by

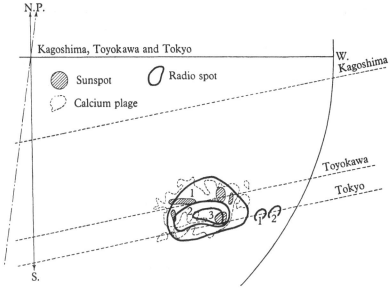

Fig. 3. Comparison between radio and optical observations. Dotted lines indicate the paths of the northern limb of the moon at three stations.

Table 2. (*Unit*: 10^{-22} w.m.$^{-2}$ (c./s.)$^{-1}$)

Source	Research Institute of Atmospherics (3750 Mc./s.)	Tokyo Astronomical Observatory (3000 Mc./s.)
Radially symmetric disk	62	66
Strong radio spot	41	45
Main part	38	42
Small patch	3	3
Eastern part	4	4
Other area	21	20
Total	128	135

Dr S. B. Nicholson. The close agreement between the radio spot and the calcium plage is very remarkable.

Besides the discussion on the radio spot it was also possible to estimate the contributions to the total flux by the different regions over the solar disk. This estimation was done almost independently by two ways as is shown in Table 2: one by combining the flux observations at Kagoshima and Tokyo and the other by combining two observations at Toyokawa. It is found that these two sets of estimations came out very closely to each other. It is to be emphasized that the brightness distribution of the so-called basic radiation cannot be radially symmetric over the solar disk, but only two-thirds of the basic radiation is radially symmetric. (Curves B in Fig. 1 are the eclipse curves reconstructed on the assumption that the whole basic radiation is radially symmetric.) The remaining one-third is denoted in Table 2 as 'other area'. 'Eastern part' in the table means the contribution by less enhanced regions in the eastern part as judged from the drift curves. The open circles in Fig. 1 show the reconstructed eclipse curve based on the present conclusion.

Details will be published elsewhere [1, 2].

REFERENCES

[1] Hatanaka, T., Akabane, K., Moriyama, F., Tanaka, H. and Kakinuma, T. *Rep. Ionosphere Research, Japan*, **9**, 195, 1955.
[2] Hatanaka, T., Akabane, K., Moriyama, F., Tanaka, H. and Kakinuma, T. *Publ. Astr. Soc. Japan*, **7**, 161, 1955.

A SURVEY OF SOVIET OBSERVATIONS OF THE RADIO EMISSION FROM THE SUN DURING SOLAR ECLIPSES

B. M. TCHIKHATCHEV

Crimean Station, Physical Institute of the Academy of Sciences, Moscow, U.S.S.R.

Soviet scientists working in the domain of radio astronomy carried out a number of observations of the radio emission from the sun during solar eclipses. The first observation was obtained by Prof. S. E. Khaikin and the author during the eclipse of 20 May 1947 at a wave-length of 1·5 metres. The solar eclipse of 25 February 1952 was observed in Archman, Turkmenian S.S.R., on wave-lengths of 3·2 and 10 cm. by V. S. Troitzky, and on 1, 1·5, 2 and 2·6 metres by the author and by V. V. Vitkevitch. The eclipse of 30 June 1954 was observed in Novomoskovsk, Ukrainian S.S.R., on 3·2 and 10 cm. by V. S. Troitzky, and on 10 and 23 cm. by V. V. Vitkevitch. It was also observed in the Caucasus on a wave-length of 3·2 cm. by A. P. Moltchanov. All the above observations were carried out in places located in the vicinity of the central path of the totality. The eclipse of 30 June 1954 was observed also outside the path of totality on the southern shore of the Crimea by a group of Dr Vitkevitch's colleagues on wave-lengths of 1, 1·5 and 3·5 metres. The maximum phase in this place was 92 %.

The following table shows the minimum values of the intensity of solar radio emission obtained during these eclipses, expressed as a percentage of the intensity of radio emission from the uncovered sun observed on the day of the eclipse.

	λ (cm.)							
Eclipse	3·2	10	23	100	150	200	260	350
20 May 1947	—	—	—	—	40	—	—	—
25 February 1952	5·2	15	—	32	36	40	50	—
30 June 1954	0·71	5·94	9·9	20*	25*	—	—	43*
	0·98	6·9						

Two values are given for 3·2 and 10 cm. for the eclipse of 30 June 1954. The upper figures were obtained by V. S. Troitzky, the lower figure on

3·2 cm. by A. P. Molchanov and on 10 cm. by V. V. Vitkevitch. The figures obtained outside the path of totality are marked with asterisks.

A quite obvious conclusion which might be made on examining the table is that in the course of these several years a compression of the solar corona and the chromosphere has taken place. Such a compression may be considered as a manifestation of periodic variations of solar activity.

It may be noted that the observation of the solar eclipse of 1947 gave the first experimental proof of the coronal origin of solar radio emission in the metre wave-lengths.

The minimum intensities of radio emission from the sun at the solar eclipses of 1952 and 1954 were observed at moments close to the maximum phase of the eclipse. The only exceptions were the 2 and 2·6 metre waves, on which the minima were observed 18^m and 30^m later, respectively. The eclipse curves for these waves showed a strong asymmetry with respect to the time of total phase. An analysis of the curves makes it possible to conclude that the intensity of radio emission from the coronal region above the large filament, which was located in the western part of the sun on 25 February 1952, was considerable. The curve for the eclipse on 2·6 metre wave-length also shows an appreciable radio emission of coronal rays, which were partly covered by the moon in the course of the eclipse.

Discussion

Hagen: Perhaps I may point out that I share with Khaikin and Tchikhatchev the honour of having made the first radio observation of a solar eclipse in 1947.

RESULTS OF OBSERVATIONS OF THE SCATTERING OF RADIO WAVES BY THE ELECTRONIC INHOMOGENEITIES OF THE SOLAR CORONA

V. V. VITKEVITCH

Crimean Station, Physical Institute of the Academy of Sciences, Moscow, U.S.S.R.

I

A new method for the investigation of the solar corona, suggested by us (Vitkevitch, 1951) [1], consists of observing the radio source identified with the Crab nebula (NGC 1952; $\alpha = 05^h\ 31^m\ 40^s$, $\delta = 22°\ 10'$) when it is covered by the solar corona. This occurs every year on 14–15 June.

Although the angular dimension of this source is smaller than the radio diameter of the sun, it is possible to resolve the radiation of the Crab nebula from the radiation of the sun with the aid of the interferometer method.

The first observations aimed at investigating the propagation of radio waves through the solar corona were carried out in 1951 at a wave-length of 4 metres by means of a sea interferometer. These observations did not give any positive results, as from 9 to 22 June the solar radiation was disturbed. These first observations showed, however, that during the time when the sun is quiet the interference picture of the Crab nebula might be seen quite clearly on the background of the sun.

Observations of 1952, confirmed by observations of 1953, showed that when the source of radio emission is covered by the solar corona the amplitude of the interferometer records from the source is appreciably decreased, even when the angular distance between the sun and the source is fairly large. The influence of the solar corona upon the propagation of radio waves of 3·5 metres begins to be perceptible at a distance of about 10 $R_\odot$. The influence of the corona at wave-lengths of 6 metres is perceptible at a distance of 15 $R_\odot$.

It was then suggested that the observed effect is caused by the scattering of radio waves by the electronic inhomogeneities of the outermost parts of the solar corona, which will be designated as the solar 'supercorona'.

Observations in 1954 by means of two interferometers made it possible to measure the angular dimension of the radio source and to establish how much this dimension increases when covered by the solar corona. The angular dimensions of the source may reach 18' for 3·5 metre wave-lengths and 27' for 5·8 metre waves (Vitkevitch, 1955) [2]. Thus it was confirmed that the observed effect is actually an effect of scattering by the electronic inhomogeneities of the solar corona. Analogous conclusions were also made by Hewish at the same time (1955) [3].

The results of our observations, and also Hewish's data, permit us to conclude that the 'supercorona' is, like the corona, somewhat asymmetric, its dimensions being somewhat larger towards the equator.

2

Let us now apply the results obtained to some problems of solar radio emission. As the electronic inhomogeneities of the 'supercorona' surround the sun, the solar radio emission is determined not only by the character of the generation of the radio emission, but also by the 'supercorona' which influences the waves passing through it. The effect of scattering by the 'supercorona' affects the observed angular dimensions of individual local sources of radio emission, the distribution of the radio emission from the 'quiet' sun, and the duration of individual short periodic radio emission bursts (peaks).

A formula given by Chandrasekhar (1952) [4] may be applied in our calculations. This formula determines the root-mean-square value of the scattering angle ϕ_c in an inhomogeneous medium:

$$\phi_c = 2 \sqrt[4]{\pi} \Delta n \, l^{-1/2} \sqrt{z} = \psi \sqrt{z}$$

where Δn is the root-mean-square deviation of the refractive index from unity, l is the dimension of the inhomogeneities and z the length of the path.

Writing for $z=0$: $\phi_c = \phi_0$, we have

$$\phi^2(z) = \phi_0^2 + \psi^2 z.$$

If Δn and l are slowly varying functions of z, we obtain

$$\phi^2(z) = \phi_0^2 + \int_0^z \psi^2(x) \, dx.$$

Applying this expression to the experimental data reported above, the following values of $\psi^2(r)$ were obtained, where r is the distance from the

314

centre of the sun and the values of ϕ, ψ, and r are expressed in minutes of arc:

$$\lambda = 5 \cdot 8 \text{ metres}; \ \psi^2(r) = 7 - 0 \cdot 027r,$$
$$\lambda = 3 \cdot 5 \text{ metres}; \ \psi^2(r) = 5 \cdot 5 - 0 \cdot 033r.$$

These values of $\psi^2(r)$, obtained by us, refer to $r > 4R_\odot$, but in calculating the scattering effect on radio spots we extrapolated these expressions to smaller values of r.

It is seen that the values $\psi^2(r)$ increase with decreasing r. This is natural, as with the approach to the photosphere a greater influence of inhomogeneities upon the scattering of radio waves may be expected. Let us point out that the values $\psi^2(r)$ for different wave-lengths should satisfy the following relation:

$$\psi^2(r, \lambda_1) = \psi^2(r, \lambda_2) \frac{\lambda_1^4}{\lambda_2^4}.$$

In our case this relation is not fulfilled. The reason apparently is the observational errors. In the present paper we do not intend to obtain the real numerical data, and shall therefore examine the functions $\psi^2(r, \lambda)$ for two wave-lengths of 5·8 and 3·5 metres independently.

On the basis of the values of $\psi^2(r, \lambda)$ obtained above we may calculate the minimum dimensions of the radio spots caused by the scattering in the 'supercorona'. The results of these computations are tabulated in Table 1.

Table 1

	$\lambda = 3 \cdot 5$ metres		$\lambda = 5 \cdot 8$ metres	
	$\phi = 0°$	$\phi = 90°$	$\phi = 0°$	$\phi = 90°$
$r_2 = 25'$	18'	21'	25'	29'
$r_2 = 40'$	16'	21'	27'	29'

Here r_2 is the distance between the source generating the radio waves and the centre of the sun, ϕ is the angle between the direction towards the earth and the source of radio emission. It is seen that the minimum dimensions of the radio sources are 16' to 21' on the 3·5 metre waves and 25' to 29' on the 5·8 metre waves. The influence of scattering upon the dimensions of the radio-emission sources is thus very important.

3

Let us pay attention now to some experimental results. The diameters of 'radio spots' obtained at the Crimean Station of the Physical Institute of the U.S.S.R. Academy of Sciences are the following: 7'·6 and 9'·8 on the

1·5 and 2 metre waves, respectively (data by Tchikhatchev), and 20′ on the 3·5 metre waves (Vitkevitch).

Let us attempt to compare these experimental results, supposing that the true dimensions of the source for two neighbouring wave-lengths are equal, and that the differences in the apparent dimensions are caused by scattering. Denoting by ϕ_n the measured angle of the observed radio spot, by ϕ_0 the proper dimensions of the radio source, and ϕ_p the angle of scattering, and assuming that the effect of scattering should be combined with the proper dimension by adding the squares, we obtain the following table:

Table 2

	Comparison of 1·5 and 2 metre waves		Comparison of 2 and 3·5 metre waves	
	1·5 metres	2 metres	2 metres	3·5 metres
ϕ_n	7′6	9′8	9′8	20′
ϕ_0	5′8	5′8	7′7	7′7
ϕ_p	4′4	7′8	6′	18′

We see that for 1·5 and 2 metre waves the angular dimension of the radio spot equals 5′8 and the scattering effect is of the same order. On 3·5 metre waves the effect of scattering equals 18′, which exceeds appreciably the dimensions of the source, equalling 7′7. It is seen besides that there may be some increase of the dimension of the radio source with increasing wave-lengths.

One deviating observation may be mentioned. On 11 July 1955 a case was registered on 3·5 metres wave-length, when the dimension of the radio spot was much smaller than 20′. Consequently, either the 'supercorona' above the spot was much less intense during that day, or the source was located very far from the photosphere, which is less probable.

4

The above calculations may be applied not only to local radio spots, but also to the radio emission of the 'quiet' sun.

Comparison of experimental data on the distribution of radio brightness with calculations based on a model solar corona, obtained from optical data, has shown considerable discrepancies up to the present. The radio diameter of the sun obtained experimentally for the metre wave-lengths is appreciably larger than that obtained theoretically. For instance, Machin (1951) [5], who compares the theoretical and experimental data of the distribution of solar radio brightness at 81·5 Mc./s., finds that theoretically

316

the radiation must be absent for values $r > 2R_\odot$, while experimentally the fall of radiation takes place at distances of r exceeding $3R_\odot$. The divergence is considerable and equals 20'. A correction of this order of magnitude might be expected on account of the scattering of radio waves by the electronic inhomogeneities of the supercorona.

The author uses the opportunity to express his gratitude to V. A. Udalzov and J. I. Alekseev for their valuable help in the measurements.

REFERENCES

[1] Vitkevitch, V. V. *Dokl. Akad. Nauk U.S.S.R.* **77**, 585, no. 4, 1951.
[2] Vitkevitch, V. V. *Dokl. Akad. Nauk U.S.S.R.* **101**, 429, no. 3, 1955.
[3] Hewish, A. *Proc. Roy. Soc.* A, **228**, 238, 1955.
[4] Chandrasekhar, S. *M.N. R.A.S.* **112**, no. 5, 1952.
[5] Machin, K. E. *Nature*, **167**, 889–91, 1951.

Discussion

Owren: What are the linear dimensions of scattering elements?

Vitkevitch: There is not a unique answer. A possible combination of values is $l = 100$ km., $\Delta N_e = 1300$, at $R = 5R_\odot$; $\Delta N_e = 600$, at $R = 10R_\odot$; $\Delta N_e = 280$, at $R = 15_\odot$.

Hewish: Observations on the occultation of the radio star in Taurus have been carried out in Cambridge from 1950 onwards, but useful results were not obtained until 1952 owing to the enhanced radiation from sunspots. Detailed measurements made in 1953, using the large radio-telescope in addition to three interferometers of different spacing on 3·7 and 7·9 metres, were reported at the previous Jodrell Bank symposium two years ago. These showed that the apparent reduction of intensity, which could be detected to a distance of 20 solar radii, was caused by multiple scattering in the outer corona. By the application of a diffraction theory it was possible to make deductions about the scale and density of the coronal irregularities (Hewish, 1955) [1].

Similar observations carried out in 1954 and 1955 showed good agreement with the previous measurements, indicating that the irregularities are a permanent feature of the outer corona. In addition, the results have all shown a slight systematic asymmetry, the scattering being greater when the radio source recedes from the sun. Because the sun's axis is not exactly perpendicular to the path of the radio source this could be explained if the contours of constant scattering were ellipses, extended in the equatorial plane, having an axial ratio of about 0·6.

REFERENCE

[1] Hewish, A. *Proc. Roy. Soc.* A, **228**, 238, 1955.

PART V

THE ACTIVE SUN

SPECTRAL OBSERVATIONS OF SOLAR ACTIVITY AT METRE WAVE-LENGTHS

J. P. WILD

Radiophysics Laboratory, Sydney, Australia

This paper summarizes some recent results obtained with the solar radio-spectroscope at Dapto (Sydney), which has been in operation since 1952. The basic instrument is a receiver which sweeps through the frequency range 40–240 Mc./s. every half a second. Each spectrum is displayed as an intensity-modulated line, and successive spectra are recorded photographically on continuously moving film to give a dynamic record of the spectrum.

In describing these results reference will be made to the three distinctive types of burst previously recognized. These are depicted in the idealized sketches of Fig. 1, all being shown on the same time-scale.

I. DERIVED VELOCITIES OF CORPUSCULAR STREAMS FROM THE SUN

We have previously described the characteristic frequency drift in bursts of spectral types II and III. The drift is believed to be due to ionized matter streaming up through the solar atmosphere, exciting plasma oscillations at both fundamental and second-harmonic frequencies; as the matter passes into successively more rarefied regions of the corona, so these frequencies gradually decrease. Thus the frequency scales in Fig. 1 have been shown also as a scale of height above the solar photosphere. This scale is based on electron densities of the undisturbed corona given by Allen and van de Hulst. The scale is intended to refer only to radiation at the fundamental frequency of emission, and is not applicable to the second harmonic.

Velocities derived from the frequency drift of a number of bursts of types II and III are given in the upper histogram of Fig. 2. (In some cases type III drifts 'stop' before reaching the 40 Mc./s. level; these have been excluded from the histogram.) It is seen that the velocities lie in the range 300 to 700 km./sec. for type II bursts, and 2×10^4 to 2×10^5 km./sec. for type III bursts.

Most type II bursts and certain type III bursts occur at the time of solar flares, and the suggestion has been made that the two classes of stream correspond respectively to the geomagnetic storm particles which arrive at the earth 1 to 3 days after large flares, and to the cosmic rays which arrive about 1 hr. after large flares. The lower histograms of Fig. 2 show tentative

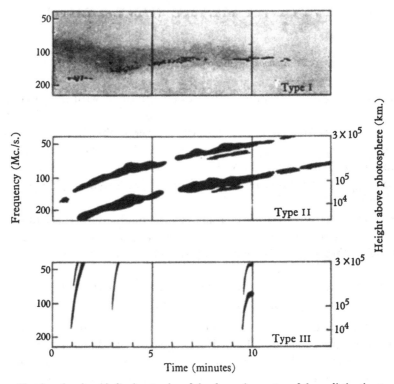

Fig. 1. Sketches showing *idealized* examples of the dynamic spectra of three distinctive types of solar radio disturbance. The scale given on the right-hand side gives the approximate height of the plasma-frequency level in the solar atmosphere. For types II and III the second harmonic is radiated as well as the fundamental; the height-scale refers to the latter only.

data on the velocities of these solar-terrestrial effects, inferred from published sun–earth time delays after flares.

The histogram indicates that any earth-bound corpuscles ejected from a flare with type III velocities would reach the earth at about the time when the cosmic-ray increase begins. On the other hand the energy of corpuscles with type III velocities (~ 100 MeV if the corpuscles are protons) is considerably less than the cosmic-ray energies. The possible association between the two phenomena is therefore obscure, and detailed simultaneous observations of cosmic-ray intensities and type III bursts will be required to discover whether any connexion exists.

The type II velocities are distinctly slower than the velocities of known geomagnetic streams. However, it is important to note that the latter velocities apply only to the 'great' flares of importance 3⁺, while the limited sample of type II bursts so far observed correspond to smaller

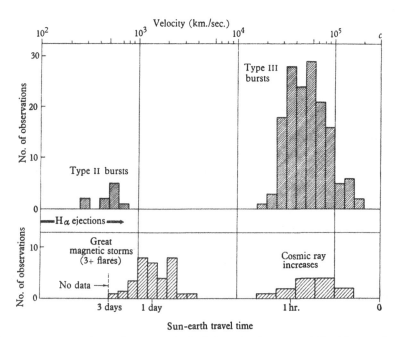

Fig. 2. Histogram (*above*) of velocities derived from the frequency drift of bursts of spectral type II and III. These velocities are compared (*below*) with those inferred from the sun–earth time delays of terrestrial magnetic storms and cosmic-ray increases, and with H_α ejections from flares. The magnetic-storm histogram is that published by H. W. Newton and W. Jackson, and the cosmic-ray data were collected by E. P. George (private communication).

flares. The type II velocities also correspond to the fastest observed H_α ejections from flares. Hence the type II streams may correspond either to the geomagnetic particles or H_α ejections (the latter two phenomena may themselves be related).

2. TWO CLASSES OF SOLAR RADIO STORM

We now recognize two distinct classes of solar radio storm (i.e. prolonged periods of disturbed radio emission lasting hours or days). One kind consists of a continuous sequence of the polarized, narrow-band type I bursts. These are often clustered within restricted but variable frequency bands and superposed on a slowly changing background continuum (Fig. 3c). The other consists of an intermittent sequence of the broad-band type III bursts occurring irregularly at a rate of the order of 10–100 per hour (Fig. 3b). Both kinds of storm appear to be linked with sunspot activity. One kind can occur without the other, and both can occur together.

3. THE MACROSCOPIC PATTERN OF TYPE I STORMS

In an attempt to gain an understanding of type I storms, we have recently reproduced records on a highly compressed time-scale, so that, while individual bursts lose their identity, the 'macroscopic' features of a storm are clearly revealed. These records have been derived from normal-speed records using a photo-mechanical process.

Fig. 3d shows an example of part of a type I storm which was restricted to the higher frequencies above about 150 Mc./s. In addition to irregular features, emission bands are seen to drift repeatedly from high to low frequencies. Assuming as before that the emission takes place near the plasma or zero-refractive-index level, we infer that the sources of emission repeatedly make movements upwards through the solar atmosphere at speeds of the order of 100 km./sec. Inspection of other records not yet analyzed in detail indicate occasional downward movements into the sun. Such behaviour is reminiscent of ascending and descending prominence material.

4. COMBINED OBSERVATIONS OF SPECTRUM AND POLARIZATION

A simple technique has been devised by which the polarization of the received waves is determined at a number of frequencies across the spectrum. Two mutually perpendicular aerials are connected to the swept-frequency receiver by unequal lengths of cable. When polarized radiation is incident upon the aerial system, the signals coming from the two elements interfere, causing the true spectrum to be modulated sinusoidally. The frequency interval between adjacent maxima of this modulation is determined by the difference in the two cable lengths, and the phase of the modulation by the type of polarization (left-hand or right-hand circular, linear, elliptical, etc.). The results largely confirm what had previously been inferred—that type I activity (bursts and background continuum) tend to be strongly polarized, while type II and type III bursts tend to be

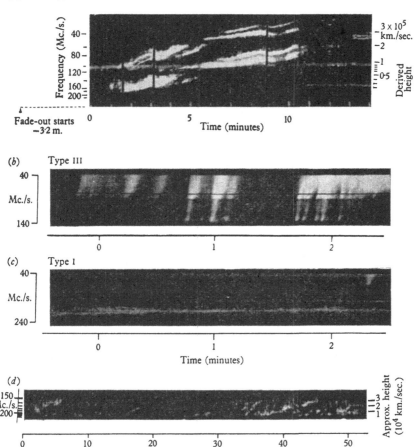

(a) Type II burst following flare

(b) Type III

(c) Type I

(d)

Fig. 3. Dynamic spectrograms of solar activity: (a) Type II burst accompanying a solar flare, showing two harmonic bands with internal fine structure—21 November 1952, 23.49 U.T. The horizontal streaks (e.g. at 120 Mc./s.) are instrumental in origin. (b) An intense part of a prolonged storm of type III bursts (in this case the intensification occurred at the time of a flare)—4 October 1952, 01.58 U.T. (c) Part of a prolonged storm of type I bursts (an isolated type III burst is present near the right-hand edge)—27 April 1953. (d) One hour's record of the same type I storm as in (c), shown on a highly compressed scale of time. Systematic frequency drifts are evident in this case.

Note on conversion of time scales

Figs. 3b and 3c are direct prints from the original records. In Figs. 3a and 3d, however, the time scale of the original record has been compressed by a large factor. This was achieved by projecting the record on an opaque screen in which was cut a narrow slit parallel to and extending across the projected frequency scale. The slit was covered by a ground-glass screen capable of being photographed with a recording camera placed behind the screen. In operation the original record was passed continuously through the projector while fresh film was passed through the recording camera at a slower speed determined by the degree of compression required.

unpolarized. However, some type III bursts observed early in the morning and late in the evening were found to be polarized, suggesting a terrestrial influence in these cases.

BIBLIOGRAPHY

Wild, J. P. and McCready, L. L. *Aust. J. Sci. Res.* A, **3**, 387, 1950.
Wild, J. P., Murray, J. D. and Rowe, W. C. *Aust. J. Phys.* **7**, 439, 1954.
Wild, J. P., Roberts, J. A. and Murray, J. D. *Nature*, **173**, 532, 1954.

RELATION BETWEEN OPTICAL SOLAR
FEATURES AND SOLAR RADIO EMISSION

H. W. DODSON

McMath-Hulbert Observatory, University of Michigan, U.S.A.

During the last five years, we at the McMath-Hulbert Observatory have been attempting to determine the characteristics of radio-frequency radiation associated with certain of the transient solar features that we normally observe on monochromatic spectroheliograms or in integrated light. We have had the privilege of using the original 200 Mc./s. records at Cornell University and the 2800 Mc./s. records at the National Research Council in Ottawa.

It is well established that transits across the solar disk of certain centres of activity coincide in time with periods of enhanced emission at radio frequencies. In general, centres of activity include sunspots, the surrounding bright plage, and superposed regions of red and green coronal emission. The relative roles of these three aspects of solar activity with respect to emission at radio frequencies are hard to determine.

There are certain observations that lead us to believe that calcium plages are better indicators of the regions emitting 2800 Mc./s. radiation than are spots. First, Covington's detailed study of 2800 Mc./s. radiation on 16 October 1951 showed three emissive regions on the solar disk [1]. For the same day there were three similarly located maxima in the east–west distribution of calcium plage intensity, but sunspots were observed in only two of the three regions [2]. Secondly, a plot of measures of areas of calcium plages from 1947 to 1953 versus Covington's measurements of the basic daily flux at 2800 Mc./s. for the same period, gives a single curve for the entire period. In this plot, there is no indication of a change in the zero intercept from year to year (see Fig. 1 for 1947 and 1952 data). Consequently, when calcium plages are used as indicators of the slowly varying component at 2800 Mc./s. it is not necessary to postulate a variation in the 'quiet sun' from year to year. Thirdly [3], studies of the formation and development of a spot and plage between 20 and 27 August 1954 on the otherwise feature-less solar disk, indicate that the basic daily flux at 2800 Mc./s. followed more

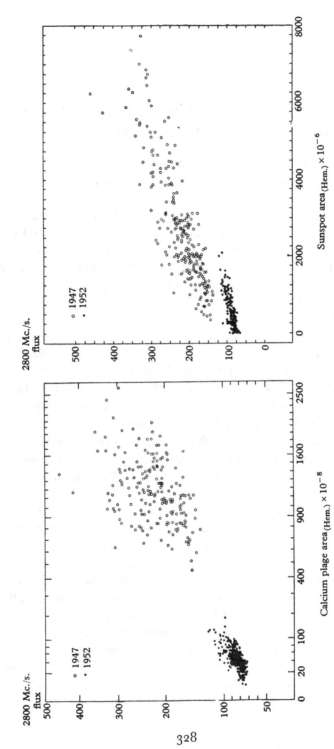

Fig. 1. Plot of daily values of 2800 Mc./s. flux for 1947 and 1952 versus area of calcium plages and area of sunspots.

closely the growth of the plage than that of the associated spot. The area of the spot began to diminish before the area of the plage and radio flux showed significant diminution (see Fig. 2).

Detailed changes within an active centre are often associated in time with detailed features of the radiation at radio frequencies. The short-lived changes in the centre of activity fall into two main categories, (1) active prominences (or 'active dark flocculi' when observed in projection on the disk) and (2) the flare-brightenings. Active dark flocculi often occur with flares, generally during the post maximum phase.

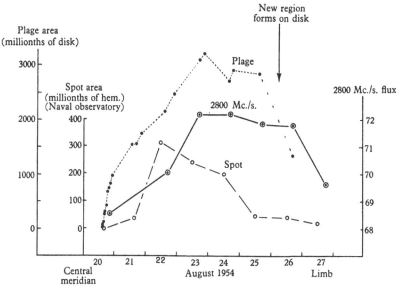

Fig. 2. Comparison of 2800 Mc./s. flux with area of sunspots and calcium plage 20 to 26 August 1954.

The 200 Mc./s. records show a small number of time coincidences between the appearance of active dark flocculi and distinct radio bursts. Most active flocculi occur without associated bursts at the frequencies we have studied. Furthermore, information on the absence or presence of the post-maximum dark flocculi has not been of assistance in distinguishing between flares with and without associated bursts at 200 and 2800 Mc./s. Except for the very high velocity ejections observed in the early stages of some flares near the solar limb [4], our studies have provided only very limited associations between prominence activity and distinctive events at radio frequencies.

The pattern and time relationships between enhancements of 200 and

2800 Mc./s. radiation and the brightenings in H$_\alpha$ flares have been thoroughly described in the past [5, 6]. The double aspect of the flare-associated radiation at both of these frequencies has been pointed out.

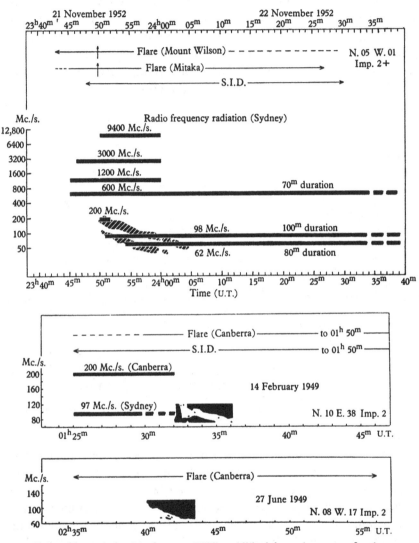

Fig. 3. Time relationships between Wild's published dynamic spectra of outbursts (type II) and observed H$_\alpha$ Flares.

Furthermore, published 97 Mc./s. records from Sydney show that the same general form and time relationships occur also at this much lower frequency. However, the relationship between these flare patterns and the dynamic spectrum or frequency sweep data for outbursts (Wild's type II burst) is not clear in the literature. For the published cases for which we can deduce the circumstances of the flares, the spectrum drift data refer only to the later stages of the flare and to the second or late part of the radio-frequency pattern (see Fig. 3).

If the position of flares associated in time with distinctive radio events be used to locate the radio emitting source, the presently limited inter-ferometric data can be supplemented considerably. Interesting relation-

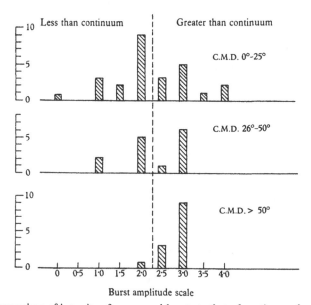

Fig. 4. Comparison of intensity of superposed bursts to that of continuum for 200 Mc./s. noise storms associated with flares at different distances from central meridian.

ships emerge from the use of this type of information. For example, position measurements are available for flares associated in time with the onsets of fifty-three noise storms on the 200 Mc./s. Cornell records. When the flares were within 25° of the central meridian the amplitude of the superposed bursts was generally less than the intensity of the continuum. Between 25° and 50° C.M.D. there were as many cases with the burst amplitude greater than the continuum as the reverse. When the flares were more than 50° from the central meridian the amplitude of the bursts was greater than the intensity of the continuum in more than 90% of the cases (see Fig. 4). This variation from centre to limb in the relative intensity of storm base-level and burst amplitude suggests that the two aspects of the noise storm are separate phenomena and perhaps originate at different levels in the solar atmosphere.

There is growing evidence that the flare mechanism is in some way intimately associated with much of the transient activity at radio frequencies. Our studies indicate that at 2800 Mc./s. there are distinctive or outstanding events *only* when a flare is in progress on the sun. At 200 Mc./s· both outbursts and the distinct onsets of noise storms are flare-associated. Furthermore, intercomparison of the Cornell records from 1950–55 with optical data indicates that 200 Mc./s. noise is often recorded from active centres *having flares* even when the region is at the very limb of the sun (8 May 1951). Our study of the development of the active centre in August 1954 on the otherwise featureless solar disk shows that the noise storm at 200 Mc./s. began when the region started to have flares, even though this did not take place until the active area was about 45° from the centre of the solar disk. Either flares themselves or a stage of development in which flares can occur may be a necessary circumstance for the emission of the greatly enhanced radio radiation at low frequencies.

REFERENCES

[1] Covington, A. E. and Broten, N. W. 'Solar Disk at 10·3 cm.', *A.J.* **119**, 569, May 1954.
[2] Dodson, H. W. 'Intensity Measures of Calcium Plages for Comparison with 10·3 cm. Solar Radiation, 16 October 1951', *A.J.* **119**, 564, May 1954.
[3] Dodson, H. W. and Hedeman, E. R. 'Detailed Study of the Development of an Active Solar Region, 1954 August 20–27', *M.N.R.A.S.* (in the Press).
[4] Dodson, H. W., Hedeman, E. R. and Chamberlain, J. 'Ejection of Hydrogen and Ionized Calcium Atoms with High Velocity at the Time of Solar Flares', *A.J.* **117**, 66, January 1953.
[5] Dodson, H. W., Hedeman, E. R. and Owren, L. 'Solar Flares and Associated 200 Mc./s. Radiation', *A.J.* **118**, 169, September 1953.
[6] Dodson, H. W., Hedeman, E. R. and Covington, A. E. 'Solar Flares and Associated 2800 Mc./s. (10·7 cm.) Radiation', *A.J.* **119**, 541, May 1954.

Discussion

Wild: With reference to Miss Dodson's Fig. 3, I remember vividly the case of the outburst of 14 February 1949. We failed to record the first part simply because the recording camera jammed!

Hey [1]: An analysis of solar radio bursts at 4 metres wave-length during 1946–51 has shown that the bursts occur more frequently in association with flares on the eastern half of the solar disk than on the west. The most likely explanation is that it is due to absorption and refraction in an asymmetrical east–west structure in the solar atmosphere above the flare-active region. When regions associated with high geomagnetic activity are analyzed there appears to be a dip in radio-burst activity near central meridian. Attenuation in corpuscular streams may be responsible for this. Similarly, there is asymmetry in the numbers of flares observed in H_a and this may be explained in a similar way in terms of H_a absorption.

Biermann: From what we have just heard from Dr Hey, it is evident how important new observations of the east–west symmetry of the enhanced radio frequency radiation of the sun would be. At the present time, these observations seem to provide one of the most effective means of studying solar corpuscular radiation. Apart from the direct information about the emission by the sun, it appears that this information may have some bearing on the problems of the mechanism of production of the radiation in the radio-frequency range.

REFERENCE

[1] Hey, J. S. and Hughes, V. A. *Nature*, **173**, 771, 1954.

SUNSPOTS: RADIO, OPTICAL AND GEOMAGNETIC FEATURES

P. SIMON

Observatoire de Paris, Section d'Astrophysique, Meudon (Seine-et-Oise), France

Noise storms on metre wave-lengths originate high in the corona over particular sunspots. Generally, the enhanced radiation comes to a maximum when these sunspots cross the central meridian. If the radiation exceeds a fixed intensity, the sunspots are called noisy (R sunspots); the others are called quiet (Q sunspots). Unfortunately until now only a few interferometer measurements enabled us to know these 'noisy sunspots', but the strong directivity of this emission can be used to determine these sunspots in a statistical manner. We have done this for the sunspots having an average area bigger than 100 millionths of the sun, from 1947 to the middle of 1951. We have thus studied the statistical features for 350 sunspots, 160 of which are noisy and 190 quiet.

I. OPTICAL FEATURES

One research aim was to see if there is some correlation between radio and optical features. A sunspot has a better chance to be noisy the bigger it is, but of the biggest spots one-quarter are quiet. In the Zürich evolution types, the F sunspots are significantly more often noisy than quiet. Strangely enough, the spots with the most important flares or the ones with an abnormal number of flares are not particularly liable to be noisy. In conclusion we think there is no special correlation between radio and optical features.

2. GEOMAGNETIC FEATURES

Another research aim was to find the correlation between the central meridian passage of the sunspots and the geomagnetic activity. The method of superimposed epochs was applied to the K_p figures. The results were:

(*a*) The central meridian passage of the noisy or R sunspots is correlated with an increase of geomagnetic activity during the first days following this passage; the central meridian passage of the quiet or Q sunspots is correlated with a decrease of geomagnetic activity.

(*b*) As far as the central meridian passage of the spots with the most important flares or of the ones with an abnormal or subnormal number of flares is concerned, we detected that the geomagnetic activity is more closely correlated with the radio features than with the optical features.

Studying the solar enhanced radiation appears to be a new and very interesting method of investigating solar terrestrial relations. The solar radio observations during the next years, especially during the International Geophysical Year, will be very useful in making these relations more explicit, particularly those between the solar activity and the geomagnetic activity or ionospheric disturbances.

A detailed account of this investigation has been published [1].

REFERENCE

[1] Simon, P., *Ann. d'Astrophys.* **19**, 122 (no. 3), 1956.

335

OPTICAL EVIDENCES OF RADIATIONAL AND CORPUSCULAR EMISSION FROM ACTIVE SOLAR REGIONS

W. O. ROBERTS

High Altitude Observatory, Boulder, Colorado, U.S.A.

The detailed physical processes involved in centres of activity still remain obscure. Optical and radio noise observations enable us to state in increasing detail the phenomena that any comprehensive theory must explain. At times it seems that the complexity of the picture provided by these new data grows more rapidly than our ability to invent adequate theoretical models.

From the optical data it is clear that the solar atmosphere in outstanding active centres is characterized by very high temperatures, relatively high densities, and by rapid motions on a large scale. The temperature evidence comes from several directions. First, we know that in pronounced active centres the yellow coronal line is often strongly emitted. This line arises from Ca xv, with the highest ionization potential of any known solar spectrum lines [1]. Spectrum lines of the corona correspond to very small optical depths and nevertheless exhibit very large line-breadths over many active regions, strongly implying [2] that the temperatures lie in the range 2×10^6 to $6 \times 10^6 °$ K. Active regions of pronounced character generally display not only stronger line emission than other regions for lines of all ionization potentials, but also relatively stronger coronal line emissions from higher ionization lines (for example, $\lambda 5303$ of Fe xiv compared to $\lambda 6374$ from Fe x). These facts also indicate their high temperature [3].

Active-centre prominences show line-breadth effects that appear to be thermal, since they depend on atomic weight [4]. These suggest temperatures of the order of $10^5 °$, which is high compared with prominence temperatures elsewhere, but not compared with the corona. Line breadths from flares cannot be interpreted as thermal, because of self-absorption effects [5], and because of the excessive temperature values to which such determinations would correspond. But there is ample evidence that in association with solar flares very high temperature effects occur in the solar atmosphere.

Active regions in the solar atmosphere are not only hotter, but they are denser and probably patchier than the quiet solar atmosphere. The electron corona is generally brighter, eclipse photographs show, over active regions [6], implying also that the electron density is greater. Sunspot-type prominences are generally abundant in such regions, and they too are probably much denser than quiescent prominences. They are certainly denser than the surrounding coronal space. The frequent surge prominences are undoubtedly very dense relative to prominences in general.

Goldberg, Dodson and Müller [5] have shown that flares are not only optically thick in H_α emission, but their line width is satisfactorily explained as radiation damping, if their density is considered to be high (number of H atoms on the second quantum level in line of sight = 10^{15} to 10^{16} per cm.²). The increased 5303 A and 6374 A emission of coronal lines from active regions, though the ratio is temperature dependent, also implies a definite increase over active regions of densities in the coronal line-emitting atoms [7], though the absolute coronal densities remain very low [8].

Active centres are characterized also by rapidly moving material in prominences [9]. This is particularly noticeable in looking at routine line spectrograms of the corona and prominences. Strong active regions have prominences with knots showing Doppler displacements corresponding to large average motions in the line of sight. The average velocities of such regions are well above the averages from prominences as a whole. One of the paradoxes, however, is that coronal lines rarely exhibit gross Doppler effects corresponding to such motions [10].

The trajectories of prominences over active regions often suggest a field of motion that is homogeneous throughout a very large volume (radius of the order of 10^5 km.). For example, trajectories of the outstanding active region of 26 February 1946 have been traced and analyzed by Dr Malcolm Correll and his assistants Miss Martha Hazen and Mr John Bahng at High Altitude Observatory this summer (Fig. 1). The trajectories suggest a magnetic field in the solar atmosphere like that of a dipole buried 0·2 solar radii below the photosphere, with the axis of the dipole approximately radial. If such fields do exist, the fact is of great significance in solar radio noise studies.

Before we can fully understand active regions, we must probably understand the 'normal' chromosphere. The behaviour of plages (chromospheric faculae) in active centres suggests that the active-region chromosphere may differ in many respects from the normal, and in particular that it is hotter. We have analyzed one such region and found it not only extended in height, but also hotter at given heights [7,11]. The close connexion of coronal line emission and plages suggests this, too, and ultimately we may have to conclude that the abnormal chromosphere found at centres of activity is responsible for the coronal line emission maxima, or vice versa.

The trend towards higher temperatures in active centres emphasizes the relatively greater importance of the far ultra-violet and X-ray emission of

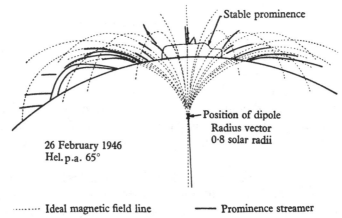

Fig. 1. Comparison between prominence streamers and magnetic lines of force.

the active regions. Specifically, we speculate that the role of the Lyman-like spectrum of ionized helium may be more important, relative to hydrogen emission, than previously recognized.

Direct optical evidence of solar corpuscular emission is scanty. The case for corpuscular solar emission is patched together from many fields, ranging from solar physics through cometary study, to auroras and geomagnetism; it is, however, a good case. Thus far, all efforts to prove directly the existence of ion streams by their absorption in the H and K lines of calcium have ended dubiously or negatively [12]. We are left with direct observations of surge prominences from active centres, but these surges have velocities well below those expected in order to fit geomagnetic evidence (600 to 3000 km./sec.), and with rather indirect speculations from the shape and spectrum of the electron corona above active centres. Above active centres, as mentioned above, there is usually heightened continuous emission from the electron corona. Eclipse photographs reveal that still higher in the solar corona there is a tendency towards a rayed structure, suggesting a diverging cone-like bundle of coronal streamers directed to space. Day-to-day observations of the electron corona are most important in establishing the orientation and directions of these streams, but do not hold out much hope for direct observation of the ejection velocities.

In the study of corpuscular emission the field of radio astronomy has much to offer. Among the high priority studies, as I see it, should be observations of polarization of radio noise bursts at metre wave-lengths. These should assist us to understand the magnetic field structure well up in the solar atmosphere. They should, therefore, be closely co-ordinated with the analysis of solar prominence motion for the same regions and times, and with studies of photospheric magnetic fields at the same times by techniques similar to those of the Babcocks [13] for the weak-field regions of the solar disk.

REFERENCES

[1] Pecker, C., Billings, D. E. and Roberts, W. O. *Ap. J.* **120**, 509, 1954; Layzer, D. *M.N.R.A.S.* **114**, 692, 1954.

[2] Billings, D. E., Pecker, C. and Roberts, W. O. *A.J.* **59**, 316, 1954.

[3] Waldmeier, M. *Zs. Ap.* **30**, 137, 1952.

[4] Billings, D. E. and Zirin, H. *A.J.* **60**, 155, 1955; Zirin, H. *Ap. J.* **123**, 536, 1956.

[5] Goldberg, L., Dodson, H. W. and Müller, E. A. *Ap. J.* **120**, 83, 1954.

[6] Bugoslavskaya, E. J. *Structure of the Solar Corona*, 1949.

[7] Athay, R. G. and Roberts, W. O. *Ap. J.* **121**, 231, 1955.

[8] See also M. Waldmeier: *Convegno Volta*, p. 171, 1953.

[9] Dolder, F. P., Roberts, W. O. and Billings, D. E. *Ap. J.* **120**, 112, 1954.

[10] Roberts, W. O. *Ap. J.* **115**, 488, 1952; Dollfus, A. *Convegno Volta*, p. 221, 1953.

[11] Athay, R. G., Evans, J. W. and Roberts, W. O. *Observatory*, **73**, 244, 1953.

[12] Smyth, M. J. *M.N.R.A.S.* **114**, 503, 1954; Richardson, R. S. *Ann. Report of Mt Wilson Observatory*, 1943–44; Brück, H. A. and Rutllant, F. *M.N.R.A.S.* **106**, 130, 1946.

[13] Babcock, H. D. and Babcock, H. W. *Ap. J.* **121**, 349, 1955.

THE ENHANCED RADIATION FROM SUNSPOT-REGIONS

M. WALDMEIER

Swiss Federal Observatory, Zurich, Switzerland

The so-called enhanced radiation appears at all frequencies in connexion with sunspot-regions. For $\lambda < 25$ cm. it may be interpreted satisfactorily as thermal radiation of the coronal condensations (Waldmeier, 1955) [1]. For $\lambda > 50$ cm. the mechanism responsible for the production of this emission is still unknown. The enhanced radiation from 62 to 200 Mc./s. shows a strong concentration in the radial direction and, therefore, becomes observable only when the source of radiation is close to the centre of the sun's disk. For $\sin \theta = 0.3$, i.e. at an angular distance of $\theta = 17°$ from the centre of the solar disk, the intensity has dropped to 50% of that for $\theta = 0°$ and for $\sin \theta = 0.5$ to about 25%. For $\sin \theta = 0.8$ the radiation intensity is certainly $< 10\%$. This property has recently been investigated again by Machin and O'Brien (1954) [2] as well as by H. Müller [3], using the entire material of the years 1947–53 published in the *Quarterly Bulletin on Solar Activity*. According to Müller the limb darkening is very similar for the frequencies 62 to 200 Mc./s., the radiation intensity having decreased at $\sin \theta = 0.29$ to half of its value for $\sin \theta = 0$. A diminution to 10% may be found for 62 Mc./s. at $\sin \theta = 0.54$, for 80 Mc./s. at $\sin \theta = 0.59$, for 98 Mc./s. at $\sin \theta = 0.62$, for 175 Mc./s. at $\sin \theta = 0.72$ and for 200 Mc./s. at $\sin \theta = 0.77$. For 600 and 1200 Mc./s. the limb darkening is much weaker, the intensity decreasing to 50% at $\sin \theta = 0.57$ and 0.75 respectively.

As the chief sources of the enhanced radiation the coronal condensations have to be considered again. If a point source and spherical distribution of the electron density in the corona are assumed, the calculated limb darkening comes out much weaker than the observed one, no matter how high above the photosphere the coronal condensations are assumed to be. Several authors have expressed the opinion that the enhanced radiation is strongly concentrated to the radial direction already in the emission process itself. In my opinion the strong concentration is produced by a special structure of the corona over the spot-region, which has a marked deviation from the spherical symmetry hitherto assumed.

This interpretation may be illustrated by an example (Fig. 1). On the occasion of the solar eclipse of 25 February 1952 an active spot-region of type E was on the western limb, over which was a well-developed coronal condensation, i.e. a region of increased electron density. This condensation

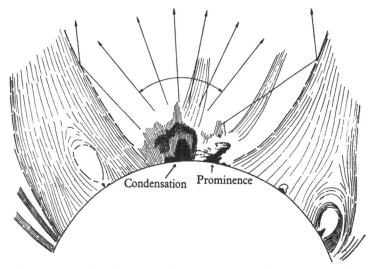

Fig. 1. The structure of the solar corona above a coronal condensation observed at the solar eclipse of 25 February 1952. The arrows indicate the directions in which the enhanced radiation may escape from the corona.

lay at the base of a relatively dark cone, i.e. a region of reduced electron density. The enhanced radiation produced in the condensation can leave the corona through this region of low density. Rays emitted at large angles to the radial direction have to penetrate the dense regions of the corona, whereby they are absorbed. The sharp boundary of the dark region in the shape of a giant parabola corresponding to high gradients of the density is likely to have an additional directional effect.

REFERENCES

[1] Waldmeier, M. *Z. Ap.* **36**, 181, 1955.
[2] Machin, K. E. and O'Brien, P. A. *Phil. Mag.* **45**, 973, 1954.
[3] Müller, H. *Z. Ap.* **39**, 160, 1956.

ON THE ORIGIN OF SOLAR RADIO NOISE

D. H. MENZEL AND M. KROOK

Harvard College Observatory, Cambridge, Mass., U.S.A.

The radio-frequency emission observed in solar bursts cannot reasonably be interpreted as thermal radiation. Its origin is to be sought for rather in terms of co-operative behaviour of systems of charged particles. In any case, we cannot avoid having to examine the physical consequences that arise from such co-operative behaviour.

Violent disturbances, involving considerable amounts of kinetic energy, are frequently observed in the lower layers of the solar atmosphere. These disturbances, initially localized, tend to be propagated through the atmosphere. Determination of the mode of propagation will depend on a study made within the framework of the dynamical theory of ionized gases. In particular, we must treat the material and the radiation fields on an equal footing as coupled systems. In general, the disturbance is propagated through the agency of the electro-magnetic field as well as through material motions. To account for noise bursts, we must discover physical processes that efficiently convert the available mechanical energy, by way of the propagation process, into electro-magnetic energy capable of leaving the boundary of the atmosphere as a radio wave.

We must, therefore, expect to treat the fundamental problem as dependent upon initial and boundary values in the dynamical theory of ionized gases. In constructing models for bursts, we expect the following factors to play essential roles:

(1) The non-linearity of the fundamental equations.

(2) The gradients of density, temperature, etc. in the solar atmosphere.

(3) The possible presence of magnetic fields.

We are still far from having a solution to any definite problem with specified initial and boundary values. The non-linearity of the equations leads to considerable complication in the theory. For this reason progress has practically been confined to investigating a few simple problems with a linearized form of the equations. In general, however, a disturbance, even when it has small amplitude in regions of high density, may grow to large amplitude in regions of low density. We can then no longer describe the phenomena even in terms of cross-modulation between solutions of the linearized equations. The non-linearity of the equations then becomes a very important part of the physical picture, as for example in the generation of shock waves.

It is, however, worth while to explore the consequences of the linearized theory as completely as possible. In view of the great extent of the solar atmosphere, we have to consider oscillations over the whole range from high to low density, i.e. from conditions where collisions predominate to conditions where collisions are negligible. It is extremely difficult to treat this whole range in a unified way on the basis of the Boltzmann kinetic equations. However, a unified treatment, in terms of a special model introduced by Bhatnagar, Gross and Krook, satisfactorily represents the effects of molecular collisions in gases. The model leads to kinetic equations which possess the same physical features as the Boltzmann equation but are mathematically more tractable. For small-amplitude oscillations, the distribution functions can be determined explicitly over the whole density range; dispersion relations can then also be determined for the whole range. Gross and Krook have applied the model to trace the continuous transition from sound waves at high density to longitudinal plasma oscillations at low density.

In the presence of magnetic fields the situation is much more complex. The mode of propagation depends on the relations between a number of parameters, e.g. collision frequencies, plasma frequencies, gyro-frequencies, etc. In general, longitudinal oscillations and transverse oscillations are coupled. In the limit of high density, transverse plasma oscillations take on the character of magneto-hydrodynamic waves; as the density tends to zero the oscillations go over continuously into electro-magnetic waves in free space.

A mechanical disturbance in the lower layers of the solar atmosphere tends to excite longitudinal modes of oscillation of the medium. In the presence of a magnetic field, these oscillations in turn excite transverse modes. As these transverse waves penetrate into regions of lower density, their amplitudes tend to increase. At the same time the waves take on more and more the character of electro-magnetic waves. Under suitable conditions an appreciable fraction of the original mechanical energy may, in this way, eventually escape from the sun in the form of radio-frequency emission.

This research was supported in part by the Air Force Cambridge Research Center, Geophysics Research Directorate, through Contract AF 19(604)–1394 with Harvard University.

343

Discussion

Minnaert: Is there any explanation why the third harmonic has not been detected?

Menzel: Theory has not been refined to that point.

Pawsey: It would seem likely that the polarized bursts need an explanation in which a magnetic field is essential but the non-polarized bursts might have an explanation without a field.

Menzel: Our view has been that a magnetic field is necessary for transverse waves. But random fields would tend to randomize the polarization.

ON THE MECHANISM OF SOLAR OUTBURSTS

K. O. KIEPENHEUER

Fraunhofer Institute, Freiburg im Breisgau, Germany

Today it is taken for granted that the solar outbursts are produced by plasma oscillations of coronal matter, excited by a corpuscular radiation originating in the inner corona (Pawsey and Smerd, 1953) [1]. A certain component of this corpuscular radiation is obviously identical with that generating geomagnetic storms.

In favour of such a hypothesis is the fact that the order of magnitude of the velocity of the exciting corpuscles (500 to 1500 km./sec. for type II bursts) is the same as that deduced for geomagnetic storms. Also plasma oscillations are observed in the laboratory when corpuscles are shot through a gas. Adopting this hypothesis we may derive some interesting conclusions on the mechanism of outbursts and also on the properties of the corpuscular radiation. The physical interpretation of the outbursts, which belong to the most direct manifestations of solar activity, in some respects even more directly than chromospheric flares (Davies, 1955) [2], is of great importance to the solar physicist as well as to the radio astronomer.

First, we mention a few conclusions that may be drawn from the observed band-width, variation of frequency and region of frequency emitted by bursts.

Band width. It is of the order of 10 Mc./s., which is only one-tenth of the total region of frequency covered. The extent of the exciting corpuscular radiation in the direction of its motion therefore can be only a small fraction of the radial extension of the corona, probably less than 10^{10} cm. It would be better to speak about a corpuscular cloud than about a corpuscular stream.

Range of frequency. Co-ordinating the observed frequencies with the corresponding electronic coronal densities, it follows that the exciting corpuscles already reach their final velocity in the very inner corona. The process of acceleration therefore occurs along a very short distance of the order of 20,000 to 50,000 km.

Variation of frequency. Assuming a constant velocity of the exciting corpuscles, the observed variation of frequency with time corresponds to an undisturbed decrease of the density of coronal electrons. This result is reasonable if one concludes that the normal distribution of electronic density in the corona is not disturbed appreciably by the corpuscular radiation penetrating the corona.

Next, we mention a few more quantitative data about this type of solar corpuscles as they can be estimated from the analysis of geomagnetic storms (Kiepenheuer, 1953) [3]. In the vicinity of the earth the corpuscular cloud should have the form of a gaseous shell, the thickness of which ($\sim 10^{11}$ cm.) is small compared with its lateral extension ($\sim 10^{13}$ cm.). The total mass of this gaseous skin should be of the order of 10^{12} gm. (10^{-20} of the solar mass). Assuming that this mass is taken from the inner corona and originally filled the volume of a sphere of 50,000 km. diameter (= extension of a mean spot group), an initial density of 10^7 to 10^8 protons/cm.^{3} is needed. This is a density similar to that of the corona. The mass is of the order of an average prominence.

The kinetic energy of the corpuscular cloud is estimated as 10^{28} to 10^{29} ergs. The total radio emission of an outburst generally does not exceed 10^{22} ergs. It follows that only a very small part of the kinetic energy is used for the generation of electro-magnetic radiation.

I. ORIGIN OF THE CORPUSCULAR CLOUD

What mechanism can accelerate a corpuscular cloud of the described properties in the very inner corona? There is very little doubt that the kinetic energy of the cloud comes from the field energy of a local magnetic perturbation. All other stocks of energy are quite inadequate. Only by the intervention of a magnetic field can it happen that the energy of a great collection of particles is transferred to a small number of particles with a reasonable efficiency, thus giving them very high kinetic energies.

Outbursts as well as geomagnetic storms suggest that the corpuscular cloud is always produced in the very vicinity of a spot-group within the inner corona and that it reaches its final velocity—a multiple of the velocity of sound—along the short distance of 20,000 to 50,000 km. For this reason we had better speak of a coronal explosion, in which the conversion of magnetic field energy into kinetic energy is occurring very rapidly. The following proposed model of this explosion is in a certain sense the reverse of a process, which, according to the theory of Chapman and Ferraro (1931, 1932, 1933) [4], happens when the same cloud of corpuscles enters the outer parts of the magnetic field of the earth.

On approaching the earth the corpuscular cloud is exposed to an increasing magnetic field. This induces a system of currents along the surface of the cloud facing the earth, which screens the rear parts of the cloud completely against the earth's field. The thickness of this current system is of the order of $\sqrt{\tau/\sigma}$, where σ is the electric conductivity of the cloud's material and τ the time taken by the field to increase sensibly. The thickness of the current system can also be regarded as the penetration depth of an electro-magnetic wave of the period τ into a conductor of the conductivity σ. It is extraordinarily small compared with the dimensions of the cloud. The formation of the current system and the deformation of the earth's field resulting from it is made at the expense of the kinetic energy of the cloud. The braking force of the induced current system per unit area is $H^2/2\pi$ and affects only the thin current-bearing layer, forming a strong compression along the top front of the cloud. The deformation of the earth's field resulting from this current system is, according to Chapman and Ferraro, recorded on the earth's surface and is supposed to represent the initial phase of a geomagnetic storm.

So much for Chapman and Ferraro's model of a geomagnetic storm. Let us now try to picture the reverse of this process in the solar atmosphere, where, on the contrary, a solar magnetic storm should set in motion a swarm of corpuscles. H. D. and H. W. Babcock (1955) [5] have shown that such storms occur in the solar photosphere. They observed field changes of several gauss in a few minutes within regions of more than 10,000 km. Much stronger changes of field are to be expected in the centres of activity, i.e. in the vicinity of spot-groups. Unfortunately no reliable measurements of these rapid field changes are available up to now. There can be little doubt, however, that the main cause of the group of phenomena occurring in a solar centre of activity is of magnetic nature. Especially chromospheric flares and the violent surge activity sometimes coinciding with a flare should be caused by a local magnetic perturbation rising from greater depth. It is known that the photospheric granulation behaves rather normally in these centres of activity. For this reason, it seems clear that the disturbing field is transported into the photosphere by the regular convection observed in the form of granulation. This is also to be expected because in this region the mean kinetic energy (> 100 ergs/cm.³) probably exceeds the magnetic field energy (about 5 ergs/cm.³ for 10 gauss). Observational results are not yet quite conclusive, but outside the penumbra of spots in general no fields greater than 10 gauss are observed. All solar magnetic fields as observed by the Zeeman-effect are located in a photospheric layer which is even greater in depth than the layers seen in a spectrohelioscope.

Almost nothing is known about how these observed photospheric field variations are transferred into the chromosphere or to the corona. The region between the depth of line formation and the height where chromospheric structures penetrate into the inner corona or where variable magnetic fields could be injected into the corona is *terra incognita*, a region where probably strong sound-, shock- and hydrodynamic-waves occur. There is no doubt that the observed surge activity, represented by chromospheric matter injected into the corona with velocities up to 700 km./sec., is one of the consequences of such deeper-seated magnetic field variations and that this injected matter carries with it magnetic fields of the order of 1 gauss or more. For this reason we have to expect rather rapid field changes in these parts of the inner corona. The existence of such field changes can be inferred with certainty from the observed rapid motions and changes of form of the so-called active filaments in the vicinity of sunspots. Without going into details, we may represent such a field change by a perturbation of the dimension $2a_0$ with a 'magnetization' H_0, which emerges into the corona within the time τ.

The arrival of this perturbation produces in the adjacent coronal matter a system of induction currents, the depth of which is again of the order $\sqrt{\tau/\sigma}$. Putting $\tau = 200$ sec., and $\sigma = 2 . 10^{-5}$ e.m.u., a depth of the current system of only 3000 cm. results. This means only that the thickness of the current-bearing layer is very small compared with the dimension of the disturbance. The actual layer will have at least the thickness of several free-paths.

This current-bearing skin is set in motion by the magnetic pressure $H^2/2\pi$, that is, by the repulsing force between the inducing and the induced current system. The moving layer pushes the coronal matter above it. The detachment of the induced current system from the primary perturbation in its first stage is difficult to understand. It might occur easily, however, when the density of the perturbation (chromosphere) is much greater than that of the initial current-bearing layer (corona). In other words, it will start preferably in regions with a steep density gradient. After a small detachment is accomplished, the separation will grow rapidly.

The magnetic disturbance, therefore, repels the adjacent coronal matter in such a way that an almost empty bubble is formed, the inner surface of which becomes the bearer of the induced current system. By this system the region outside this bubble is screened completely against the primary disturbance. If the disturbance has a dipole character, then the pressure on the unit area will decrease as H^2 or as $(a_0/a)^6$, where a is the radius of the bubble, a_0 that of the perturbation. The acceleration,

348

therefore, has to be accomplished along a distance of the order of $a_0/2$. The amount of mass collected by the current-bearing layer along this distance per unit area will be of the order of

$$m \approx \tfrac{1}{2} a_0 \rho_0,$$

where ρ is the mean density of the inner corona. The acquired final velocity $\dot{r}_\infty$ can be estimated from an approximated energy balance

$$\frac{m}{2}\, \dot{r}_\infty^2 \approx \frac{a_v}{4}\, \rho_0 \dot{r}_\infty^2 \approx \frac{1}{2\pi} \int_{a_0}^{\infty} H^2 dr,$$

and thus is

$$\dot{r}_\infty^2 \approx \frac{H_0^2 a_0}{2\pi m} \approx \frac{H_0^2}{\pi \rho_0}.$$

The total mass being accelerated to this velocity is of the order of

$$M \approx 2\pi \left(\frac{a_0}{2}\right) m \approx \frac{\pi}{2} \rho_0 a_0^3,$$

and the total kinetic energy gained becomes

$$E_{\text{kin.}} \approx a_0^3 \rho_0 \frac{\dot{r}_\infty^2}{2}.$$

As can easily be verified, the kinetic energy gained by this process equals the total magnetic energy of the perturbation.

This high efficiency is not real, however, because two effects may, under certain circumstances, reduce it drastically. First of all, the reaction of the induced on the inducing current system has been neglected in our simplified model. This reaction, however, cannot seriously alter the magnitude of the resulting acceleration. The second objection concerns the nature of the primary field perturbation. If this perturbation is represented by a rigid body carrying a magnetic field with it, then the building up of its external field will be a very slow process. Because of the 'skin effect' only the currents in the very outer parts of this magnetized body will contribute. This skin effect, however, can be annulled almost entirely by internal turbulence, which brings in succession the internal fields to the surface, as long as the field is small enough not to prevent the turbulent motion.

The acceleration was estimated on the assumption that the emerging time τ of the disturbance is smaller than the time of acceleration t, which is

$$t \approx \frac{a_0 \sqrt{\rho_0}}{H}.$$

Taking plausible values for a_0, ρ_0 and H an acceleration time of the order of a few seconds seems quite possible. The question is, whether appreciable field changes are to be expected in this short interval of time. The emerging time of a magnetic disturbance transported into the inner corona by a chromospheric surge will be of the order of a_0/v, where a_0 is the diameter and v the velocity of the surge. For a velocity of the surge of 100 km./sec. and a diameter of 1000 km. an emerging time of about 10 sec. results. Assuming that the field increases linearly in this time, a field change of

$$\frac{dH}{dt} \approx \frac{v}{a_0} H_0$$

will be produced, which can amount easily to 5 gauss/sec. in the vicinity of active spot-groups.

As the observations of Wild, Roberts and Murray (1954) [6] have shown, type II bursts (~ 500 to 1500 km./sec.) are often preceded by type III bursts ($> 10{,}000$ km./sec.) of short duration. One is tempted to believe that the fast corpuscles exciting the type III bursts originate in the same coronal explosion as depicted above. But why, then, is not the entire spectrum of velocities between 500 and 50,000 km./sec. observed? To answer this question would require a deep look into the mechanism of the explosion. But even without going into too much detail, it can be shown that there must be at least two groups of velocities.

Let us assume for a moment that the magnetic perturbation leaking into the inner corona is surrounded by a vacuum and that there is no other field than that of the perturbation. Then the lines of force of the developing field move outward with a velocity which satisfies the condition

$$\left(\frac{\partial H}{\partial t}\right)_r + \dot{r}\frac{\partial H}{\partial r} = 0, \quad \text{so} \quad \dot{r} = \frac{\partial H}{\partial t}\bigg/\frac{\partial H}{\partial r}.$$

Assuming a dipole field, we find

$$\dot{r} = \frac{a_0}{3}\frac{\dot{H}_0}{H_0}.$$

This means that in the very first moment, when the field breaks through into the corona and H_0 is still very small, the lines of force will fly into space with a very high velocity. But how can this field develop in the corona with its high electric conductivity? It is known that the material is bound to the lines of force. The initial rapid outward motion of the lines of force must be accompanied therefore by a mass motion. This will be true only as long as the magnetic drag force exceeds the forces of inertia acting on accelerated material. That is, as long as $H^2/2\pi \gg m\ddot{r}$ ($m =$ the total mass per cm.², set into motion by the field). This condition is best fulfilled for the type II bursts, which require an acceleration of about 2×10^6 cm./sec.^{2} For the type III bursts, however, accelerations up to about 10^9 cm./sec.^{2} are needed.

Assuming fields of the order of a few gauss it can be seen easily from the above conditions that magnetic pressure and inertia become equal for an accelerated mass of 10^{-10} gm./cm.^{2}, that is, for a coronal column of about 10 km. length.

The conclusion may be drawn that the swarms of fast corpuscles producing the type III bursts originate in those low layers of the corona which are hit first by the emerging magnetic perturbation. As shown elsewhere (Kiepenheuer, to be published) [7], it also seems reasonable to explain the acceleration of cosmic-ray particles on the sun in terms of such a mechanism. The radio observation of solar corpuscles up to velocities of 0·2 the velocity of light favours strongly such an hypothesis.

2. DETACHMENT OF THE CORPUSCULAR CLOUD AND ITS MOTION THROUGH THE CORONA

When the accelerating force of the magnetic disturbance has ceased because of the increasing distance of the induced current system, the protons and electrons are exposed only to gravitation and friction. Gravitation can be ignored for velocities exceeding the velocity of escape, which is about 600 km./sec. on the sun's surface. The friction is represented by the interaction of protons and protons, the collisional cross-section being

$$q \approx 8\pi \, \frac{c^4}{m_p^2 v^4} \, \log\left(\frac{1}{\theta_{\min}}\right),$$

where m_p is the protons mass, v the relative velocity of the colliding protons and $\theta_{\min}$ is the smallest angle of deviation, which still means a collision. It turns out, that protons with $v \gtrsim 4 \cdot 10^7$ cm./sec. are capable of escaping from the inner corona without collision. For this reason the fast protons, responsible for type III bursts, after being blown into the corona during the initial phase of the explosion, will detach themselves and penetrate the corona unimpeded while the main explosion is going on.

The bulk of exploding material will continue to heap up and to compress coronal matter, as long as its velocity relative to the surrounding corona is sensibly less than 400 km./sec. and as long as the magnetic pressure lasts. When the critical velocity and distance from the primary disturbance is reached, the shell of collected coronal material flies freely through the corona, steadily increasing its dimensions but without disturbing the shape of the corona.

351

It remains to discover whether the corpuscular cloud flying through the corona almost without friction can transfer enough energy to the corona to excite plasma oscillations of the required intensity. For a corpuscular velocity of 1000 km./sec. the cross-section per proton amounts to 5×10^{-21} cm.^{2}, the cross-section of the total corona to about 5×10^{-3} cm.^{2}/ cm.^{2}. Therefore about five per mille of the kinetic energy of the cloud, that is 5×10^{25} to 5×10^{26} ergs, are transferred to the corona. This is about 10^4 times the radio energy transmitted by an outburst. The required efficiency of the mechanism of excitation of plasma oscillations, therefore, turns out to be of the order of 10^{-4}.

H. D. and H. W. Babcock have also proposed a model for the acceleration of solar corpuscles [5]. They believe that by the occasional encounters of photospheric turbulent bodies magnetic fields and matter can be squeezed into a tube-like volume. The compressed matter can then escape only along the lines of force parallel to this tube. According to the steep gradient of density in the solar atmosphere the velocity of expansion would become very high in the upper parts of this tube. The material thus accelerated is identified by the authors with the solar corpuscular radiation. This simple mechanism, which transfers the kinetic energy of large photospheric masses to the ejection of a few particles by the intervention of a magnetic field, is tempting but is subject to serious objections. In particular, it is difficult to imagine what forces are able to maintain the form of the postulated tube formed by the lines of force. This tube has to survive a drop of density of the order $1:10^5$ to permit the desired mechanism of acceleration to work. We think that the field fluctuations produced by photospheric or chromospheric turbulence manifest themselves in the explosional process described in this paper.

In order to come to a quantitative estimate of the accelerating force in a coronal explosion, the mechanism proposed by us has been very much simplified and idealized. Certain details which might be of great importance in a thorough treatment of the process have been ignored completely, such as shock-waves and hydromagnetic-waves. Also very important observational details have been omitted in order to get a clear picture of the explosion as such. It is my hope that the wrong as well as the right parts of my mechanism might be a stimulus to the experts in this complicated field.

It is evident that a very close co-operation between solar physicists (especially those who can measuer weak magnetic fields on the sun) and radio astronomers can contribute greatly to a better understanding of this type of coronal phenomena.

REFERENCES

[1] Pawsey, J. L. and Smerd, S. F. in G. P. Kuiper, *The Sun*, pp. 508ff., Chicago, 1953.
[2] Davies, R. D. *M.N.R.A.S.* **114**, 74, 1955.
[3] Kiepenheuer, K. O., in G. P. Kuiper, *The Sun*, pp. 449ff., Chicago, 1953.
[4] Chapman, S. and Ferraro, V. C. A. *J. Terr. Magn. Atm. El.* **36**, 77, 1931; *ibid.* **37**, 421, 1932; and *ibid.* **38**, 79, 1933.
[5] Babcock, H. W. and Babcock, H. D. *Ap. J.* **121**, 349, 1955.
[6] Wild, J. P., Roberts, J. A. and Murray, J. D. *Nature*, **173**, 532, 1954.
[7] Kiepenheuer, K. O. Report of the Cosmic Ray Conference, held at Guanajuato (Mexico) in September 1955, to be published.

Discussion

Ellison: We have as yet no optical evidence for the expulsion of particles from the flare regions at speeds of 60,000 km./sec., but there is considerable evidence for the emission of hydrogen atoms at speeds of the order of 500 km./sec. [1, 2].

REFERENCES

[1] Ellison, M. A. *Pub. Roy. Obs. Edin.* **1**, 110, 1952.
[2] Ellison, M. A. *Relations entre les phénomènes solaires et terrestres* (C.I.U.S.). Huitième rapport (1954), p. 33.

REMARKS ON THE ENERGY OF THE NON-THERMAL RADIO-FREQUENCY EMISSION

L. BIERMANN AND R. LUST

Max Planck Institute of Physics, Göttingen, Germany

It is proposed to discuss the role of the radio-frequency emission in the whole set of the non-thermal emissions of the sun, which originate in the solar corona and the uppermost regions of the chromosphere. Of these, the radiative emissions give probably an amount of between 10^4 and 10^5 ergs/cm.^{2} sec., to which the radio-frequency region contributes only very little (even during an intense outburst, when the total radiative emission is much larger, not more than 10^0–10^2 ergs/cm.^{2} sec.); the contribution of the lower chromosphere, however, is not yet well known. The corpuscular emissions under normal conditions seem to require $\gtrsim 10^5$ ergs/cm.^{2} sec. (but again much more in active, e.g. 'M', regions), and to constitute a normal feature of the outer solar corona. These emissions, we propose, are maintained by the same supply of mechanical energy which secures the thermal and radiative equilibrium of the inner corona. That is to say, some part of the flux of acoustic energy originating in the hydrogen convection zone and, according to this theory, heating the upper chromosphere and the inner corona, is believed always to reach the outer corona, where the radiative loss is smaller, and this part is then at least comparable with that dissipated in the inner corona.*

It is evident that the travelling waves transmitting this energy outwards, and the mass motions constituting the corpuscular emission, are greatly affected by the solar magnetic fields. If one assumes 1 gauss as an estimate for the large-scale magnetic field intensity on the sun, the magnetic stress becomes comparable with the ordinary gas pressure just in the transition region from the chromosphere to the corona. The magnetic field will cause the acoustic energy to flow along the magnetic lines of force rather than at right angles to it. These questions have been discussed in some detail by

---

* For a more detailed discussion, with a somewhat different emphasis, and for references, see the contribution of one of us [1] to the Joint Discussion on Turbulence in Stellar Atmosphere, I.A.U. General Assembly in Dublin, 1955.

one of us [2]. The mechanism by which the mechanical energy, after heating the regions considered, causes the acceleration of the outer regions of the corona, will be considered in the communication by Arnulf Schlüter.

From this point of view, the radio-frequency radiation of the chromosphere and the corona ultimately is of completely non-thermal origin, although the emission mechanism of the normal radiation, and probably also that of the slowly varying component, is microscopic. The coherent oscillations causing the more violent types of radiation are considered as produced in the processes mentioned before, for instance directly by the pressure gradients connected with the shock waves, or in the boundary regions of the corpuscular streams. These streams, however, should not necessarily be regarded as something physically different from the outer corona.

REFERENCES

[1] Biermann, L. *Trans. I.A.U.*, **9**, p. 750, 1957.
[2] Lüst, R. *Z. Naturf.* **10**a, 125, 1955; see also *Z. Ap.* **37**, 67, 1955.

SOLAR RADIO EMISSION AND THE ACCELERATION OF MAGNETIC-STORM PARTICLES

A. SCHLUTER

Max Planck Institute of Physics, Göttingen, Germany

The shift of the emitted frequencies towards lower frequencies during a solar outburst is usually interpreted as due to a progressive rarefaction of the emitting gas. If one assumes that the emitted frequency is identical with the plasma frequency and furthermore that the density of the emitting plasma is similar to the density of the solar corona at the location of the radiating material, then it follows that this material is subject to an acceleration throughout the solar corona which compensates or exceeds the effect of the gravitational field of the sun.

The existence of the general magnetic field of the sun, as observed by the Babcocks, seems to present a twofold difficulty to this interpretation. First, during the initial part of an outburst the radio-frequency emission is usually unpolarized, whereas it would be expected to be polarized in the presence of a magnetic field. Secondly, motion of ionized material across magnetic lines of force is impossible since the electrical conductivity is almost as high in the corona (due to the high temperature) as it is in ordinary metals; the so-called reduction of the conductivity across the magnetic lines of force (or more appropriately the Hall effect) does not affect this conclusion appreciably, as I have shown earlier. Therefore, if the ejected material is penetrated by the 'general' magnetic field of the sun, it has to drag this field along and overcome the Maxwell tensions which counteract any lengthening of the lines of force. Even if the accelerating mechanism were much stronger than required to compensate the gravitational field of the sun, ejection of material to infinity would seem almost impossible.

Both difficulties are circumvented if one assumes that the ejected material is *not* magnetized and moves *between* the magnetic lines of force, punching holes into the field. The material then behaves like a diamagnetic body and the Maxwell tensions of the magnetic field produce an accelerating force ('melon seed' mechanism). If one neglects the gas pressure of the *un*accelerated coronal gas and assumes a constant temperature of the ejected material, the magnitude and the direction of this force are proportional to $-\mathrm{grad}\,\log H^2$, where H denotes the magnitude of the solar magnetic field that would exist at the considered place if no 'diamagnetic' matter were present. The logarithmic dependence is caused by the expansion of the ejected matter when it enters regions of smaller magnetic field strength. Therefore, the diamagnetic force is smaller than would be exerted upon a rigid body. (Only the latter possibility is usually considered in text-books.)

A closer analysis shows that the energy of acceleration stems from the heating mechanism of the corona while the magnetic field provides only the forces needed to balance gravitation and the change of momentum. All non-magnetized material, which is heated to a few million degrees (the exact value of the temperature depends on the structure of the magnetic field but not on its strength), will be accelerated and will reach the vicinity of the earth with a speed comparable to the velocity of escape at the point where the acceleration started. The speed of the particles which cause the ordinary magnetic disturbances on earth is usually thought to be of the order of the escape velocity near the surface of the sun. This seems to confirm the present theory. The increased speed of the particles during large magnetic storms could be explained by the assumption that these particles originate in locations on the sun where the magnetic field is considerably stronger than the 'general' magnetic field of the sun.

Discussion

Menzel: You do not mean to say that the dipole field can support everything? The force involved must be expressed as the divergence of the magnetic tensor, not as the gradient of the magnetic pressure.

Schlüter: I agree. The distortion component causes the support.

POLARIZATION OF SOLAR RADIO BURSTS

T. HATANAKA

Tokyo Astronomical Observatory, Mitaka, Tokyo, Japan

I. TIME-SHARING RADIO POLARIMETER

The polarization observation of solar radio emission at 200 Mc./s. was started in December 1954 at the Tokyo Astronomical Observatory with a new type of radio polarimeter [1, 2]. With a pair of crossed dipoles at the focus of a 10 metre paraboloid of equatorial mounting, it gives almost simultaneously six components of polarization, two circular and four linear, on a time-sharing basis by using electronic switching. Fig. 1 shows the block diagram of the circuit and the combination of modulators and demodulators. 6AS6 pentodes are used as the time modulators. The duration of each gate pulse is 1/1600 sec. and one set of observations is obtained in 1/200 sec. One of the principal features of the present scheme is to use a common receiving system for all the components in order to avoid inevitable errors due to differences or changes in the gain and in the central frequencies of different receivers. A new type of square-law detector is employed in the present system [3]. It has rapid response time and excellent stability.

Six pieces of information are fed to a six-pen recorder made specially for our purpose. At the same time six Y.E.W. recorders, which are quite similar to Esterline-Angus recorders, are used in order to get a wider latitude in intensity. The overall time-constant is 1/2 sec. for the former and 1/5 sec. for the latter. The speed of tape for the former recording is 60 mm./min. in active periods. Examples of the observed records are shown in Fig. 2.

2. SIX COMPONENTS OF POLARIZATION

It is known that four independent pieces of information are necessary in order to determine completely the state of polarization. It seems, however, that no such observation has been carried out in radio astronomy. It has been customary to observe two circularly polarized components and at

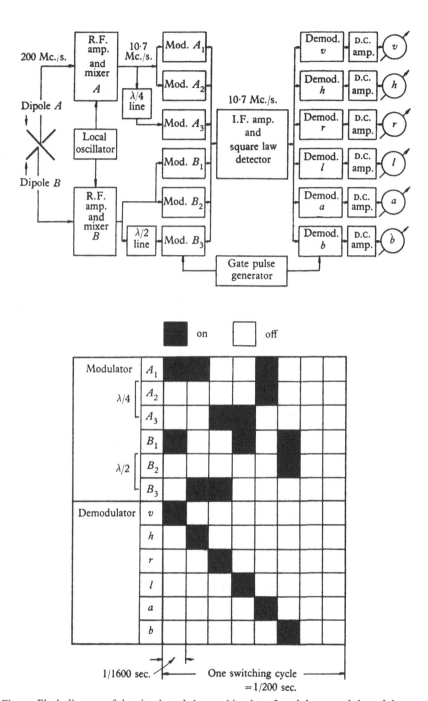

Fig. 1. Block diagram of the circuit and the combination of modulators and demodulators.

359

most to add one linearly polarized component [4]. The polarization of solar radio bursts, therefore, has been said to be either random or circular or at most a mixture of circular plus random.

The present system is to observe six components of polarization. This means that our system gives two checks besides necessary information.

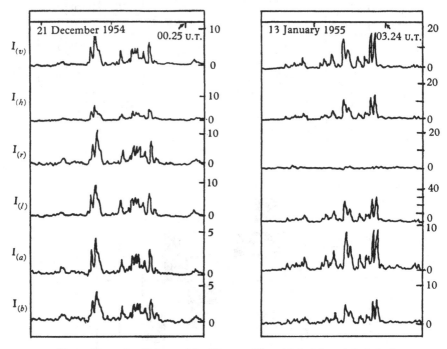

Fig. 2. Examples of the records of bursts. (One-third of the original size.)

This system has been adopted also because of the simplicity in the reduction, since the differences between two complementary components give directly three of four Stokes' parameters, i.e. Q, U and V [2].

3. POLARIZATION OF TYPE I BURSTS

A preliminary analysis reveals the following points on the polarization of type I bursts.

(i) The radiation is a mixture of two components: one is elliptically polarized and the other random. In other words the polarization of type I bursts is to be described by the most general state of polarization.

(ii) The ellipticity varies from nearly 100 %, i.e. circular, to 10 %, i.e. nearly linear. The variation from day to day is, on the average, similar to the ellipticity of a circle put tangentially on the sun's surface at the position of the source and viewed from the earth. The situation is illustrated in Fig. 3. The curves in the figure give the expected ellipticity according to the hypothesis mentioned above. The dotted part means that the source was not active during the period as judged from the interferometer observation at Cornell.

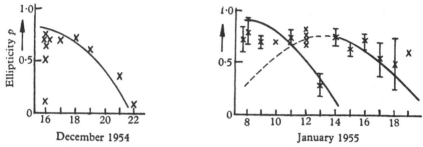

Fig. 3. Variation of the ellipticity with the position of the source.

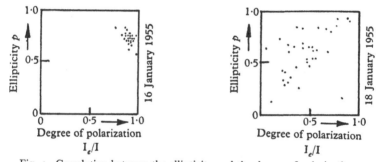

Fig. 4. Correlation between the ellipticity and the degree of polarization.

(iii) The degree of polarization, or the fraction of the elliptically polarized component, is more than 90 % on most days, but sometimes drops below 50 % and even down to 10 %.

(iv) The ellipticity and the degree of polarization of bursts are usually the same in one day within the accuracy of the determination, but sometimes they show large scatter. Examples are shown in Fig. 4. It seems that large scatter in such a diagram is observed either on the day the source becomes suddenly active, or when the source is very close to the limb.

(v) The tilt angle of the ellipse is almost constant during the observing time in a day, but is different on different days.

4. FARADAY EFFECT ON THE POLARIZATION OBSERVATION

It is pointed out that the tilt angle of the polarization ellipse is rotated to some extent because of the Faraday effect in the earth's atmosphere. The amount of the rotation is estimated to be several radians at 200 Mc./s. for a model ionosphere. It is to be remarked that if the solar radio emission is observed in such a direction that it passes the ionosphere almost transversely to the earth's magnetic field a change in the shape of ellipse besides the rotation of the axis is to be observed. The frequency of 200 Mc./s. is, however, too high to observe this 'transverse magnetic field effect'. Such an observation at much lower frequency, say at 50 Mc./s., will be of interest [5].

The Faraday effect is also expected to be present in the solar atmosphere. If we adopt a model solar corona and the general solar magnetic field of a dipole type with the polar field of a few gauss the amount of rotation of the polarization ellipse becomes so large that the amount of differential rotation within the receiver band-width becomes conceivable. This will imply a change in the ellipticity and the degree of polarization by passing through the solar corona. This 'finite band-width effect' will be of importance in the theory of the emission and propagation of bursts and also for the structure of the solar atmosphere.

Part of the present research has been carried out during the author's stay at the School of Electrical Engineering, Cornell University.

REFERENCES

[1] Hatanaka, T., Suzuki, S. and Tsuchiya, A. *Proc. Japan Academy*, **31**, 81, 1955.
[2] Hatanaka, T., Suzuki, S. and Tsuchiya, A. *Publ. Astr. Soc. Japan*, **7**, 114, 1955.
[3] Suzuki, S. *Publ. Astr. Soc. Japan*, **7**, 121, 1955.
[4] Hatanaka, T. *Research Report EE*-179, Cornell University, 1953.
[5] Hatanaka, T. *Research Report EE*-257, Cornell University, 1955. Also Hatanaka, T. *Publ. Astr. Soc. Japan*, **8**, 73, 1956.

Discussion

Wild: Recent records of noise storms at Sydney confirm that at times there is a prolonged period of linear polarization.

DISTURBED RADIO EMISSION FROM THE SUN AS A SUM OF SMALL MONOCHROMATIC PEAKS

V. V. VITKEVITCH

Crimean Station, Physical Institute of the Academy of Sciences, Moscow, U.S.S.R.

I

Observations of the radio emission from the sun carried out during recent years at the Crimean Station of the Physical Institute of the U.S.S.R. Academy of Sciences showed that the occurrence of spots appreciably increases the intensity of the solar radio emission in the range of metre wave-lengths. This increase of intensity has two components. The first (*S*-component) changes comparatively slowly with time. The second (*P*-component) consists of individual brief bursts (of the order of a second and less) of small amplitudes (10–100 % of the intensity of the quiet sun). The *P*-component is manifested most clearly in the emission connected with spots of small areas, when the general increase of the intensity is insignificant. Such a situation has been utilized for the study of the spectrum of individual small peaks.

Observations were carried out as follows [1]. Two radio receivers, adjusted for frequencies differing by 4 Mc./s., were connected successively to the aerial with ninety-six half-wave dipoles with a reflector. As the band-width of the aerial equalled 4–5 Mc./s., radio waves were received reliably by the two receivers. The intensity of the radio emission from the sun for the two wave-lengths was recorded at the output terminals of each of the two receivers. The intensities of the peaks that appeared simultaneously on the two frequencies were compared. A statistical method was applied in order to determine the effective width of the spectra of individual peaks. It was found that the band-width is on the average 6 Mc./s. $= 0 \cdot 029 f_0$ for peaks of small amplitude and 11 Mc./s. $= 0 \cdot 054 f_0$ for peaks of large amplitude. This corresponds to values of $Q = f_0/\Delta f = 35$ and 18, respectively.

These preliminary mean values characterize the radio emission spectrum of the peak component of the sun for 1·5 metres wave-length. It is seen that the bands in which the peaks are generated are extremely narrow. The radio emission of separate peaks thus is of a monochromatic nature. Similar results were obtained by other authors [2].

2

A number of observations of the radio emission from the peaks was carried out by means of a multi-channel radio-spectrograph. The multi-channel radio-spectrograph has the advantage over the Australian radio-spectrograph of discrete bands for the reception of the radio waves. Records are obtained at all channels simultaneously. The sensitivity of our radio-spectrograph is much higher than that of the Australian so that the separate peaks of radio emission from the sun are observed quite distinctly.

A comparison of the intensities of the peaks occurring in neighbouring wave-lengths in the range of 2–4 metres, in a similar manner as was used for the 1·5 waves, showed that the band-width of the peaks is again extremely narrow, of the order of 4–7 Mc./s.

Thus it may be concluded that the narrow band in which a peak is generated is a common feature for radio waves in the range of metre wave-lengths. The duration of small peaks for radio waves of 3–4 metres is somewhat longer than for radio waves of 1·5 metres.

Observations have shown that, if peaks occur continually at some wave-length of the metre range, they are also observed at other wave-lengths.

Cases were met when peaks were not observed on a wave-length λ_0 and all larger wave-lengths, while they were noticed in shorter waves. Thus, the peaks seem to break off abruptly somewhere in the region of shorter wave-lengths, being absent on larger wave-lengths. The value λ_0 often lies in the range of 2·5–3 metres. A detailed investigation of the range of wave-lengths where the peaks are generated is extremely important and should be carried out in detail.

3

Daily observations of the radio emission from the sun on 1·5 metres have been carried out by us since 1953. Observations were carried out during 1 or 2 hr. per day in the periods when the radio emission was of a quiet nature, and in addition during some hours without interruption, when the radio emission had a disturbed character.

During days of disturbed radio emission, a general increase of the radio emission from the sun was always accompanied by the occurrence of small peaks. Their duration averaged 1–2 sec. We paid particular attention to the relation between the P- and S-components. We shall define the value S as the increase of the intensity of the continuous radio emission, and the value P as the mean intensity of the peaks. It is convenient to express both values in units of the radio emission from the quiet sun during subsequent days. It was discovered that the P-component may occur in the absence of the S-component. But the S-component is never present unless the P-component is there. It follows that a rise of the intensity is always connected with the presence of peaks.

A further study of this question showed that whenever the S-component is present the peaks follow each other without interruption and no cases when peaks were absent in the presence of the S-component were observed. As soon as one peak disappears another one, superimposed upon the foregoing peak, originates. It is possible to conclude from this fact that the component of the disturbed radio emission from the sun connected with spots and flocculi is the sum of monochromatic peaks of radio emission, unresolved in time.

An increase of the intensity of radio emission not connected with the occurrence of peaks was noticed only once in the course of the entire period of observation. Such cases must apparently be considered as exceptional and their observation is of particular importance.

REFERENCES

[1] Vitkevitch, V. V. *Dokl. Akad. Nauk U.S.S.R.* **101**, 229, 1955.
[2] Blum, E. *Ann. Univ. Paris*, **23**, no. 1, 136, 1953.

Discussion

Smith: Very fast recordings with a time constant of 0·1 sec. made in Cambridge at $\lambda = 1\cdot4$ metres, in 1949 and 1950, showed the presence of a steady component with superimposed bursts often well separated in time.

Owren: Recordings with a time constant of 0·01 sec. made at Cornell University at 200 Mc./s. since 1950 fully confirm this.

SOME PROPERTIES OF SOLAR RADIO-TRANSIENTS ON FAST 200 MC./S. RECORDS

C. DE JAGER AND F. VAN 'T VEER

Astronomical Observatory, Zonnenburg, Utrecht, Netherlands

Some 6 km. of high-speed (Brush) records of the solar radio radiation at 200 Mc./s. made at Cornell University by Ch. L. Seeger in 1949 and 1950 have been kindly put at our disposal. They were made with a paper speed of 0·5 cm./sec., sometimes with 2·5 cm./sec. The response time of the recording pen was $0^s \cdot 02$. Some records have been made simultaneously at 205 Mc./s. and at 200 Mc./s. This paper gives some results of a discussion

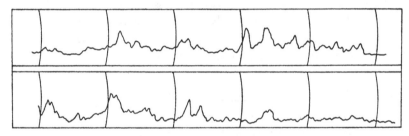

Fig. 1. Typical specimen of a record made on 200 (lower part) and 205 Mc./s. (upper part). Time increases from right to left. The vertical lines are 10 sec. apart. Date: 30 April 1950.

of these records. The properties that were studied are: the clustering tendencies among the radio-pips, their band-widths, the echo-phenomena and the duration of the pips.

A typical example of the records is given in Fig. 1. The figure suggests that the basic phenomenon of the burst activity of the sun consists of radio-pips of very short duration with, in the great majority of the cases, a duration shorter than 1 sec. Such pips are hardly detectable on records obtained by means of a slower recording instrument. As a 'pip' we consider every measurable transient phenomenon on the records; in order to be measurable it appears that the top intensity must be greater than about 0·5 mm. We note for comparison that the base-level intensity of the quiet sun is generally of the order of 5 to 10 mm.

A glance at Fig. 1 suggests that the pips cluster together. Quantitatively this can be shown by dividing the record into equal time-intervals and by making a histogram showing the frequency of occurrence of time intervals containing n pips (see Fig. 2; discontinuous curve). In all our counts the unit time-interval has been chosen equal to 30 sec. The smooth curve in the same figure shows the probability distribution for the same record computed according to Poisson's law, which corresponds to a random distribution of the pips. The observed curve differs from the computed one

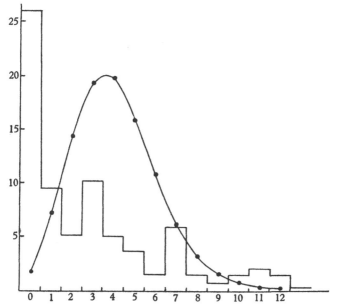

Fig. 2. Example of a histogram of the number of pips counted in time intervals of 30 sec. The smooth curve gives the corresponding Poisson distribution. Date: 15 May 1950.

in two respects: the number of intervals with many pips is much greater than follows from Poisson's law and the same is true of the intervals without any pips.

The difference can be explained by assuming that the pips occur in groups. We assumed: (*a*) that all pips occur in groups, the number of pips per group being distributed according to Poisson's law; (*b*) that the number of groups per unit time-interval is also defined by Poisson's law. A statistical formula was derived, based on these hypotheses. The discussion of a number of records showed that the average number of pips per group remains practically constant and has the mean value of 2·2.

Interpreting this formally with Poisson's law, this means that about ten per cent of all pips occur singly and the others in groups. Two comparisons with observed distributions are shown in Fig. 3. Close examination of this figure shows, however, that deviations still occur for large values of n. This probably means that apart from groups consisting of 2 or 3 members, there occurs also another kind of group consisting of a much larger number of members.

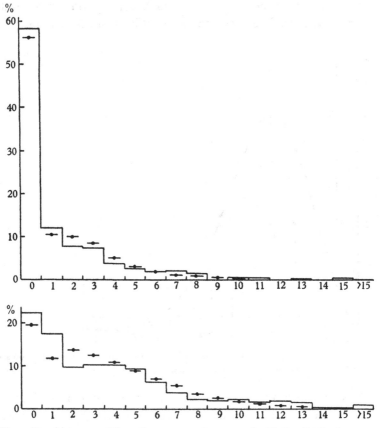

Fig. 3. Two histograms (discontinuous curves) compared with the distribution computed from a formula based on the assumption of clustering. Date: 5 May 1950.

2. BAND-WIDTH

Fig. 1 gives two simultaneous records; one made at 200 Mc./s. and the other at 205 Mc./s. Although most pips can be identified on both records, it is also clear that the ratio between the intensities of the same pip, observed in each of these frequencies, changes considerably from one pip to another. We interpret this observation by the obvious assumption that the pips have a small band-width.

Computations of the statistical distribution of the intensity ratios of the pips observed on both frequencies were made, assuming a gaussian spectrum of the pips. The best agreement with the observed results was obtained for a band-width of about 7 Mc./s.

3. ECHO PHENOMENA

If the burst of radio noise is emitted by gas masses above the layer of zero refractive index for the frequency of our observations, we can expect to observe two signals: the direct and the reflected one. The time difference between the reception of both signals may be up to some seconds. It depends on the location of the emitting source above the reflecting surface. A short computation shows that the intensity of the reflected signal will, in general, not be negligible compared to the direct signal (De Jager and Van 't Veer, in the Press) [1]. In order to examine whether any such phenomena occur on our records we have selected all isolated and well-pronounced pips. These pips, which have been called by us central pips, have been selected in such a way that another pip with an intensity of only 10 % of the 'central pip' should still be measurable on our records. The total number of the pips selected in this way was about 600.

We have looked carefully in all available records for all pips in an interval of time from 5 sec. before to 5 sec. after the central pip.

There seems to be no indication that the number of pips following the central one would be greater than that preceding it. In the following table we give the counted numbers of pips in equal time-intervals around the central pip.

	Interval in seconds, before or after central pip		
	0·25–1·75	1·75–3·25	3·25–4·75
No. of pips before central pip	233	228	184
No. of pips after central pip	263	224	189

Echoes are most likely to occur in the first seconds after the central pip. The slight preponderance observed in the first interval after as compared with the first interval before the central pip does not seem to be significant. However, since echoes will principally be produced by limb sources it may be interesting to repeat this investigation with a selected material consisting of pips produced by limb sources.

Our conclusion must be that these records do not show the echo effect.

This may lead to the conclusion that the pips are produced in the layer where the plasma frequency is equal to the frequency of the pips.

4. THE DURATION OF THE PIPS

On sixteen of our records the 'e-widths', defined as the width of a pip at $1/e$ times the top intensity, were measured for all measurable pips. For each of these records a histogram has been constructed showing the distribution of the measured e-widths. We were struck by the fact that the histograms thus obtained may vary considerably from one day to the other, a fact that has already been studied earlier by Dr Seeger (personal communication). In general, maxima in the e-width distributions are observed at values of $0^s.4$ and $0^s.6$ and sometimes at $1^s.0$. In a few other cases we observed very broad pips having an e-width of the order of 2 sec. There are indications that these pips of long duration may be associated with the first stage of the flare-associated radio phenomena, but this observation is based on too small a number of data and needs further confirmation.

At any rate Reber's (1955) [2] observation that the half-width of radio-pips expressed in seconds should be equal to the wave-length in metres is not confirmed by our observations: Reber's law would predict a mean half-width of 1·5 sec. while we observed only in very few cases an e-width of about 2 sec. of time. The great majority of the pips at 200 Mc./s. has an e-width of $0^s.5$, and a still smaller half-width.

Note added in proof.

Solar noise measurements by T. de Groot during 1956 at the Dwingeloo Radioastronomical Observatory at 400 Mc./s. yielded an average half-width of $0^s.18 \pm 0^s.03$. The sample consisted of about 500 pips. There was no indication of a second maximum in the histogram.

REFERENCES

[1] Jager, C. de and Veer, F. van 't. *B.A.N.* (in the Press).
[2] Reber, G. *Nature*, **175**, 132, 1955.

A PECULIAR TYPE OF SCINTILLATION
OF SOLAR RADIO RADIATION

A. D. FOKKER

Section 'Ionosphere and Radio Astronomy' of the Netherlands P.T.T., The Hague

Some authors (Payne-Scott and Little (1952) [1]; Owren (1952) [2]) have mentioned a phenomenon in the enhanced solar radio emission which they call 'non-selective fading'. The present paper is meant to call attention to another, rather peculiar, type of scintillation in the radio emission of the sun, which differs from the non-selective fading in some important respects. This scintillation has been observed since 1952 by the division 'Ionosphere and Radio Astronomy' of the Netherlands Telecommunications Service in the course of a continuous survey of solar radio radiation. It has been found at the radio frequencies 140, 200 and 545 Mc./s.

The main characteristics of this type of scintillation are the following:

(1) Compared with ordinary scintillation of point sources it has a more fading-like character. Sometimes the intensity fluctuations are of an oscillatory type, but the enhancements never surpass the decrements.

(2) The duration of the fades ranges between 0·3 and 4 minutes, being mostly of the order of one minute.

(3) A high degree of correlation exists between the intensity fluctuations on different frequencies. Most of the fluctuations that are found simultaneously at 200 and 545 Mc./s. correspond in detail.

(4) At 200 Mc./s. the fractional depth of the fadings often reaches values of some 20–40 %; in exceptional cases they may be so strong as to extinguish almost the entire solar radiation. The average amplitudes at 140 and 545 Mc./s. are 1·8 and 0·3 times those at 200 Mc./s.

(5) The phenomenon may be present during periods of quiet solar radio emission as well as during disturbed periods.

(6) Solar scintillation occurs with greatest strength and frequency during the months May–August.

(7) Its occurrence is rather sporadic, being sometimes limited to just one depression or to one oscillation. The fraction of time during which solar scintillation has been observed amounts to 1 % of the total observing

time in the months May–August of the years 1953, 1954 and 1955. The dates on which it has occurred show some tendency to cluster.

(8) The frequency of occurrence is evenly distributed over the main part of the day, until about 17ʰ local time. After 17ʰ the frequency is four to five times greater.

Many arguments make it extremely improbable that the phenomenon is of an instrumental nature. The only cause that might affect the two receivers in use simultaneously could be situated in the mains stabilizer, which is common to both receivers. The fading should then also be present when the receivers are connected to the dummy load. That this is not the case is clearly demonstrated by those few cases in which a fade happened to be in progress when the receivers were switched to the dummy load. In all these cases the level corresponding to the dummy load had remained unchanged.

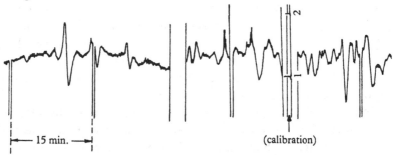

15 min.   (calibration)

Fig. 1. Two examples of solar scintillation at 140 Mc./s. (21 July 1953).

No mention of this kind of solar noise scintillation has yet been made in the literature. However, E. J. Blum, G. Eriksen and R. A. J. Coutrez have noticed a kind of scintillation that is probably the same (private information).

Unlike the non-selective fading, which by Owren (*loc. cit.*) is attributed to the sun, the solar scintillation seems to be controlled by terrestrial, probably ionospheric, conditions because of its seasonal and diurnal dependence. One might consider sporadic *E*-clouds in some way responsible. However, no correlation has been found between the occurrence of the fadings and the presence of sporadic *E*. We certainly are confronted with a puzzling phenomenon.

REFERENCES

[1] Payne-Scott, R. and Little, A. G. *Aust. J. Sci. Res.* **5**, 40, 1952.
[2] Owren, L. *Radio Astr. Rep. of Cornell Univ.* no. 15, 74, 1954.

PART VI

METEORS AND PLANETS

PAPER 72

SOME PROBLEMS OF METEOR ASTRONOMY*

INTRODUCTORY LECTURE BY

F. L. WHIPPLE

Harvard College Observatory, Cambridge, Mass., U.S.A.

We are now in an era of remarkable progress in the study of meteors. The electronic techniques developed at Jodrell Bank by Lovell, Clegg and others, and now by Davies, have culminated in giving us the power to observe the equivalent of 8th or 9th magnitude meteors for velocities, radiants and orbits. And certainly the limit has not yet been reached. The Harvard Super-Schmidt meteor cameras represent nearly the limit of current photographic-optical techniques; they approach close to the visual limit for very slow meteors, and with great precision, to 0·1 % in velocity and radiant.

I cannot take the time to review the subject of meteoric astronomy because too much has happened. I shall only mention the state of the art in certain areas of interest and point out some of the problems that I feel we should make special efforts to solve. These problem areas are as follows:

1. Meteor orbits and their generic significance.

2. The physics of persistent meteor trains and ionization.

3. The physics of the meteoric processes with special emphasis on the determination of meteoroid masses.

4. Possible atmospheric effects in the occurrence of radio, photographic and visual meteors.

5. Problems of faint radio meteors and micro-meteorites, including the Zodiacal Light and possible correlations with rainfall, the earth's magnetic field, etc.

* This research was supported by the U.S. Office of Naval Research, Contract N5ori-07647, and the Geophysics Research Directorate of the Air Force Cambridge Research Center under Contract No. AF 19(122)–482. Reproduction in whole or in part permitted for any purpose of the U.S. Government.

I. METEOR ORBITS

We may now accept as proven the fact that bodies moving in hyperbolic orbits about the sun play no important role in producing meteoric phenomena brighter than about the 8th effective magnitude. In the radio region, McKinley[1] at Ottawa and Lovell and his colleagues at Jodrell Bank have proved that the hyperbolic component lies below the 1% level. The Harvard photographic programme provides a similar demonstration for the brighter visual meteors. Among the fireballs and meteoritic falls, the evidence presented first by H. A. Newton[2] and more recently by Wylie[3] and by Whipple and Hughes[4] is not quite conclusive but is very convincing.

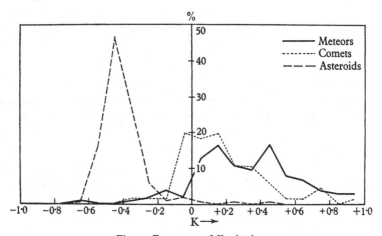

Fig. 1. Frequency of K criterion.

The major remaining difficulty at the moment, however, lies in discriminating between the meteoric contributions by comets and those by asteroids. The photographic orbits by Whipple[5] appear to be mostly (90%) of cometary character, as judged by the arbitrary K criterion

$$\left[\log_{10} \frac{a(1+e)}{1-e} - 1\right],$$

determined by the aphelion orbital velocity. Fig. 1 presents the frequency distribution of this quantity (for $K < +1$) for short-period comets, meteors and asteroids. Only two of the seven asteroids passing within the earth's orbit show positive values of K while all other asteroids except Hidalgo show negative values. Comets and meteors show largely positive values, with a similar distribution.

376

The majority of meteors, whether photographic or radio, show a strong preference for direct motion near the ecliptic (or the invariant plane). Hawkins has shown this in a radio survey of sporadic radiant points [6] and Almond, Davies and Lovell [7] have shown with their apex and antapex experiments that the mean helio-centric velocity of radio meteors is about 34 km./sec., corresponding to an aphelion distance of 3 a.u. Jupiter's dominance is clearly indicated in the uniform distribution of aphelion distances of photographic meteor orbits between 3·0 and 6·5 a.u.

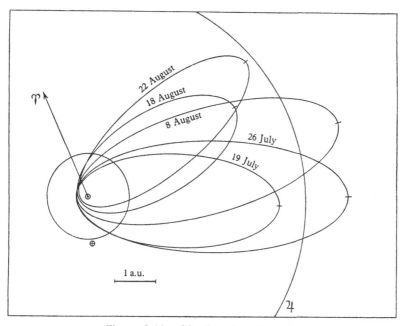

Fig. 2. Orbits of Southern Iota Aquarids.

Significantly, only one out of some 300 Harvard meteors has aphelion within the orbit of Mars (0·987 a.u.!, unpublished). This fact tends to support Öpik's conclusion that the earth has swept away any remnants of primitive asteroidal material crossing its orbit. Unfortunately the theory of the meteoric processes is still inadequate to distinguish chemically or physically between cometary and asteroidal meteoroids travelling in orbits with aphelia in the asteroid belt. A small fraction of the sporadic photographic orbits may well be of asteroidal origin.

Among the recognized meteor streams, however, the photographic evidence appears to rule out an asteroidal origin except possibly for the Geminids and the daytime o Cetids [8]. As shown in Table 1 and Figs. 2,

377

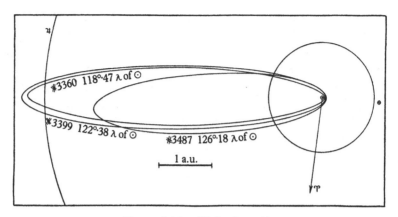

Fig. 3. Orbits of Delta Aquarids.

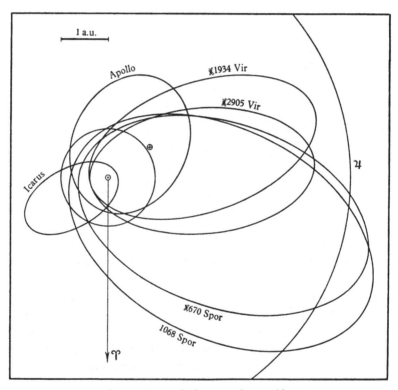

Fig. 4. Orbits of Meteors and Asteroids.

Table 1 presents a summary of the mean photographic orbital elements of all meteor streams determined from photographic data at present available at Harvard, along with the radiants for the streams, the date of maximum, in Universal Time, when possible, and the duration of the showers. The streams identified by Roman numerals are generally based on too few meteors to be immediately accepted as real, although the probability of such similarities in orbits is rather small (see Whipple [5]).

Table 1. *Mean orbital elements of meteor streams*

Stream	No. of orbits	V_∞ km./sec.	V_G km./sec.	V_H km./sec.	ϖ (°) 1950·0	Ω (°) 1950·0	i (°) 1950·0	π (°) 1950·0	a (a.u.) $1/(1/a)$	e	P (yr.)	q (a.u.) $a(1-e)$	q' (a.u.) $a(1+e)$	Elong. (°)	U.T. date at max.	Duration (days)	Corr. radiant α (´50)	δ
Quadrantids	(1)	44·1	42·4	39·3	167·9	282·2	73·8	90·1	3·42	0·715	6·3	0·974	5·87	62·9	3 Jan:		230°	+48°
'I'	(2)	14·7	9·8	37·2	172·2	354·4	11·1	166·6	2·20	0·550	3·3	0·990	3·41	131·8	(14–17 Mar.)		57°	+69°
Virginids	(3)	30·8	28·9	38·3	285·8	353·7	5·2	279·5	2·82	0·857	4·7	0·403	5·24	81·4	(5–21 Mar.)		183°	+4°
'II'	(2)	15·2	10·6	37·8	187·1	27·3	11·0	214·4	2·67	0·626	4·4	0·999	4·34	134·4	(14–21 Apr.)		157°	+56°
Lyrids	(3)	48·4	47·0	41·6	213·9	31·8	79·9	245·6	29·6	0·969	161	0·918	58·3	61·0	21 Apr.		270°	+33°
ι Aquarids	(4)	35·75	33·82	37·93	127·5	311·0	6·0	78·5	2·88	0·920	4·9	0·230	5·52	73·4	(19 July–22 Aug.)	39:	338°	−14°
δ Aquarids	(5)	42·98	41·54	37·40	154·7	302·9	29·3	97·6	2·60	0·976	4·2	0·062	5·14	60·9	30 July	27	339°	−17°
α Capricornids (a)	(10)	25·53	23·02	37·45	270·5	132·8	4·0	43·3	2·57	0·779	4·3	0·568	4·57	89·9	1 Aug.	37	308°	−10°
α Capricornids (b)	(5)	25·05	22·50	36·97	270·5	122·5	7·2	33·0	2·35	0·755	3·6	0·576	4·12	90·0	(July α Capricornids)			
α Capricornids (c)	(4)	25·96	23·47	38·01	269·6	143·6	0·6	53·2	2·91	0·804	5·0	0·570	5·25	90·1	(Aug. α Capricornids)			
Perseids	(11)	60·44	59·30	41·29	151·2	138·1	113·7	289·3	20·8	0·955	95	0·936	40·7	39·8	12 Aug.	27	46°	+58°
κ Cygnids	(4)	26·6	24·2	39·2	204·2	144·3	37·0	348·4	4·09	0·762	8·3	0·973	7·21	93·4	(19–22 Aug.)		Incomplete	
'IV'	(2)	40·4	38·8	37·4	146·1	331·9	21·0	123·8	2·51	0·958	4·0	0·105	4·91	64·8	(21–29 Aug.)		0°	−7°
'V'	(2)	22·2	19·4	39·8	203·8	192·9	26·4	36·6	4·66	0·794	10·1	0·960	8·36	106·2	(4–9 Oct.)		307°	+48°
So. Arietids	(1)	31·4	29·5	39·5	122·2	27·2	6·0	149·5	1·91	0·845	2·6	0·206	3·52	75·3	(15–17 Oct.)	14	42°	+10°
Orionids	(2)	66·5	65·5	40·8	86·8	29·8	163·2	116·5	7·70	0·930	21	0·539	14·86	26·1	22 Oct.		94°	+16°
So. Taurids	(5)	30·2	28·1	37·4	111·9	45·1	5·4	156·9	2·30	0·835	3·5	0·380	4·22	79·9	1 Nov.	32	51°	+14°
No. Taurids	(3)	31·3	29·5	37·0	298·4	221·8	3·2	160·2	2·14	0·849	3·1	0·323	3·96	77·1	17 Oct.–2 Dec.	47	52°	+21°
'VI'	(2)	21·3	18·5	39·0	65·3	43·4	6·0	107·8	3·34	0·776	6·1	0·748	5·93	104·4	(6–7 Nov.)		30°	+6°
Leonids	(5)	72·0	70·8	41·5	173·7	235·0	162·5	48·7	12·76	0·924	46	0·970	24·6	10·2	17 Nov.	6	152°	+22°
'VIII'	(2)	30·4	28·4	38·0	289·2	257·2	1·6	186·5	2·49	0·846	3·9	0·383	4·60	80·7	(9–10 Dec.)		86°	+24°
'IX'	(2)	24·5	21·9	37·4	88·2	79·0	5·0	167·1	2·22	0·740	3·3	0·577	3·86	90·4	(10–13 Dec.)		79°	+16°
'VII'	(3)	64·5	63·4	42·2	264·5	259·1	135·7	163·7	574·	0·997		0·172		34·5	(8–13 Dec.)		151°	+33°
'X'	(2)	30·7	28·4	38·7	105·4	79·8	3·0	185·2	2·92	0·859	5·0	0·412	5·43	82·4	(10–14 Dec.)		88°	+20°
Monocerotids	(2)	44·0	42·4	42·6	128·2	81·6	35·2	209·9	−84·4	1·002		0·186		69·4	(13–15 Dec.)		103°	+8°
Geminids	(19)	36·5	34·7	34·1	324·3	261·2	24·0	225·6	1·39	0·899	1·6	0·140	2·64	62·8	14 Dec.	6	113°	+32°
Ursids	(1)	35·2	33·4	40·6	212·2	264·6	52·5	116·8	5·91	0·845	14·4	0·916	10·90	79·2	17 Dec.:		206°	+80°

: = doubtful.

V_∞ : Velocity in the atmosphere relative to the station after correction for atmospheric resistance.
V_G : Velocity relative to the centre of the earth after correction for diurnal rotation and the earth's attraction.
V_H : Velocity relative to the sun after correction for earth's motion and attraction.
Orbital elements: i, a, e, P, q and q' denote, as usual, the inclination, semi-major axis, eccentricity, period, perihelion and aphelion; while ϖ, Ω and π give the angle from the ascending node to the perihelion point, measured along the orbit in direction of motion, the celestial longitude of the ascending node as seen from the sun, and where π is the sum of ϖ and Ω, all referred to ecliptic and equinox of 1950·0.
Elongation is the angle between the corrected radiant and the apex of the earth's motion.

3 and 4 the October Arietids (daytime ζ Perseids), Taurids (daytime β Taurids), α Capricornids, ι Aquarids, Virginids, δ Aquarids and κ Cygnids all have their aphelia near Jupiter like the short-period comets. Hoffmeister's [9] use of the term 'ecliptic currents' for such streams has suggested an asteroidal generic connexion which the best orbital data do not support. We still require, however, objective physical criteria to eliminate all possibility of asteroidal origin for one or two of the recognized streams. Jacchia's [10] discovery that the Geminid meteoroids have either greater densities (2·5 times) or greater luminous efficiencies (10 times) than do average meteoroids leaves some element of doubt as to their true nature. It is difficult to find a mechanism whereby a comet could have attained the small aphelion distances of the Geminids or of the daytime ο Cetids.

The writer, however, still prefers the working hypothesis that all meteor streams and almost all fainter meteors are of cometary origin. But we are badly in need of observational or theoretical methods whereby we can check this hypothesis critically.

2. THE PHYSICS OF PERSISTENT METEOR TRAINS AND IONIZATION

To date we have not even a rudimentary theory to explain the long persistence of radiation in a meteor trail. We can conclude only that the energy for the persistent radiation is derived from the meteoric process and not parasitically from the atmosphere by the introduction of foreign meteoric atoms. This conclusion follows from the well known fact [11] that faster meteors are much more efficient than slower meteors in producing long-enduring trains. Since slower meteors are more massive than faster meteors of the same brightness they should be better train producers than the faster meteors, if the parasitic hypothesis were correct.

Although it seems likely that active nitrogen stores the meteoric energy for the required seconds or minutes, nevertheless we need a detailed substantial theory of meteoric trains. Numerous checks on the theory are already available: the velocity dependence mentioned above, the strong variation in train-decay rates with altitude, knowledge of the physical parameters of the high atmosphere and even spectral information from Millman's [12] observation of a short-lived train between the shutter breaks of a meteor spectrum.

At Harvard we shall soon have much more statistical data on meteor trains and their height-decay rates [13], and hope to photograph train spectra directly. Relevant laboratory studies are badly needed to improve the theory. There is much to be done in this respect. I feel that a detailed understanding of persistent luminous trains will also involve a much improved understanding of the electron decay and diffusion observed in radio meteors. Millman[11] finds a strong correlation between the persistence of train luminosity and the persistence of electrons in meteor trails. The excellent foundation in this theoretical field by Lovell and Clegg[14], Herlofson[15], Kaiser and Closs[16], Greenhow and Hawkins[17] and others at Jodrell Bank must be extended and integrated with a theory of meteor trains to give a comprehensive understanding of the physical process occurring after a meteoroid has passed.

3. THE PHYSICS OF THE METEORIC PROCESSES

I have arbitrarily separated the present subject from the previous one for two reasons: (a) An important symposium of several days length on the Physics of the Meteoric Process occurred at Jodrell Bank last year, so there is no need to repeat or to condense these more prolonged discussions. (b) The problems of meteor-train physics appear more readily soluble than those of the general meteoric phenomena.

From radio and photographic observations we have now found numerical relationships, for meteors of a given velocity and brightness (or ionization), between the quantities meteoroid mass (m), meteoroid density (ρ_m), luminous efficiency and ionization efficiency. If any one of these four quantities can be determined, by any means whatsoever, the other three can be derived from our store of observations as a function of meteor velocity and brightness (or ionization). This peculiar situation arises from the nature of the 'drag equation' representing the observed decrease in velocity of the meteoroid caused by atmospheric resistance. When the other physical factors such as the drag coefficient (Γ), a dimensionless shape factor (A_0), and the atmospheric density ρ are approximated, observed, or derived theoretically we find that the measures of velocity (v) and deceleration $\left(\dfrac{dv}{dt}\right)$ give us the quantity $m^{1/3}\rho_m^{2/3}$ by the equation

$$m^{1/3}\rho_m^{2/3} = -\Gamma A_0 v^2 (dv/dt)^{-1} \rho. \tag{1}$$

Had we even a semi-adequate theory for the production of light or of ionization in the meteoric process, we could determine the mass and hence, from equation 1, the density of a meteoroid. But no such theory exists. From the Harvard photographic data and the limiting assumption that *all* the kinetic energy is converted to radiation, I found[18] that the density of photographic meteors must be smaller than the density of stony meteorites. Since the luminous efficiency must certainly be much smaller than unity the densities must also be reduced. Jacchia[19] has found observational evidence for the fragility of meteoroids and Opik[20] has presented theoretical evidence for their low densities.

Recently, A. F. Cook and I have found a more direct method for determining meteoric masses. In measuring high-altitude winds by the multiple photography of persistent luminous trains from two stations, we found an example in which the complete wind vector could be determined. The nearly vertical component down the meteor trail exceeded 40 m./sec., comparable to the horizontal wind velocities. There could be no doubt that this motion measured the transfer of momentum from the meteoroid to the surrounding air mass. With a suitable diffusion theory for calculating the momentum transfer Cook has calculated the mass, and, from equation (1), the density of the meteoroid. The density turns out to be 0·05 gm./cm.^{3}, on the basis of the most likely constants and theory. Its upper possible value, obtained by pressing all the uncertainties in the proper sense, appears not greater than 0·3 gm./cm.^{3}.

A few other examples of photographed trains that may give total velocity vectors are now under reduction. The present result should not be accepted as typical until we have had opportunity to confirm it thoroughly. Nevertheless, such a low meteoric density appears not inconsistent with other meteor theory and observation. The density of 0·05 gm./cm.^{3} leads to meteoroid mass nearly two orders of magnitude greater than those derived by the early Öpik theory. A zero (visual) magnitude meteor of velocity 28 km./sec. would have a mass of about 25 gm. The luminous and ionization efficiencies in terms of energy would be of the order of 10^{-4}. Such low values are quite reasonable when one considers the small atomic cross-sections for excitation or ionization by encounter.

Furthermore, a mass discrepancy discovered by van de Hulst[21] would be eliminated if small meteoritic bodies have such low densities. He calculated that the zodiacal cloud of small particles, if sufficiently extensive to produce the Zodiacal Light, should shower the earth by some 10^4 times the mass rate estimated by Watson[22] on the basis of Öpik's meteor theory. The increase in the meteoritic mass striking the earth discussed above will account for a factor of some 10^2. But van de Hulst determined the dimensions of the zodiacal particles, not their masses directly, so he had to assume meteoritic densities. Hence the density decrease of some two orders of magnitude in his calculation completes the removal of the 10^4 times discrepancy.

Cometary debris of extremely low mean density and of extremely fragile character is entirely consistent with the author's theory [23] for the icy comet model. That a simple cometary theory based on present-day estimates of the cosmic abundances of elements leads to a mean density for meteoroids of 0·3 instead of 0·05 gm./cm.³ cannot be considered a discrepancy. It seems that cometary debris must be made of imperfect crystals, full of holes at all dimensions from molecular to macroscopic.

The search, however, for independent methods of determining masses, densities, luminous efficiencies and ionization efficiencies for meteors must be pressed to the limit in order to clarify the basic problem of meteor physics.

4. POSSIBLE ATMOSPHERIC EFFECTS

A long-outstanding meteor problem was presented by Poulter [24] during the second Byrd polar expedition (1933–5). Poulter and a trained group of meteor observers found, near the South Pole of the earth, that with binoculars they could count some sixty times the normal frequency of meteors as compared with corresponding observations made at moderate latitudes. No explanation for this result has yet proved convincing. During the International Geophysical Year the observations will be repeated and perhaps the puzzle can be solved.

A new meteoric problem has recently come to light in the comparison of the radio and photographic ratios of meteor occurrence. Figs. 5 and 6 show, respectively, the mean hourly rates of night and daytime radio meteors averaged over each month for two years for sporadic meteors, with the major showers superimposed. The data are from Jodrell Bank, recorded by Hawkins [6] and Aspinall, at 72 Mc./s., uncorrected for astronomical theory but reduced to the total rate over the entire sky. The shower rates, averaged over the entire day, do not overwhelm the sporadic background. Both night and day rates show a maximum in June, July and August and a minimum in January, February, March and April. The summer/winter ratio is greater than three, exclusive of the major showers.

For comparison, the corresponding hourly rates of meteors doubly photographed by the Super-Schmidt meteor cameras in New Mexico are shown in Fig. 7. The averages for each lunation represent approximately 100 meteors, so that the statistical fluctuation is appreciable but not excessive. The seasonal variation is relatively small, certainly not exceeding a factor of 1·5 in the summer/winter ratio. Now, the photographic data are subject to variable discovery rates near the film limits due to changes in personnel. Also the blackening of the film by the Milky Way tends, somewhat, to reduce the sensitivity during the summer months. We are investigating these effects quantitatively but it seems worthwhile to present the preliminary raw data at this time, even though they are subject to some later correction.

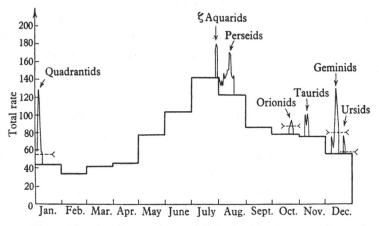

Fig. 5. Mean hourly rate of meteors for each night, showers and sporadic.

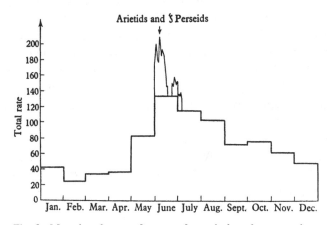

Fig. 6. Mean hourly rate of meteors for each day, showers and sporadic.

Unless the effects of Milky Way fogging or some other observational error are excessive, the yearly photographic meteor rates vary much less over the year than the radio rates. Three possibilities are obvious:

(1) The fainter radio meteors in space have a different orbital distribution from that of the brighter photographic meteors.

(2) Variations in the high atmosphere affect the radio rates seasonally,

in the sense that greater ionospheric activity may increase the ionization efficiency, reduce ion diffusion rates or reduce electron decay rates, etc.

(3) Meteors of different velocity groups may encounter the earth at seasonally variable rates and the differences in sensitivity as a function of meteoric velocity between radio and photographic methods may account for the observed seasonal difference.

One prefers to avoid the first possibility until the others have been thoroughly explored, first on basic principles and secondly because of the

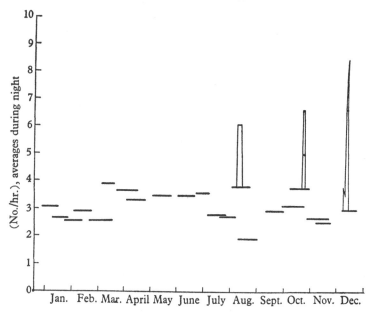

Fig. 7. Photographic meteor frequencies (no./hr.) for twenty-one lunations.

observation by Almond, Davies and Lovell [7] that the velocity distribution of radio meteors is independent of limiting magnitude.

The second possibility, an effect arising from seasonal changes in the ionosphere, could be readily checked by comparable radio observations from the southern hemisphere. The seasonal effect in meteor rates should remain, but be reversed with respect to the calendar months. Weiss [25], indeed, has published such radio meteor rates as observed in Australia, but the counts are not sufficiently numerous or well enough distributed over the year to be definitive. They appear, however, to show no change or possibly a slight increase during the southern summer as compared to the southern winter.

That the third possibility appears to operate in the required direction has been shown tentatively in the Harvard photographic studies. McCrosky [26] has completed the measurements, by a rapid graphical method, of about 1000 velocities for photographic meteors. The radio technique in meteor detection appears to be relatively more efficient than the photographic for meteors in the velocity range 25–50 km./sec. [27]. McCrosky's data show that photographic meteor rates in this velocity range do indeed show a seasonal variation like that of the Jodrell Bank radio meteors, but of smaller amplitude. Hawkins [6] has suggested that the seasonal effect may arise from a greater concentration of cometary orbits near the earth's orbit during the summer months, following Hoffmeister's statistics in this regard.

Whether or not the observational data are yet sufficiently firm to support the above arguments, we are in duty bound to clarify the matter. Systematic measurements of meteor rates from southern latitudes are essential. Furthermore we must all make every effort to ensure the validity of our statistics and the uniformity of measurement. Some important and surprising results with regard to the high atmosphere, to meteor physics or to meteor orbits may be in the offing.

5. PROBLEMS OF FAINT RADIO METEORS AND MICROMETEORITES

E. G. Bowen [28] has introduced a most fascinating problem in the area of micro-meteorites by his arguments that micro-meteorites produce condensation nuclei to trigger off unusually heavy rainfall about thirty days after certain meteor showers. Although statistical arguments against Bowen's theory have been presented by D. F. Martyn and others, Bowen has collected independent geophysical evidence for his hypothesis such that we in meteor astronomy cannot ignore his proposals. Millman [29] has outlined some of the meteoric difficulties involved and it is clear that some radical changes in our general concepts of meteoric astronomy are demanded by Bowen's hypothesis. We must investigate carefully to be certain whether these changes do or do not, in fact, violate our observations. Our theories, of course, have no intrinsic merit except as they integrate the observations.

In the meteor-rainfall correlations, certain meteor showers, such as the Geminids and the Quandrantids, must produce condensation nuclei in numbers that are markedly greater than in the sporadic background. But we see from Hawkins's data (Figs. 5 and 6) and the photographic data

(Fig. 7) that few ordinary meteor showers at maximum intensity exceed the sporadic background rate. Nor is there any indication that the shower/sporadic rate increases with decreasing particle size; in fact, Davies [30] finds just the reverse. Probably the radio meteor showers are less conspicuous against their sporadic background than the photographic showers of larger meteoroids. But below some particle dimension the showers must become overwhelmingly strong with respect to the sporadic rates if Bowen's hypothesis is true.

This postulate of powerful, as yet unobservable, streams of micrometeorites is further required by Bowen's [31] correlations of heavy rainfall with the Bielid and Draconid showers, which are only occasionally observed by visual, photographic or even radio techniques.

It appears that the meteor-rainfall hypothesis requires that relatively 'young' meteor streams carry with them an unobserved and excessive amount of fine dust, which, in space, does not survive to accompany statistically the sporadic meteoroids isolated or detached from meteor streams. Careful and detailed theoretical investigations will be required to ascertain the extent to which the above requirement is consistent with the observations of the Zodiacal Light, van de Hulst's [21] and Allen's [32] theory for it, and Whipple's [23] discussions of the dynamics of particles in the zodiacal cloud. The requirement of short-lived dust appears to be best explained by the erosive action of corpuscular radiation from the sun. Removal of dust from the streams by the Poynting-Robertson effect or the corresponding effect of corpuscular radiation appears not to be a solution to the problem. Whether or not the concentration of fine dust in meteor streams alone would produce such spatial irregularities as to cause *measurable* variations in the brightness or position of the Zodiacal Light requires further study. The meteor-rainfall hypothesis certainly leads to the necessity for variations in the Zodiacal Light; the only question concerns the amplitudes to be expected.

The above considerations, although sketchy and preliminary, indicate the need for new and more extensive attacks on the problem of micrometeorites. The methods range through (a) dust collection at all possible altitudes including the deep-sea oozes, (b) observations of optical scattering by dust in space and in the atmosphere, (c) more sensitive radio-meteor detection mechanisms, (d) geophysical measurements during meteor showers of possible concomitant effects in the ionosphere, in the earth's magnetic field or in other parameters, (e) laboratory and theoretical studies of the effects of corpuscular radiation on small particles, and (f) more extensive studies of the dynamics and physics of the zodiacal cloud. Whether or not Bowen's hypothesis is correct, these studies are needed to supplement the extensive research conducted on the larger meteoric particles.

Bumba's [33] interesting correlations between variations in the earth's magnetic field and the occurrence of meteor showers indicates the type of frontiers that may be opened up by such research.

Finally, another area of new possibilities is suggested by Davies' current measurement of orbits for faint radio meteors. Whereas the photographic meteor orbits appear representative of cometary orbits, as though the large photographic meteoroids had been injected by comets into these orbits, the faint radio meteors show a much greater preponderance of smaller, more circular orbits. If these radio orbits represent a systematic change in character from the 'injection' orbits, the cause most probably lies in the action of corpuscular solar radiation, which produces an effect similar to the Poynting-Robertson [34] spiralling effect. Since the larger meteoroids appear to be eliminated by collisions with zodiacal material [23], a knowledge of the meteoroid dimension at which spiralling occurs provides a measure of the mean momentum carried by corpuscular radiation. Meteoroid densities, of course, are also required in this solution.

Thus many problems of the interactions of radiation, corpuscular radiation and meteoritic material in the solar system appear to be soluble in the near future.

I am particularly indebted to Richard E. McCrosky, Gerald S. Hawkins, Luigi G. Jacchia and John G. Davies, for the use of new observational material in advance of publication, and to Miss Frances W. Wright for compiling Table 1.

REFERENCES

[1] McKinley, D. W. R. *Ap. J.* **113**, 225, 1951.
[2] Newton, H. A. *Am. J. Sci.* **36**, 1, 1888.
[3] Wylie, C. C. Private communication, 1954.
[4] Whipple, F. L. and Hughes, R. F. *J. Atmos. Terr. Phys.* 'Special Supplement', Pergamon Press, Ltd. 1955.
[5] Whipple, F. L. *A.J.* **59**, 201, 1954.
[6] Hawkins, G. S. Doctoral Thesis, Manchester Univ., England, 1952.
[7] Almond, M., Davies, J. G. and Lovell, A. C. B. *M.N. R.A.S.* **113**, 411, 1953.
[8] Almond, M. *M.N. R.A.S.* **111**, 37, 1951.
[9] Hoffmeister, C. *Meteorströme*, Sonneberg Obs., 1948.
[10] Jacchia, L. G. Har. Col. Obs., *Tech. Rep.* no. 10, 1952.
[11] Millman, P. M. *J. R.A.S. Can.* **44**, 209, 1950.
[12] Millman, P. M. *Nature*, **172**, 853, 1953.

[13] Liller, W. and Whipple, F. L. *Rocket Exploration of the Upper Atmosphere*, ed. Boyd and Seaton, Pergamon Press, London, 1954, pp. 112–130.
[14] Lovell, A. C. B. and Clegg, J. A. *Proc. Phys. Soc.* **60**, 491, 1948.
[15] Herlofson, N. *Rep. Phys. Soc. Progr. Phys.* **11**, 444, 1947.
[16] Kaiser, T. and Closs, R. L. *Phil. Mag.* **43**, 1, 1952.
[17] Greenhow, J. S. and Hawkins, G. S. *Nature*, **170**, 355, 1952.
[18] Whipple, F. L. *A.J.* **57**, 28 (abstract), 1952.
[19] Jacchia, L. G. *Ap. J.* **121**, 521, 1955.
[20] Öpik, E. *J. Atmos. Terr. Phys.* 'Special Supplement', Pergamon Press, Ltd., 1955.
[21] Hulst, H. C. van de. *Ap. J.* **105**, 471, 1947.
[22] Watson, F. G. *Between the Planets*, The Blakiston Co., Philadelphia, 1941.
[23] Whipple, F. L. *Ap. J.* **121**, 750, 1955 *b*.
[24] Poulter, T. C. *Meteor Observations in the Antarctic Byrd Expedition II, 1933–35*, Stamford Res. Inst., California, 1955.
[25] Weiss, A. A. *Aust. J. Phys.* **8**, 148, 1955.
[26] McCrosky, R. E. Doctoral Thesis, Harvard Univ., 1955.
[27] Whipple, F. L. *Ap. J.* **121**, 241, 1955 *a*.
[28] Bowen, E. G. *Aust. J. Phys.* **6**, 490, 1953.
[29] Millman, P. M. *J. R.A.S. Can.* **48**, 226, 1954.
[30] Davies, J. G. These symposium proceedings, Paper 73, p. 390.
[31] Bowen, E. G. Private communication, 1955.
[32] Allen, C. W. *M.N. R.A.S.* **106**, 137, 1946.
[33] Bumba, V. *Bull. Astr. Inst. Czech.* **6**, 1, 1955.
[34] Robertson, H. P. *M.N. R.A.S.* **97**, 423, 1937.

Discussion

Lovell: Whipple's conclusion that meteors have such large diameters (25 cm. for zero magnitude) will have considerable repercussions on the theory of the scattering of radio waves by meteor trails. These theories are based on the idea that the diameter of the ionized trail is very small compared with the radio wave-length both at the instant of formation and for some time afterwards.

Whipple: But the radio meteors are much fainter; their diameters do not exceed 1 cm.

Greenstein: Would not all the peculiar variations in the distribution of the radio meteors have been smoothed out by the Poynting-Robertson effect if they had been going around so long?

Whipple: A good percentage at any time ought to be newly injected.

Gold: A density < 0·1 is almost impossible to achieve with ordinary materials. It would have to be a body composed of long thin needles or threads. More compact pellets stuck together even with a lot of interspaces would result in a higher density. It is interesting to note that needles are also required for other purposes in interstellar space.

Whipple: I think indeed of an extremely fragile body from which all the ice has evaporated.

ORBITS OF SPORADIC METEORS

J. G. DAVIES

Jodrell Bank Experimental Station, University of Manchester, England

A survey of sporadic meteor orbits has been made using a spaced receiver technique. The apparatus consists of three receivers spaced on the ground by about 4 km. The received signals are brought together by radio links and displayed on a single film. In this way both diffraction pattern and the time delay between the diffraction patterns received at the different stations can be measured. From this the velocity and the direction of motion of the meteor can be calculated. The range of the echo and the local sidereal time are also recorded on the film.

From May 1954 to April 1955 twelve 24-hr. periods have been analyzed at the rate of one per month, thus yielding a general picture of the distribution of sporadic meteor orbits. The number of meteors measured in 24 hr. varied from 120 in February to 330 in July, the total for the year being approximately 2400. The aerial system used was designed to provide approximately equal sensitivity to meteors from all parts of the sky north of declination $-10°$ in the course of 24 hr., and the sensitivity of the equipment was such that most meteors measured were of magnitude $+7·5$ to $+8$.

Not one of the 2400 orbits measured is hyperbolic, except for a few close to the parabolic limit which may be ascribed to the effect of experimental errors on long-period orbits.

Orbits with aphelion distances between 3 and 10 a.u., about 35 % of the total, show considerable concentration in the plane of the ecliptic. In the case of aphelion distances greater than 10 a.u. there is a slight concentration towards the ecliptic. Since the accuracy is often insufficient to distinguish between an orbit with aphelion distance 10 a.u. and a parabolic orbit, this may in fact represent errors of measurement on medium-period orbits.

Over 50 % of the orbits have aphelion distances less than 3 a.u. Of these orbits 40 % are nearly circular (eccentricities less than 0·5) and show concentrations in inclinations 60° and 120°. Inclinations less than 30° and close to 90° are rare in this class of orbit.

The proportion of low eccentricity orbits increases towards fainte magnitudes. At magnitudes +6 and +7 the low eccentricity orbits are more concentrated at high inclinations than at magnitude 8.

A few meteors of magnitude +4·5 and brighter have also been observed. The low eccentricity group is absent, and most orbits have aphelion distances greater than 5 a.u.

The Geminid shower was observed in 1954 to assess the accuracy of the experiment. The results indicate clearly that accuracy of velocity measurement is limited by atmospheric deceleration of the meteors to about ±2 km./sec., and the accuracy of direction measurement to ±3° by distortion of the meteor trail by non-uniform winds in the upper atmosphere. The mean correction for deceleration to be applied to these faint Geminid meteors was found to be about four times greater than the value inferred from Harvard photographic observations on bright meteors. This correction has been applied in the experiment on sporadic meteors described above.

The method and results of the observations on the Geminid stream are described in detail elsewhere [1], and a further paper relating to the sporadic meteor observations is in preparation. It is planned to continue the work on sporadic meteors, and to make a study of shower meteors.

REFERENCE

[1] Gill, J. C. and Davies, J. G. *M.N.R.A.S.* **116**, 105, 1956.

THE DISTRIBUTION OF METEOR MASSES

I. C. BROWNE, K. BULLOUGH, S. EVANS AND T. R. KAISER

Jodrell Bank Experimental Station, University of Manchester, England

Kaiser (1953) [1] has shown how it is possible, from statistical studies of radio echoes from meteor trails, to obtain values of the exponent s in the assumed equation for the number of meteors $n(\alpha)$ of line density α

$$n(\alpha_m) \ d\alpha = \text{const.} \ \frac{d\alpha}{\alpha_m^s},$$

where α_m is the maximum electron line density (number of electrons per cm.) produced in the trail of a meteor. If the simple theories of the meteor ionization process are correct, α_m is proportional to the meteor mass m, so that from radio echo studies it should be possible to obtain a measure of the meteor mass distribution:

$$n(m) . dm = \text{const.} \ \frac{dm}{m^s},$$

where $n(m)$ is the number of meteors with mass in the range from m to $m + dm$.

This paper presents new work carried out during the last three years by three independent methods giving information about different magnitude ranges, namely:

Mag. $M = +7, +3$: distribution of echo amplitudes,
$+2 > M > 0$:   distribution of echo durations,
$+7 > M > +5$:   distribution of echo heights.

The table below shows weighted mean values for sporadic meteors and various meteor showers, together with visual results for comparison. The probable errors are indicated by the number of significant figures in the values of s. It can be seen that radio and visual results are in close agreement.

A value of $s = 2$ corresponds to constant total meteoric mass per magnitude range. Values of s greater than 2 correspond to distributions in which the mass per magnitude range increases with magnitude, and vice versa.

It will be seen that for all showers, with the exception of the Arietids, s is less than 2 for faint meteors, but greater than 2 for bright meteors, suggesting that the mass per magnitude range for these showers has a maximum value around $M = +3$.

Values of s for sporadic and shower meteors

Shower	Visual $M = +10$	Radio $M = +7$	Radio $7 > M > 5$	Radio $M = +3$	Radio $2 > M > 0$	Visual $M < +3$
Sporadic	2·0[1,2]	1·98	—	2·4	—	2·0[1,2]
Quadrantid	—	—	1·78	—	1·8*	2·5[2]
Perseid	—	1·38	1·59	2·0	2·0	2·1[2,3]
Arietid	—	—	2·7	—	1·8*	—
Geminid	—	1·45	1·62	2·24	2·3	2·3[2]

* These values are lower limits to s.
[1] Watson (1941)[2].
[2] Levin (1955)[3].
[3] Mean value from analysis of measurements by Öpik (1922)[4], Levin (1955)[3], Kresak and Vozarova (1953)[5], Hruška (1954)[6] and Ceplecha (1952)[7].

Detailed results, including a discussion of the variation of s with solar longitude during the Geminid and Perseid showers, are being published elsewhere (Browne, Bullough, Evans and Kaiser, 1956) [8].

REFERENCES

[1] Kaiser, T. R. *Advances in Physics (London)*, **2**, 495, 1953.
[2] Watson, F. *Between the Planets*, Blakiston, 1941.
[3] Levin, B. J. *Meteors* (ed. Kaiser, T. R.), Pergamon Press, p. 131, 1955.
[4] Öpik, E. J. *Publ. Astr. Obs. Tartu*, **25**, no. 4 (Quoted by Lovell, A. C. B. *Meteor Astronomy*, Oxford, 1954, p. 384), 1922.
[5] Kresak, L. and Vozarova, N. *Bull. Astr. Inst. Czech.* **4**, 139, 1953.
[6] Hruška, A. *Bull. Astr. Inst. Czech.* **5**, 13, 1954.
[7] Ceplecha, Z. *Bull. Astr. Inst. Czech.* **3**, 65, 1952.
[8] Browne, I. C., Bullough, K., Evans, S. and Kaiser, T. R. *Proc. Phys. Soc.* B, **69**, 83, 1956.

Discussion

Whipple: Bowen's rainfall theory should require many more weak meteors than suggested by Browne's results.

Browne: Indeed, by extrapolation to the 10th magnitude one would expect the showers to be hardly visible against the sporadic background

JUPITER AS A RADIO SOURCE

B. F. BURKE AND K. L. FRANKLIN

Department of Terrestrial Magnetism, Carnegie Institution of Washington,
Washington, D.C., U.S.A.

Records obtained at a declination of about $+22°$ during the first quarter of 1955 with the 22·2 Mc./s. Mills Cross of the Carnegie Institution of Washington occasionally exhibited an interference-like event which was apparently a function of sidereal time. A plot of the right ascensions of beginning and end of each event (duration roughly fifteen minutes) against the date of occurrence revealed a smooth change of Right Ascension corresponding, initially, to a westward motion (Fig. 1). This pre-

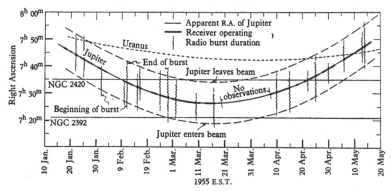

Fig. 1. Apparent positions of variable radio source observed at 22·2 Mc./s.

cluded correlation of the event with passage through the pencil beam of fixed objects like the galactic cluster NGC 2420 and the planetary nebula NGC 2392 which otherwise would have been candidates. The retrograde motion suggested a planet, and a canvass of the solar system uncovered only Jupiter and Uranus as possibilities. Of these, Jupiter exhibited the same position and the same change of position as did the event recorded, while Uranus was well out of the pencil beam much of the time. It was therefore concluded that the source of the radio emission was associated with Jupiter.

The phenomenon itself occurred on twenty out of sixty-eight transits during the observation period, for an average occurrence rate of about one out of three days. The limited material was insufficient to show any trustworthy correlation with central Jovian meridian for either system I or II at the time of observation, (January–May 1955), indicating that there was, apparently, more than one active region, if the events were associated with localized occurrences on the surface of the planet.

The characteristic appearance of each event was that of a succession of intense, sharp bursts, occasionally being completely off the scale of the ecorder. A typical record is shown in Fig. 2. It may be noticed that many

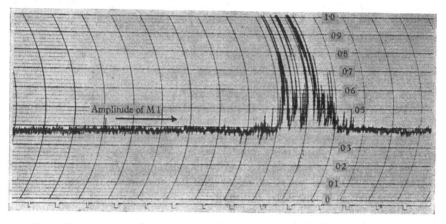

Fig. 2. Typical record of bursts from radio source associated with the planet Jupiter.

of the bursts are more than five times as strong as Taurus A. Extreme events reach or may surpass the intensity of Cygnus A or even Cassiopeia A. Comparisons with the burst intensity of local lightning strokes indicate that the intrinsic radio intensity of the events at Jupiter, assuming an isotropic radiator, are often stronger by a factor of at least 10^9 than terrestrial lightning.

An audio tape-recording was obtained during one of the active transits of Jupiter in April 1955. The individual bursts appear to have a duration of the order of $0^s.5$, longer than typical bursts due to local lightning. The Jupiter bursts do not usually have abrupt rises and falls; it is estimated that the rise to maximum intensity may have occupied about one-tenth of the burst duration.

On 6 June 1954, during observations of the solar occultation of Taurus A, a strong 'temporary source' was observed passing through the lobes of a 22·2 Mc./s. interferometer. This source exhibited large, irregular fluctuations in intensity while it was present. Simultaneous records obtained with an interferometer at 38 Mc./s. failed to show any trace of this source. Since the identification of Jupiter as a radio source, a re-examination of the interferometer record showed that Jupiter could account for this 'temporary source'. Recently, equipment operated by H. W. Wells at the Carnegie Institution at 27 Mc./s. exhibited a record which could be correlated with a simultaneous record of Jupiter at 22 Mc./s. C. A. Shain, in Sydney, has records showing Jupiter at 18·3 Mc./s. while F. G. Smith at Cambridge, England, had been unable to observe Jupiter at 81 Mc./s. Thus it is possible to state that this phenomenon is one occurring over the frequency range 18 to 27 Mc./s., but is presently unobservable at 38 Mc./s. and higher. Closer specification of the spectrum is not possible at the present time.

Discussion

Oort: Have you observed Saturn, or Venus?

Burke: Any effect on Saturn is at least five times weaker; we have not looked at Venus.

Smith: In various runs of one week we found nothing on Jupiter with a big aerial at 80 Mc./s. Any bursts must have less than 1/700 of the intensity of Taurus A. This means that the spectral index must have a large negative value, at least $-5\frac{1}{2}$. This does not at all agree with the mechanism of a gas discharge over a long path like terrestrial lightning.

Dewhirst: Analyzing Burke's data of the first half of 1955 we did not find any fixed point rotating with the period of the equatorial regions, but the material was very limited.

LOCATION ON JUPITER OF A SOURCE
OF RADIO NOISE

C. A. SHAIN

Division of Radiophysics, Commonwealth Scientific and Industrial Research Organization, Sydney, Australia

A search of old records of cosmic noise, taken in the Radiophysics Laboratory at 18·3 Mc./s., has confirmed the discovery, by Burke and Franklin [1], of radio radiation from Jupiter, and has further shown that in 1951 the radio radiation came from a very localized region of the planet.

The records studied fall into two series. The first series comprised records, using an aerial with beam-width 17°, taken at intervals during the period October 1950 to April 1951. Radiation from Jupiter appeared on about one-half of the days on which records were suitable (correct aerial direction, no obvious interference, etc.), and it was clear that the Right Ascension of the source changed with that of Jupiter. For some of these records more accurate direction-finding was possible using a split-beam technique, and these records proved that the position of the source was within ± 1° of Jupiter.

The second series of records had been taken with the aerial modified in such a way that the beam was still narrow in declination but was very broad in hour angle. Because of its broad beam it was capable of receiving signals from an extra-terrestrial source for nearly eight hours per day—a time which would cover nearly one complete rotation of Jupiter. This series of records, from 15 August to 2 October 1951, revealed Jupiter radiation on twenty-seven out of the thirty days on which there were suitable records, although on all days the bursts of radiation came in groups only an hour or two long.

A most interesting new fact coming out of the examination of these records was the very close relation between the times of occurrence of bursts and the rotation of Jupiter. For each record the times of activity were noted. Fig. 1 shows the longitudes of the central meridian during the times when bursts were received. The equatorial regions of the visible surface of Jupiter rotate more rapidly than the remainder of the planet and two conventional systems of zenocentric longitude are used. System I, which applies to the equatorial regions, is based on a rotation period of about $9^h 50^m$; system II on a rotation period of about $9^h 55^m$. It will be seen that the lines indicating times of occurrence of bursts are almost directly under each other when plotted against system II longitudes and show a steady drift towards increasing longitudes in system I. Closer examination shows that there is actually a small negative drift in the system II diagram. These drifts imply that the source of radiation had a rotation period slightly shorter than that adopted for the calculation of system II longitudes; the period of rotation is found to be $9^h 55^m 13^s$, with an estimated probable error of $\pm 5^s$.

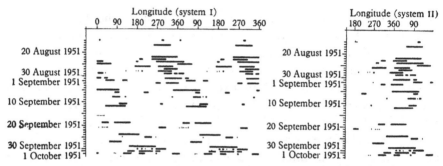

Fig. 1. Periods of occurrence of 18·3 Mc./s. radiation from Jupiter plotted against longitude of the central meridian at the time of observation.

After allowing for the slight drift in longitudes, all the lines in Fig. 1 were superimposed to give a histogram of the frequency of occurrence of the noise for 5° intervals of central meridian longitude. This is shown in Fig. 2, the longitudes being system II longitudes on 14 August. This figure shows that for a band of longitudes centred on 67° and extending from about 0° to 135°, the frequency of occurrence was much greater than outside this band. This suggests an origin in a very localized source on Jupiter. Since there were about 120 rotations of Jupiter during the period of the observations on which the figure is based, the probability that the effect observed is due to chance is extremely small.

Fox [2] has given a summary of the observations of Jupiter at this time which were communicated to the Jupiter Section of the British Astronomical Association. A disturbance at the boundary between the South Temperate Zone and the South Temperate Belt, which was observed for several months, had an observed rotation period of $9^h 55^m 13^s$, that of the radio source. All the other belts which move with system II were either faint, with no certain markings, or moved slightly more slowly than system II. Fox gave a sketch by E. J. Reese which depicted the visually most active region of the South Temperate Belt at the end of November 1951. The radio observations ceased on 2 October, but, allowing for a continuing drift in longitude, the longitude of the radio source on 30 November would have crossed the middle of one of the prominent white markings in Reese's sketch. Therefore, although the identification is not proved beyond all doubt, it seems very probable that this visually dis-

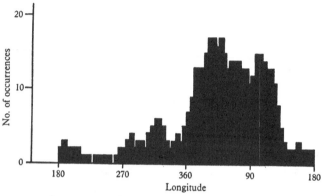

Fig. 2. Frequency of occurrence of 18·3 Mc./s. radiation from Jupiter for intervals of 5° in longitude (based on a rotation period of $9^h 55^m 13^s$). The longitudes are system II longitudes on 14 August 1951.

turbed region was responsible for the radio radiation, the most intense source lying in one of the white spots at the southern edge of the South Temperate Belt.

The mechanism of origin of the radiation is still unknown. Observations during the last few months have confirmed a result suggested by the 1950–51 observations that the individual bursts have durations ranging from fractions of a second to many seconds. An origin in electrical discharges analogous to terrestrial atmospherics is not unlikely. The recent observations are also consistent with the hypothesis of a localized source of the radiation.

A full discussion of the 1950–51 observations will be given in the *Australian Journal of Physics.* [3]

REFERENCES

[1] Burke, B. F. and Franklin, K. L. *J. Geophys. Res.* **60**, 213–17, 1955.
[2] Fox, W. E. *J. Brit. Astr. Ass.* **62**, 280–2, 1952.
[3] Shain, C. A. *Aust. J. Phys.* **9**, 61–73, 1956.

POSSIBLE PROOFS OF THE LUNAR
ATMOSPHERE

F. LINK

Astronomical Institute of the Academy of Sciences, Prague, Czechoslovakia

In the period 1955–57 a number of occultations of two radio sources 05·01 Taurus and 06·01 Gemini by the Moon takes place [1]. They will give an excellent opportunity to examine not only the exact position and the shape of these sources but possibly also some traces of the lunar atmosphere [2].

If we adopt as upper limit of density the optical determination by Dollfus [3], who gives 10^{-9} of the terrestrial density at sea level, and if we go still three orders further to 10^{-12}, we meet analogous conditions to those in the terrestrial atmosphere on the top of F region at altitude of some 400 km. There the direct determinations by rockets lead [4] to an electron density of about $N = 10^5$ cm.$^{-3}$. The density of this order can perfectly well be traced by the method based on the radio propagation theory.

In the first approximation we may assume the validity of the Chapman formula, which gives in the uppermost part of the ionized region the relation

$$N = N^* \exp\left(-h/2H\right), \tag{1}$$

where N^* is the electron concentration on the Moon surface, from which we count the altitude h, and H is the scale height (about 10^2 km.).

The total deviation of radio waves of the frequency f will then be

$$\omega = 4\cdot03 \cdot 10^{-5} \frac{N}{f^2} \sqrt{\frac{5460}{H}} = \omega^* \exp\left(-h/2H\right) \tag{2}$$

and their intensity should be multiplied by the factor $1/s$, where

$$s = 1 - \frac{\omega}{R}\left(\frac{1736}{2H} - \frac{1}{\rho}\right) - \frac{1736}{2H\rho}\left(\frac{\omega}{R}\right)^2, \tag{3}$$

which gives the amplification of the intensity due to the convergence of the rays after the refraction in the lunar atmosphere. Here $R = 930''$ is the angular radius of the Moon and ρ is the distance at which the ray seems to

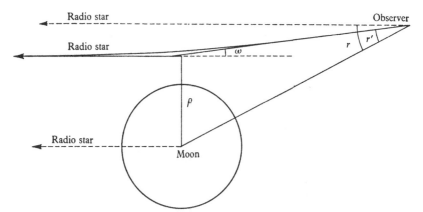

Fig. 1. Refraction by lunar atmosphere.

Table 1. *Total deviation ω and intensity 1/s of the rays at Moon surface*

N (el. cm.$^{-3}$) =	10^5	10^4	10^3	10^2	10^1
$H=$ 1365 km. $f=$ 25 Mhz.		4·6′	27″	2·7″	0·3″
		0·95	0·99	1·00	1·00
50	11·4′	1·1′	6·9″	0·7″	0·1″
	1·08	0·97	1·00	1·00	1·00
100	2·9′	17″	1·7″	0·2″	0·0″
	0·96	0·99	1·00	1·00	1·00
200	23″	2·3″	0·2″	0·0″	0·0″
	0·99	1·00	1·00	1·00	1·00
400	11″	1·1″	0·1″	0·0″	0·0″
	1·00	1·00	1·00	1·00	1·00
$H=$ 341 km. $f=$ 25 Mhz.		9·1′	55″	5·5″	0·6″
		1·02	1·17	1·00	1·00
50		2·3′	14″	1·4″	0·1″
		1·39	1·02	1·00	1·00
100	5·7′	34″	3·4″	0·3″	0·0″
	11·6	1·10	1·01	1·00	1·00
200	1·4′	8·6″	0·9″	0·1″	0·0″
	1·17	1·01	1·00	1·00	1·00
400	21″	2·1″	0·2″	0·0″	0·0″
	1·04	1·00	1·00	1·00	1·00
$H=$ 152 km. $f=$ 25 Mhz.			1·4′	8·2″	0·8″
			1·86	1·04	1·00
50		3·4′	21″	2·1″	0·2″
		3·1	1·12	1·01	1·00
100	8·6′	51″	5·1″	0·5″	0·1″
	0·56	1·27	1·03	1·00	1·00
200	1·8′	10·8″	1·1″	0·1″	0·0″
	2·7	1·05	1·00	1·00	1·00
400	32″	3·2″	0·3″	0·0″	0·0″
	1·15	1·00	1·00	1·00	1·00

pass the Moon's centre, expressed in lunar radii. In Table 1 we have calculated ω and $1/s$ for some reasonable assumptions about N^* and H. The general features of a central occultation are given in the example of Table 2.

Table 2. *Central occultation of a radio point source*

$N = 10^4$ el. cm.$^{-3}$, $f = 50$ Mhz., $H = 341$ km., $\omega^* = 137''$

ρ		$1/s$	r'	r	Δt	
1·0	137″	1·39	930″	1067″	4$^\mathrm{m}$	34$^\mathrm{s}$
1·1	106	1·28	1023	1129	6	38
1·2	82	1·20	1116	1198	8	56
1·3	64	1·15	1209	1273	11	26
1·4	50	1·12	1302	1352	14	04
1·5	38	1·09	1395	1433	16	46
1·6	30	1·07	1488	1518	19	36
1·7	23	1·05	1581	1604	29	40
2·0	11	1·02	1860	1872	31	14
3·0	2	1·00	2790	2792	62	04

ρ = Distance of the ray from the Moon's centre in Moon's radii.
r = Geometrical angle between the Moon's centre and the radio source.
r' = Observed angle between the Moon's centre and the radio source.
Δt = Time elapsed since the geometrical occultation ($r = 930''$).

From these results we can see that appreciable effects both in direction and in intensity should be expected, especially on the lower frequencies, if a trace of the lunar atmosphere were present. The deviation of 0·5″ gives in the central case a lengthening of the occultation of about 2$^\mathrm{s}$. Also an augmentation of the intensity of 10 % can be detected.

These relatively simple conditions will be complicated by the diffraction, by the shape of the source, and possibly also by irregularities in the lunar ionosphere. Exact measurements of radio flux on low frequencies are needed before we can undertake a further analysis.

REFERENCES

[1] Link, F. and Neuzil, L. *B.A.C.* 5, 112, 1954.
[2] Link, F. *B.A.C.* 7, 1, 1956.
[3] Dollfus, A. *C.R. Paris*, 234, 2046, 1952.
[4] Berning, W. W. *J. Meteorol.* 8, 175, 1951.

THE LUNAR OCCULTATION OF A RADIO STAR AND THE DERIVATION OF AN UPPER LIMIT FOR THE DENSITY OF THE LUNAR ATMOSPHERE

B. ELSMORE

Cavendish Laboratory, Cambridge, England

Observations were made at Cambridge on 26 April 1955 of the lunar occultation of the large-diameter radio source in the constellation of Gemini. This radio source, having R.A. $06^h 13^m 37^s$ and DEC. $22° 38'$ (1950·0), has been identified by Baldwin and Dewhirst (1954) [1] as the galactic nebulosity IC 443, which consists of a filamentary structure contained within a circular region of $24'.5$ radius. Baldwin and Dewhirst also succeeded in measuring the distribution of radio 'brightness' across the source using an interferometric method; their measurements indicate that the diameter of the radio source is approximately the same as that of the visible nebulosity.

The position of the moon, as seen from Cambridge on 26 April 1955, in relation to IC 443, is shown in Fig. 1. Observations were made at a wave-length of 7·9 metres during the period $16^h 05^m$ to $16^h 15^m$ U.T., using large fixed-aerial systems. At a wave-length of 3·7 metres the two pairs of the large Cambridge Radio Telescope (Ryle and Hewish, 1955) [2] were used as interferometers of low resolving power. In order to increase the period of observation, phasing cables were introduced into one pair of aerials. In this way it was possible to observe the source continuously from $16^h 08^m$ to $16^h 35^m$ U.T.

The observed relative intensity showed good agreement with that predicted from the 'brightness' distribution derived by Baldwin and Dewhirst, but the reduction was less. At $16^h 30^m$ U.T. a marked discontinuity occurred which corresponded to an increase in radio emission of about 5 %.

This rise might be due to the scattering by the moon's surface of enhanced solar radiation, but the absence of any evidence for solar activity on this day indicates that this explanation is most improbable.

At 16^h 30^m U.T. the moon's west limb is uncovering a part of the well-defined filament near η Geminorum, as shown in the figure. This indicates that the radio emission may originate mainly in regions of strong H_α emission. This hypothesis is consistent with the evidence from the smaller reduction than predicted near the time of maximum phase, since the moon was then covering relatively few areas of H_α emission.

If the identification of the increase of radio intensity with the uncovering of the bright arch is correct, then their relative times may be used to

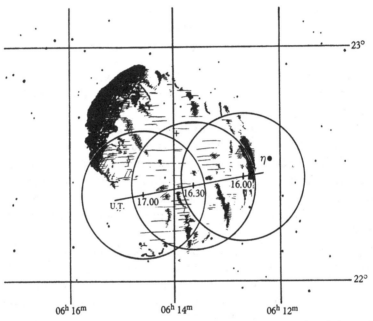

Fig. 1. The moon's position, as seen from Cambridge on 26 April 1955, in relation to IC 443.

estimate the density of the lunar atmosphere. Two effects would delay the radio occurrence: diffraction at the moon's limb, which at 3·7 metres wave-length is about one minute of time, and refraction in the lunar atmosphere.

The observations show that the total delay cannot exceed four minutes of time. Using results computed by Link [3] (see also paper 77), the electron density in the lunar atmosphere cannot exceed 10^5 cm.^{3} for a scale height between 50 and 1500 km.

It seems probable that on this sun-lit side of the moon the degree of ionization would be as great as that of the F-region of the ionosphere. On this supposition the density of the lunar atmosphere must be less than 10^{-12} of the density of the terrestrial atmosphere at sea level; a figure $1/1000$ of the previous upper limit.

REFERENCES

[1] Baldwin, J. E. and Dewhirst, D. W. *Nature*, **173**, 164, 1954.
[2] Ryle, M. and Hewish, A. *Mem. Roy. Astr. Soc.* **67**, 97, 1955.
[3] Link, F. *Bull. Astr. Czech.* **7**, 1, 1956.

Discussion

Hoyle: Would not most of the gas be on the dark side of the moon?

Elsmore: Since the age of the moon was four days, the west limb which uncovered the bright filament is sun-lit and if any gas were present there, it would be ionized. We have no information about the dark side of the moon, but it seems unlikely that the difference in temperature would cause a large difference in density.

RADIO EMISSION FROM THE MOON AND THE NATURE OF ITS SURFACE

V. S. TROITZKY AND S. E. KHAIKIN

Gorky State University and Physical Institute of the Academy of Sciences, Moscow, U.S.S.R.

A theoretical study of the integral radio emission of the moon, measured at the wave-length of 3·2 cm. (Zelinskaja and Troitzky[1]; Kajdanovsky, Turusbekov and Khaikin[2]), was carried out at the Gorky radio astronomical station 'Zimenky' and at the Physical Institute of the Academy of Sciences of the U.S.S.R. The following expression for the average radio temperature of the entire lunar disk, as a function of the lunar phase, Ωt, was obtained (Troitzky, 1954)[3]:

$$T_c(t) = 0.92(1 - k_0)\ (T_n + 0.344\Theta) + \frac{0.37(1 - k_0)\ \Theta}{\sqrt{1 + 2\delta + 2\delta^2}} \cos(\Omega t - \xi). \quad (1)$$

Here $\tan \xi = \delta/(1 + \delta)$ and $\delta = \beta/\kappa$, where β is the attenuation coefficient of the thermal wave, κ the power attenuation coefficient of the radio wave. Further, $T_m = 374°$ K. is the temperature of the subsolar point, T_n is the temperature at the lunar midnight, $\Theta = T_m - T_n$ and k_0 is the reflexion coefficient of radio waves for vertical incidence ($k_0 \approx 0.1$). The numerical coefficients in equation (1) were obtained as a result of averaging the Fresnel reflexion coefficients over the whole disk. The degree of polarization of the total radio emission was calculated and was found to be about 4%.

Two cycles of measurements of the radio emission from the moon at the wave-length of 3·2 cm., carried out in 1952 (Troitzky and Zelinskaja)[4] and 1955, yielded the result that the mean radio temperature, determined with a precision of ±5 to ±7%, does not depend upon phase. The constant component of the radio temperature was found to be:

$$T_c = 170° \pm 10° \text{ K.}$$

From this result combined with equation (1) the temperature T_n at the lunar midnight was found to be 115° K. The amplitude of the variable part of the radio temperature may be estimated from the maximum displacement of the 'centre of gravity' of the radiation relative to the geometric centre of the moon. Measurements show that no systematic displacement of the centre of emission as a function of phase exists up to $\pm$ 0·5 minutes of arc. The amplitude of the variation of the lunar radio temperature is consequently less than $10°$ K.

The experimental data of the radio emission from the moon obtained on 3·2 cm. and the data given by Piddington and Minnett (1949) [5] for 1·25 cm. were discussed on the basis of equation (1). The reduction shows that neither the one-layer model of the surface structure nor the double-layer model contradict the results obtained on these two wave-lengths. It was shown, however, that the value for the mean radio temperature of the moon obtained by Piddington and Minnett contradicts optical temperature measurements and the thermal properties of the lunar rocks (Troitzky[6]). Assuming the value $(k\rho c)^{-1/2}$ as known from optical measurements (thermal conductivity $k = 2\cdot5 \times 10^{-6}$, density $\rho = 2$, thermal capacity $c = 0\cdot2$) the electrical conductivity of the lunar rocks was found to be $\sigma_3 = 5\cdot4 \times 10^8$ cgs. (penetration depth $= 8\cdot7$ cm.) for $\lambda = 3\cdot2$ cm. and $\sigma_1 = 1\cdot4 \times 10^9$ cgs. (penetration depth $= 3\cdot3$ cm.) for $\lambda = 1\cdot25$ cm. The comparatively high values of the electric conductivity obtained suggest that the content of calcium, natrium and iron oxides in the lunar surface material is considerable. This conclusion agrees with the results of investigations carried out in reflected light (Budnikova, 1953[7]; Borissova, 1953[8]).

In order to solve the problem of the structure of the lunar surface, further precise measurements of the radio emission of the moon, particularly on 8 and 4 mm. waves, are needed.

REFERENCES

[1] Zelinskaja, M. R. and Troitzky, V. S. *Publications of the 5th Conference on Cosmogony*, p. 99, 1956.
[2] Kajdanovsky, N. L., Turusbekov, M. T. and Khaikin, S. E. *Ibid.* p. 347, 1956.
[3] Troitzky, V. S. *A.J. U.S.S.R.* **31**, 6, 511, 1954.
[4] Troitzky, V. S. and Zelinskaja, M. R. *A.J. U.S.S.R.* (in print).
[5] Piddington, J. H. and Minnett, H. C. *Aust. J. Sci. Res.* A, **2**, 63, 1949.
[6] Troitzky, V. S. *Publications of the 5th Conference on Cosmogony*, p. 325, 1956.
[7] Budnikova, N. A. *Bulletin of the Leningrad University, Math. Phys. Chem.* Series no. 8, 71, 1953.
[8] Borissova, A. N. *Ibid.* series no. 8, 89, 1953.

THE MOON AS A SCATTERER OF
RADIO WAVES

I. C. BROWNE AND J. V. EVANS

Jodrell Bank Experimental Station, University of Manchester, England

Radio echoes from the moon at 120 Mc./s. have been obtained with a pulse length of 30 msec. and a pulse interval of 1·8 sec., using a fixed aerial directed due south. Their study has given information about the nature of the scattering function of the lunar surface.

The intensity of moon echoes is proportional to the quantity $g\rho\pi a^2$, where a is the radius of the moon, ρ the mean power reflectivity of its surface for radio waves at normal incidence, and g a directivity factor, which depends on the way in which the intensity of waves scattered backwards from an element of the surface varies with the angle of incidence.

For a lunar surface in which any irregularities are very small compared with the wave-length, so that the moon is smooth, the reflexion is specular, and the directivity $g = 1$. If the irregularities are of the order of a wave-length in size, the scattering function is given by Lambert's cosine law, the moon appears limb-'darkened', and the directivity $g = 8/3$. When the irregularities are all very large compared with the wave-length, the scattering function is given by the Lommel-Seeliger law, the moon appears uniformly 'bright', and the directivity g is of the order of 5. In the optical case, as is well known, the moon roughly obeys the Lommel-Seeliger law, and the directivity $g = 5·7$.

From measurements of the echo intensity alone, it is possible to deduce only the value of the product $g\rho$. However, moon echoes show a rapid fading of intensity, with a quasi-period (at 120 Mc./s.) of the order of 0·5 sec. This, it is believed, is caused by the libration of the moon, because the changing angle of libration introduces variations into the phase-paths of the rays scattered by various parts of the moon, resulting in changes in the resultant echo signal. If this explanation is correct, the amplitudes of a series of echo pulses should follow a Rayleigh distribution, while the rate at which the echo fades should be proportional to the rate of change of the angle of libration, taking into account the librations in longitude, and in latitude, as well as the diurnal and physical librations. (The rate of fading is conveniently measured by the autocorrelation function for echo amplitude, regarding the amplitudes of echoes as a time series.) These predictions agree well with the observations made at Jodrell Bank. The form of the power spectrum of the fluctuations in echo amplitude will depend, not only upon the rate of libration, but also upon the scattering function for the lunar surface, and hence also with the directivity g. The power spectra and their associated auto-correlation functions (given by the square of the Fourier cosine transforms of the power spectra) have been computed for four models representing different possible scattering laws for the lunar surface, namely: (i) specular reflexion, (ii) Lambert's law (limb 'darkening'), (iii) Lommel-Seeliger law (uniform 'brightness'), and (iv) a model showing limb brightening. From a comparison with the observed form of the auto-correlation function, it appears that the assumption of Lommel-Seeliger scattering gives the best approximation, though in this case the fading rate is greater than predicted by some 50 %. The discrepancy has not been fully accounted for.

A value of 5·7 has accordingly been assumed for the directivity g. Assuming further that the reflectivity ρ of the lunar soil is 0·1, a value characteristic of dry terrestrial rocks, it becomes possible to compare the intensity of the observed echoes with that predicted. A comparison is made below for two days on which the slow fading of echoes (produced by the rotation of the plane of polarization of the waves during their passage through the ionosphere) permitted reliable measurements of echo intensity. The values given are averages over one minute.

	10 August 1955	17 August 1955
Observed mean signal-to-noise power of echoes	27·2 ± 1 db.	24·3 ± 2 db.
Predicted signal-to-noise power of echoes	31·8 ± 1 db.	31·0 ± 1 db.

It can be seen that the agreement is fair. It is improved by 3 db. if the assumed value of reflectivity ρ is taken as 0·05. This value is characteristic of dry dusty soils, suggesting that the moon's surface may be covered with such a layer, to a depth of several centimetres.

Full details of these measurements and their probable errors are being published elsewhere, together with an account of the effects produced by the ionospheric rotation of the plane of polarization of the radio waves during their passage to and from the moon (Browne, Evans, Hargreaves and Murray, 1956) [1].

REFERENCE

[1] Browne, I. C., Evans, J. V., Hargreaves, J. K. and Murray, W. A. S. 'Radio Echoes from the Moon', *Proc. Phys. Soc.* B, **69**, 901, 1956.

Printed in the United States
By Bookmasters